DIGITAL IMAGE PROCESSING ALGORITHMS AND APPLICATIONS

DIGITAL IMAGE PROCESSING ALGORITHMS AND APPLICATIONS

I. Pitas
Aristototle University of Thessaloniki

A WILEY-INTERSCIENCE PUBLICATION
JOHN WILEY & SONS, INC.
New York / Chichester / Weinheim / Brisbane / Singapore / Toronto

This book is printed on acid-free paper. ∞

Copyright © 2000 by John Wiley & Sons, Inc. All rights reserved.

Published simultaneously in Canada.

No part of this publication may be reproduced, stored in a retrieval system or transmitted in any form or by any means, electronic, mechanical, photocopying, recording, scanning or otherwise, except as permitted under Sections 107 or 108 of the 1976 United States Copyright Act, without either the prior written permission of the Publisher, or authorization through payment of the appropriate per-copy fee to the Copyright Clearance Center, 222 Rosewood Drive, Danvers, MA 01923, (978) 750-8400, fax (978) 750-4744. Requests to the Publisher for permission should be addressed to the Permissions Department, John Wiley & Sons, Inc., 605 Third Avenue, New York, NY 10158-0012, (212) 850-6011, fax (212) 850-6008, E-Mail: PERMREQ@WILEY.COM.

For ordering and customer service, call 1-800-CALL WILEY.

Library of Congress Cataloging in Publication Data is available.

ISBN 0-471-37739-2

Printed in the United States of America

10 9 8 7 6 5 4 3 2 1

Contents

Preface		*ix*
1	***Digital image processing fundamentals***	***1***
	1.1 Introduction	*1*
	1.2 Topics of digital image processing and analysis	*2*
	1.3 Digital image formation	*4*
	1.4 Digital image representation	*7*
	1.5 Elementary digital image processing operations	*13*
	1.6 Digital image display	*18*
	1.7 Fundamentals of color image processing	*20*
	1.8 Noise generators for digital image processing	*38*
References		*47*
2	***Digital image transform algorithms***	***51***
	2.1 Introduction	*51*
	2.2 Two-dimensional discrete Fourier transform	*52*
	2.3 Row–column FFT algorithm	*59*
	2.4 Memory problems in 2-d DFT calculations	*68*
	2.5 Vector-radix fast Fourier transform algorithm	*85*

2.6	Polynomial transform FFT	92
2.7	Two-dimensional power spectrum estimation	96
2.8	Discrete cosine transform	103
2.9	Two-dimensional discrete cosine transform	107
2.10	Discrete wavelet transform	113

References 117

3 Digital image filtering and enhancement 121

3.1	Introduction	121
3.2	Direct implementation of two-dimensional FIR digital filters	122
3.3	Fast Fourier transform implementation of FIR digital filters	125
3.4	Block methods in the linear convolution calculation	128
3.5	Inverse filter implementations	133
3.6	Wiener filters	135
3.7	Median filter algorithms	139
3.8	Digital filters based on order statistics	149
3.9	Signal Adaptive order statistic filters	156
3.10	Histogram and histogram equalization techniques	162
3.11	Pseudocoloring algorithms	166
3.12	Digital image halftoning	168
3.13	Image interpolation algorithms	174
3.14	Anisotropic Diffusion	177
3.15	Image Mosaicing	179
3.16	Image watermarking	180

References 185

4 Digital image compression 191

4.1	Introduction	191
4.2	Huffman coding	192
4.3	Run-length coding	200
4.4	Modified READ coding	203
4.5	LZW compression	205
4.6	Predictive coding	221
4.7	Transform image coding	229

4.8	JPEG2000 compression standard	235

References	239

5 Edge detection algorithms 241
   5.1 Introduction 241
   5.2 Edge detection 242
   5.3 Edge thresholding 249
   5.4 Hough transform 249
   5.5 Edge-following algorithms 257

References 273

6 Image segmentation algorithms 275
   6.1 Introduction 275
   6.2 Image segmentation by thresholding 277
   6.3 Split/merge and region growing algorithms 282
   6.4 Relaxation algorithms in region analysis 297
   6.5 Connected component labeling 300
   6.6 Texture description 303

References 319

7 Shape description 323
   7.1 Introduction 323
   7.2 Chain codes 324
   7.3 Polygonal approximations 329
   7.4 Fourier descriptors 334
   7.5 Quadtrees 336
   7.6 Pyramids 342
   7.7 Shape features 348
   7.8 Moment descriptors 352
   7.9 Thinning algorithms 356
   7.10 Mathematical morphology 361
   7.11 Grayscale morphology 369
   7.12 Skeletons 372
   7.13 Shape decomposition 376
   7.14 Voronoi tesselation 382

	7.15	Watershed transform	385
	7.16	Face detection and recognition	386

References 393

8 Digital Image Processing Lab Exercises Using EIKONA 401
   8.1 Introduction 401
   8.2 Overview 401
   8.3 Structure 402
   8.4 BW image processing 405
      8.4.1 Black-and-White 405
      8.4.2 Basic 405
      8.4.3 Processing 405
      8.4.4 Analysis 405
      8.4.5 Transforms 406
      8.4.6 Filtering 406
      8.4.7 Nonlinear filtering 406
   8.5 Color image processing 406
      8.5.1 Basic 406
      8.5.2 Processing 407
      8.5.3 Analysis 407
      8.5.4 Color Representation 407
   8.6 Modules 407
      8.6.1 Arts module 408
      8.6.2 Crack Restoration 410
      8.6.3 Watermark module 412
   8.7 EIKONA Source, Library/DLL 413
   8.8 Instructions for using the educational material 413

References 417

Index 418

Preface

Digital image processing, analysis, and computer vision have exhibited an impressive growth in the past decades, in terms of both theoretical development and applications. They constitute a leading technology in a number of very important areas, for example, in digital telecommunications and the Internet, broadcasting, medical imaging, multimedia systems, biology, material sciences, robotics and manufacturing, intelligent sensing systems, remote sensing, graphic arts and printing. This growth is reflected in the large number of papers published in international scientific journals each year, as well as in a good number of specialized books in digital image processing, analysis, and computer vision. The application tasks created the need to construct a variety of digital image processing and analysis algorithms. Most of them are scattered in the relevant scientific journals. Several of them are described in specialized books. However, most of the digital image processing and computer vision books concentrate on the theory and the applications of digital image processing rather than on its algorithmic part. The aim of this book is to present such algorithms and lab experiments in a rather systematic way. The algorithms described cover many aspects of digital image processing, analysis, and coding. However, these algorithms are only a small fraction of the total number of algorithms existing in the literature. They have been selected on the basis of their acceptance by the scientific community. Indications of their acceptance are their appearance in classical digital image processing textbooks, as well as their reference by scientists. The book is accompanied by lab exercises that are based on EIKONA, a digital image processing software developed by the author. The demo, student, and full versions of EIKONA can be found in *www.alphatecltd.com*. The book is also accompanied by transparencies in PDF format that highlight important topics covered in each book chapter. They can be used either in a traditional classroom or for self-teaching or for distance learning by videoconferencing over IP or over ISDN. Therefore, the book, together with its transparencies and its exercises, can be used as a teaching tool in a practical image processing course, either in a traditional class or in a distance learning environment. In particular, the networked version of EIKONA can be used in an Internet environment to create a virtual image processing lab, where each student can execute his lab exercises at home at his own pace. This environment has been successfully used by the author for teaching in undergraduate as well as in distance learning courses during the past two years. The accompanying multimedia material can be downloaded from $ftp://ftp.wiley.com/public/sci\_tech\_med/image\_processing$. The attachment of lab exercises and multimedia tools in this book for distance learning or self-teaching is a significant step forward compared to my earlier book on the same topic (1993). This book is a result of the active involvement of the author in digital image processing over the past decade. Therefore, it is strongly related to his research activities, especially in digital image transform algorithms, nonlinear digital image processing, and mathematical morphology. Several algorithms and programs presented in the book are by-products of the author's research and teaching activities.

The algorithms described in this book are presented in C code. This language has been chosen because of its wide acceptance by the digital image processing community. Several non-essential control commands have been removed from the body of subroutines. This increases C code readability and, at the same time, reduces software robustness, especially with respect to inappropriate subroutine parameter passing at subroutine calls. The algorithms described are compatible with each other. Together they form a digital image processing library called EIKONA. Microsoft C has been used for the development of EIKONA, because of its widespread use in IBM PC-compatible computers. The author thinks that such a personal computer equipped with a true color graphics card and the Microsoft Windows environment can be a very efficient and cheap instrument for developing digital image processing applications. The C code is ANSI compatible and can be easily linked with C programs in Unix workstations.

The algorithms are accompanied by a discussion of the related theory. In many cases, algorithm description and discussion uses advanced mathematical concepts. Thus, it is assumed that the reader already has a basic understanding of digital signal/image processing and computer vision. Such knowledge can be obtained from any classical book on digital image processing and/or computer vision. This book is not intended to serve as a substitute for these textbooks. It is, rather, a guide that can be used by the student, scientist, or engineer to deepen knowledge of digital image processing algorithms and their structure and/or to solve specific problems encountered in various applications. Advanced readers can read the material in Section 1.4 and go directly to the description of the algorithms of interest. Less advanced readers are recommended to study Chapter 1 first and to proceed to more complicated algorithms in subsequent chapters as a second step. The study of Chapter 2 is recommended before the study of Chapters 3, 4, and the study of Chapters 5, 6 is recommended before the study of Chapter 7. Those readers interested in digital image processing can focus their attention on Chapters 1, 2, 3. Those readers interested in digital image coding can study Chapters 1, 2, 4. Finally, those readers interested in digital image analysis and computer vision can study Chapters 1, 5, 6, 7.

As has already been mentioned, the material presented in this book is the result of the involvement of the author in digital image processing over the past decade in research, teaching, and the development of applications. The research has been supported by various projects funded by Greek as well as European Community research programs. The author acknowledges the help of Ph.D. students as well as Diploma Thesis students who have been involved in these projects in the past fifteen years and who helped in the preparation of the material for this book. The following colleagues, researchers, and students have contributed in preparing some parts of the manuscript of the current book version under close supervision by the author: A. Georgakis, G. Angelopoulos, A. Bors, C. Cotsakes, V. Chatzis, C. Kotropoulos, A. Nikolaidis, N. Nikolaidis, V. Solachidis, A. Tefas, S. Tsekeridou. Furthermore, the teaching of the digital image processing course at the University of Thessaloniki has given the author an excellent audience for checking the contents of this book. The author acknowledges the secretarial support of these students in preparing the manuscript of this book and thanks the Skete of Saint Andrew, Greece, the Lab of Endodontology, University of Thessaloniki, Greece and the Neuromedia Group, University of Kanterbury, England for providing permission to use their digital images in this book.

I. Pitas

Thessaloniki, October 1999

Important note: The source code presented in this book has been carefully debugged. However, the author cannot accept responsibility for any loss or damage produced by use of this code. The code cannot be used in any form (source, object, executable) in any commercial product or software package without the written permission of the author.

1
Digital Image Processing Fundamentals

1.1 INTRODUCTION

Human vision is one of the most important and complex perception mechanisms. It provides information needed for relatively simple tasks (e.g. object recognition) and for very complex tasks as well (e.g. planning, decision making, scientific research, development of human intelligence). The Chinese proverb 'One picture is worth a thousand words' expresses correctly the amount of information contained in a single picture. Pictures (images) play an important role in the organization of our society as a mass communication medium. Most media (e.g. newspapers, TV, cinema) use pictures (still or moving) as *information carriers*. The tremendous volume of optical information and the need for its processing and transmission paved the way to image processing by digital computers. The relevant efforts started around 1964 at the Jet Propulsion Laboratory (Pasadena, California) and concerned the digital processing of satellite images coming from the moon. Soon, a new branch of science called *digital image processing* emerged. Since then, it has exhibited a tremendous growth and created an important technological impact in several areas, e.g. in telecommunications, TV broadcasting, the printing and graphic arts industry, medicine and scientific research.

Digital image processing concerns the transformation of an image to a digital format and its processing by digital computers. Both the input and output of a digital image processing system are digital images. *Digital image analysis* is related to the description and recognition of the digital image content. Its input is a digital image and its output is a symbolic image description. In many cases, digital image analysis techniques simulate human vision functions. Therefore, the term *computer vision* can be used as equivalent to (or the superset of) digital image analysis. Human vision is a very complex neurophysiological process. Its characteristics are only partially known, despite the tremendous progress that has been made in this area in the past decades. Therefore, its simulation by digital image analysis and computer vision is a very difficult task. In general, the techniques used in digital image analysis and computer vision differ greatly from the human visual perception mechanisms, although both have similar goals.

Another classification of digital image processing and computer vision techniques has three distinct classes: low-level vision, intermediate-level vision and high-level vision. *Low-level vision* algorithms are essentially digital image processing algorithms: their input and output are digital images. *Intermediate-level vision* algorithms have digital images as input and low-level symbolic representations of image features as output (e.g. representations of the object contours). *High-level vision* algorithms use symbolic representations for both input and output. High-level vision is closely related to artificial intelligence and to pattern recognition. It tries to simulate the high levels of human visual perception (image understanding).

The aim of this book is to provide algorithms that are employed either in digital image processing or in digital image analysis. An overview of the topics that will be covered in this book will be given in the next section. The fundamentals of digital image representation and the description of basic image processing operations will follow in subsequent sections. Thus, this chapter will provide the algorithmic basis for more advanced techniques that will be described in subsequent chapters.

1.2 TOPICS OF DIGITAL IMAGE PROCESSING AND ANALYSIS

This section will give a short overview of the various topics of digital image processing and analysis that will be covered in this book. The description will be rather qualitative and will be treated from an algorithmic point of view. More detailed descriptions will be given in the introduction of the relevant chapters. If the interested reader needs more information or a more thorough exposure to digital image processing and analysis, he or she is referred to the excellent literature in this area [JAI89], [BAL82], [GON87], [PRA91], [SCH89], [LEV85], [ROS82].

Digital image formation is the first step in any digital image processing application. The digital image formation system consists basically of an optical system, the sensor and the digitizer. The optical signal is usually transformed to an electrical signal by using a sensing device (e.g. a CCD sensor). The analog (electrical) signal is transformed to a digital one by using a video digitizer (frame grabber). Thus, the optical image is transformed to a digital one. Each digital image formation subsystem introduces a deformation or degradation to the digital image (e.g. geometrical distortion, noise, nonlinear transformation). The mathematical modeling of the digital image formation system is very important in order to have precise knowledge of the degradations introduced.

Digital image restoration techniques concern the reduction of the deformations and degradations introduced during digital image formation. Such techniques try to reconstruct or to recover the digital image. Knowledge of the mathematical model of the degradations is essential in digital image restoration. Digital image enhancement techniques concern the improvement of the quality of the digital image. This usually involves contrast enhancement, digital image sharpening and noise reduction. In certain applications, digital image pseudocoloring and digital image halftoning are considered to belong to digital image enhancement as well. Enhancement techniques have a rather heuristic basis compared to the digital restoration techniques that have rigorous mathematical foundations.

Digital image frequency content plays an important role in digital noise filtering, digital image restoration and digital image compression. Digital image transforms are used to obtain the digital image frequency content. Thus, transform theory is an integral part of digital image processing. The transforms used are two-dimensional, because the digital image itself is a two-dimensional signal. Their computation requires a large number of numerical operations (multiplications and additions). Therefore, the construction of fast transform algorithms is a very important task.

Digital images require a large amount of memory for their storage. A color image of size 1024×1024 pixels occupies 3 MB of disk or RAM space. Thus, the reduction of the memory requirements is of utmost importance in many applications (e.g. in image storage or transmission). Digital image coding and compression take advantage of the information redundancy existing in the image in order to reduce its information content and to compress it. Large compression ratios (e.g. 1:24) can be obtained by proper exploitation of the information redundancy. Excessive image compression results, of course, in image degradation after the decompression phase. Therefore, a good compromise between fidelity and compression ratio must be found. Image compression plays an important role in several vital applications, e.g. image data bases, digital image transmission, facsimile, digital video and high-definition TV (HDTV). Therefore, intensive research has been carried out that has led to a multitude of digital image compression techniques. Some of them are already CCITT (Consultative Committee on International Telephony and Telegraphy) standards.

The first step towards digital image analysis is often the detection of object boundaries. This is performed by using edge and line detection techniques. The lines or edges detected are followed subsequently and a list of the boundary coordinates is created. Edge-following algorithms can be constructed in such a way that they are robust to noise and can follow broken edges. Special algorithms can be written to follow lines or edges having a particular shape (e.g. straight line segments or circles).

The dual problem of edge detection is region segmentation. Segmentation algorithms identify homogeneous image regions. Hopefully each of them corresponds to image objects or to image background. Regions are disjoint sets whose union covers the entire image. Region segmentation techniques can be grouped in three classes. Local techniques employ the local properties within an image neighborhood. Global techniques segment the image on the basis of global information (e.g. global texture properties). Split and merge techniques employ both pixel proximity and region homogeneity in order to obtain good segmentation results.

Object recognition is a very important task in digital image analysis. Shape description models are extensively used to attain this goal. Shape description and representation schemes have been thoroughly studied in the past two decades by researchers working in computer vision and in computer graphics. Both areas have overlapping interests: computer vision concerns the creation of object models from object pictures, whereas computer graphics concern creating digital pictures from symbolic models. Several two-dimensional shape description schemes will be described in this book. Such schemes can be divided into two classes: external and internal representations. External representations employ the object boundaries and their features. Internal representations use region descriptions and features related to the region occupied by an object. Such important object features are related to its illumination texture. Texture description schemes are of great importance in object recognition applications.

Digital image processing and analysis have exhibited a tremendous growth in the past three decades. A multitude of algorithms have been presented in the literature in all the above-mentioned areas. The aim of this book is to provide an algorithmic description of some of the established techniques and algorithms, rather than to give an extensive survey of all proposed algorithms. The algorithms that will be described will form the backbone of an algorithmic package covering all important tasks related to digital image processing and analysis.

1.3 DIGITAL IMAGE FORMATION

An image is the optical representation of an object illuminated by a radiating source. Thus, the following elements are used in an image formation process: object, radiating source and image formation system. The mathematical model underlying the image formation depends on the radiation source (e.g. visible light, X-rays, ultrasound), on the physics of the radiation–object interaction and on the acquisition system used. For simplicity, we shall restrict our description to the case of the visible light reflected on an object, as shown in Figure 1.3.1. The reflected light $f(\xi, n)$ is the optical image which is the

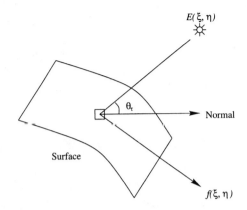

Fig. 1.3.1 Reflection of light on an object surface.

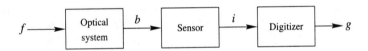

Fig. 1.3.2 Model of a digital image formation system.

input of the digital image formation system. Such a system usually consists of optical lenses, an optical sensor and an image digitizer. The model of such a formation system is shown in Figure 1.3.2.

The optical subsystem H can be modeled as a linear shift-invariant system having a two-dimensional impulse response $h(x, y)$. H is usually a low-pass system and suppresses the high-frequency content of its input image $f(\xi, n)$. Thus, its output image $b(x, y)$ is usually a blurred or unfocused version of the original image $f(\xi, n)$. Since both signals $f(\xi, n)$, $b(x, y)$ represent optical intensities, they must take non-negative values:

$$f(\xi, n) \geq 0 \tag{1.3.1}$$

$$b(x, y) \geq 0 \tag{1.3.2}$$

The input–output relation of the optical subsystem is described by a 2-d convolution:

$$b(x, y) = \int_{-\infty}^{\infty} \int_{-\infty}^{\infty} f(\xi, n) h(x - \xi, y - n) d\xi dn \tag{1.3.3}$$

The mathematical model of the sensing device depends on the photoelectronic sensor used. Several types of sensors exist, for example, standard vidicon tubes, charge injection devices (CIDs) and charge coupled devices (CCDs). In most cases, the relation between the input image $b(x, y)$ and the output

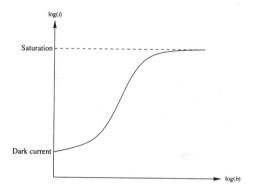

Fig. 1.3.3 Current–illumination curve of a photoelectronic sensor.

electric current $i(x,y)$ is highly nonlinear. The characteristic curve of the vidicon camera is shown in Figure 1.3.3. It is clear that the current–illumination relation is nonlinear. At the saturation region a further increase in the illumination does not affect the output current. A dark current exists, even in the absence of illumination. Even in the 'linear' part of the characteristic curve, the input–output relation is nonlinear:

$$\log(i(x,y)) = \gamma \log(b(x,y)) + c_1 \tag{1.3.4}$$

$$i(x,y) = c_2 [b(x,y)]^\gamma \tag{1.3.5}$$

Typical values of γ are 0.65 (for vidicon tubes) and $[0.95,\ldots,1]$ (for silicon vidicons).

The output $i(x,y)$ of the sensing devices is still a two-dimensional analog signal. It must be sampled and digitized before it can be processed by the computer. Sampling and digitization are performed by an A/D converter. It transforms the analog image $i(x,y)$ to a digital image $i(n_1,n_2)$, $n_1 = 1,\ldots,N$, $n_2 = 1,\ldots,M$:

$$i(n_1,n_2) = i(n_1 T_1, n_2 T_2) \tag{1.3.6}$$

The sampling (1.3.6) is performed on a rectangular grid having sampling intervals T_1, T_2. The image size is $N \times M$ pixels. Typical digital image sizes are 256×256 and 512×512 pixels. In the case of color images, sampling is performed on each channel (red, green, blue) independently, thus producing three digital images having equal sizes. The A/D converter performs quantization of the sampled image as well. If q is the quantization step, the quantized image is allowed to have illumination at the levels kq, $k = 0, 1, 2, \ldots$, as shown in Figure 1.3.4. If an image pixel is represented by b bits, the quantization step is given by:

$$q = 1/2^b \tag{1.3.7}$$

In most applications, the *grayscale* images are quantized at 256 levels and require 1 byte (8 bits) for the representation of each pixel. In certain cases,

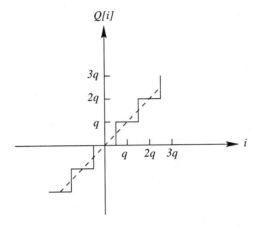

Fig. 1.3.4 Input-output curve of the quantizer.

binary images are produced having only two quantization levels: 0, 1. They are represented with 1 bit per pixel. A digital image quantized at 256, 64, 8 and 2 levels respectively is shown in Figure 1.3.5. Image quantization introduces an error term, $e(n_1, n_2)$, equal to:

$$e(n_1, n_2) = i(n_1, n_2) - Q[i(n_1, n_2)] \qquad (1.3.8)$$

where $Q[\cdot]$ is the quantization function. If P_i, P_e are the power of the image and error signals, respectively, the signal to noise ratio (SNR) for the quantizer is given by [OPP89]:

$$SNR = 10\log_{10}(P_i/P_e) = 10\log_{10} P_i + 10.79 + 6.02b \qquad (1.3.9)$$

Thus, one additional bit used in the pixel representation increases the SNR by 6 dB.

1.4 DIGITAL IMAGE REPRESENTATION

The digital image can be conveniently represented by an $N \times M$ matrix **i** of the form:

$$\mathbf{i} = \begin{bmatrix} i(1,1) & i(1,2) & \cdots & i(1,M) \\ i(2,1) & i(2,2) & \cdots & i(2,M) \\ \vdots & \vdots & & \vdots \\ i(N,1) & i(N,2) & \cdots & i(N,M) \end{bmatrix} \qquad (1.4.1)$$

The matrix elements (image pixels) are integers in the range [0,...,255] for 8 bit images. Therefore, they can be represented as characters in the C language. In certain cases, when floating point arithmetic is involved, image

8 DIGITAL IMAGE PROCESSING FUNDAMENTALS

pixels must be represented as floating point numbers in C. Character and floating point matrices of fixed size $N \times M$ can very easily be represented in C:

$$\begin{array}{ll} unsigned\ char & a[N][M]; \\ float & b[N][M]; \end{array} \qquad (1.4.2)$$

The matrices are zero-offset structures, that is, their elements have the form $a[i][j]$, $b[i][j]$, $0 \leq i \leq N-1$, $0 \leq j \leq M-1$. The storage scheme used for such arrays is of the form shown in Figure 1.4.1. This scheme corresponds to the spatial coordinates illustrated in Figure 1.4.2a. This coordinate system is equivalent to the terminal coordinate system usually employed in a computer graphics display [ANG90]. In such a system, the indices i, j of the image matrix $a[i][j]$ correspond to the axes y, x, respectively. This system is a rotated version of the coordinate system usually employed in geometry and graphics that is shown in Figure 1.4.2b.

Digital image representation by a fixed-size two-dimensional array has severe limitations. In most cases the image size needed is not known in advance.

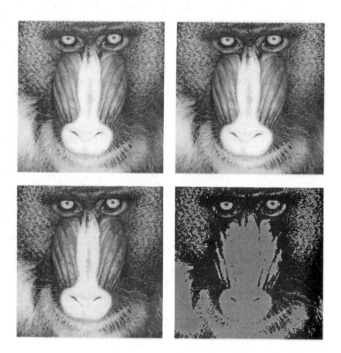

Fig. 1.3.5 A digital image quantized at (a) 256 levels; (b) 64 levels; (c) 8 levels; (d) 2 levels. *Note*: Throughout the text (a)=top left, (b)=top right, (c)=bottom left, (d)=bottom right.

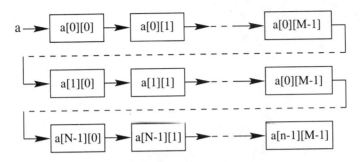

Fig. 1.4.1 Storage scheme for a two-dimensional array of fixed size.

Furthermore, the C language does not allow declarations of the form:

$$\begin{aligned}&void\ a\_routine(a, N, M)\\&float\ a[N][M];\\&\{Body\ of\ a\ routine\}\end{aligned} \quad (1.4.3)$$

Thus, the creation of routine libraries operating on variable image sizes is impossible. Fortunately, there exists an elegant solution to this problem [PRE88]. The C language has a direct correspondence between arrays and pointers. The element referenced by the expression a[i] is equivalent to the element *(a+i). Furthermore, arrays can be created dynamically by using the memory allocation procedure `malloc`:

$$\begin{aligned}&unsigned\ char*a;\\&a=(unsigned\ char*)\ malloc\ (N*size\ of\ (unsigned\ char));\end{aligned} \quad (1.4.4)$$

In this case, a one-dimensional array (vector) of unsigned characters is created dynamically. Its elements can be referred to by using a[i], $0 \leq i \leq$

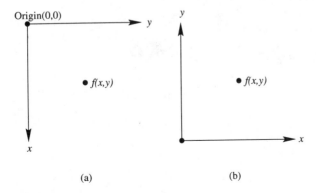

Fig. 1.4.2 (a) Spatial coordinates used in the digital image representation; (b) coordinates used in computer graphics.

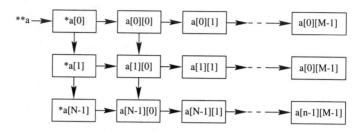

Fig. 1.4.3 Matrix representation as an array of pointers to matrix rows.

$N - 1$. A floating point vector can be constructed in a similar way:

$$float * a;$$
$$a = (float*)\; malloc\; (N * size\; of\; (float)); \quad (1.4.5)$$

The two-dimensional array can be considered as a one-dimensional array of vectors, as shown in Figure 1.4.3. The pointer **a points to the array of pointers *a[0],*a[1],...,*a[N-1] that point to the image rows. The entire matrix can be created dynamically as shown in Figure 1.4.4, for the case of unsigned char matrices.

```
unsigned char ** matuc2(i,j)
unsigned int i,j;
{
  /* Function to create a 2-d matrix of size ixj
     and of type unsigned char.
     i: number of rows
     j: number of columns */
  unsigned char ** m;   int ii;
  m=(unsigned char **)
       malloc(i*sizeof(unsigned char *));
  if(m==NULL) return(NULL);
  for(ii=0; ii<i; ii++)
     { m[ii]=(unsigned char **)
             malloc(j*sizeof(unsigned char *));
       if(m[ii]==NULL) { mfreeuc2(m,ii); return(NULL); }}
  return(m);
}
```

Fig. 1.4.4 Dynamic allocation of a two-dimensional **unsigned char** array.

The matrix elements are referenced by a[i][j], $0 \le i \le N - 1$, $0 \le j \le M - 1$. Thus, zero-offset representation results. The memory allocated by a two-dimensional array can be released, as shown in Figure 1.4.5. The rows are freed first. The column of the pointer array is freed afterwards.

```
void mfreeuc2(m,i)
unsigned char ** m;
unsigned int i;
{
/* Subroutine to free a 2-d array of type unsigned char.
   This array must have been created by either matuc2(),
   or build_image() or build_win_uc().
   m: array name
   i: number of rows                                           */
   int ii;

   for(ii=0; ii<i; ii++) free(m[ii]);
   free(m);
}
```

Fig. 1.4.5 Dynamic release of the memory allocated to an **unsigned char** matrix.

The dynamic memory allocation and release for floating point matrices is done along similar lines.

The subroutines described in this book manipulate digital image representations as just described. They perform digital image processing/analysis tasks and can be used to construct a digital image processing library. The subroutines described in this book are part of such a library called EIKONA. The subroutines of EIKONA communicate with their environment through the parameters in their definition and through the following global parameters:

NMAX: vertical image size
MMAX: horizontal image size
NWMAX: vertical window size
MWMAX: horizontal window size
NSIG: vector size

They are all of type int. It is recommended that the vector size NSIG be 256 or larger. EIKONA supports the following type definitions:

```
typedef unsigned char * * image;
typedef float * * matrix;
typedef int * * imatrix;
typedef float * vector;
```

The type **image** describes two-dimensional image buffers. It is a basic type that will be encountered in almost any subroutine to be described in this book. The type `imatrix` represents two-dimensional arrays of integers. It is used in certain cases, where the range of unsigned characters [0 ... 255] is not sufficient (e.g. in the Hough transform parameter matrix). The type `matrix` is used instead of the type `image` in those cases where floating point arithmetic is required (e.g. in the calculation of image transforms). The type `vector` describes one-dimensional arrays of floating point numbers. It will be used in the representation of digital image histograms, as will be described in Chapter 3.

Digital images can be stored on binary image files with or without a header. Two routines that can be used to load and store raw image data on a disk file are shown in Figures 1.4.6 and 1.4.7 respectively. A data buffer is used to store or load the image data row by row. No header is used and no image dimensions are stored. Thus, the image size must be known in advance in order to load the image data correctly. Several image file formats exist (e.g. TIFF, GIF, PCX, TGA) that have become almost industry standards. Their description is outside the scope of this book. The interested reader is referred to [LIN91], [RIM90] for the description of the corresponding file structures.

```
int fromdisk(imfile, a, ND1,MD1, N,M)
char imfile[];
image a;
int ND1,MD1;
int N,M;
/* Subroutine to read an image from a disk file.
   imfile: OS file name (string)
   a: image array of dimensions N,M
   ND1,MD1: starting point on the buffer a
   N,M: image dimensions in the file imfile */
{  FILE *fp;
   unsigned char * buc;
   int i,j;

   if((fp=fopen(imfile,"rb"))==NULL) return(-300);
   buc=(unsigned char *)
         malloc((unsigned int)M*sizeof(unsigned char));
   if(buc==NULL) return(-10);
   for(i=0; i<N; i++)
      { fread(buc,sizeof(unsigned char),M,fp);
        for(j=0; j<M; j++) a[i+ND1][j+MD1]=buc[j];
      }
   free(buc);   fclose(fp);
   return(0);
}
```

Fig. 1.4.6 Procedure to load an image from a disk file.

```
int todisk(a, imfile, N1,M1,N2,M2)
image a;
char imfile[];
int N1,M1,N2,M2;
/* Subroutine to store an image buffer in a disk file.
   a: buffer
   imfile: OS file name (string)
   N1,M1: start coordinates
   N2,M2: end coordinates             */
{ int k,l;
  FILE *fp;
  unsigned char * buc;
  if((fp=fopen(imfile,"wb"))==NULL) return(-300);
  buc=(unsigned char *)
          malloc((unsigned int)(M2-M1)*sizeof(unsigned char));
  if(buc==NULL) return(-10);
  for(k=N1; k<N2; k++)
     { for(l=M1; l<M2; l++) buc[l-M1]=a[k][l];
       fwrite(buc,sizeof(unsigned char),M2-M1,fp);}
  free(buc);   fclose(fp);
  return(0);
}
```

Fig. 1.4.7 Procedure to store an image on a disk file.

1.5 ELEMENTARY DIGITAL IMAGE PROCESSING OPERATIONS

Elementary digital image processing includes basic arithmetic operations, for example, image addition, subtraction:

$$c[i][j] = a[i][j] \pm b[i][j] \qquad (1.5.1)$$

and multiplication of an image by a constant:

$$b[i][j] = c \cdot a[i][j] \qquad (1.5.2)$$

A procedure for the addition of two images is shown in Figure 1.5.1. A similar subroutine can be written for image difference. Both input matrices a,b and the output matrix c are of the type image (unsigned char). Therefore, special attention must be paid so that the output image elements are in the range [0...255]. If an overflow occurs, the corresponding pixel is black on a white background. In the case of an underflow, the corresponding pixel appears white on a black background. Both overflow and underflow create visually unpleasant effects.

```
int add(a,b,c, N1,M1,N2,M2)
image a,b,c;
int N1,M1,N2,M2;
/* Subroutine to add two images.
    a,b: source buffers
    c:   destination buffer
    N1,M1: start coordinates
    N2,M2: end   coordinates    */

{ int i,j,ime;
  for(i=N1; i<N2; i++)
  for(j=M1; j<M2; j++)
    { ime=a[i][j]+b[i][j];
      if(ime>255) c[i][j]=255; else c[i][j]=ime;}
  return(0);
}
```

Fig. 1.5.1 Subroutine for image addition.

Point nonlinear transformations of the form:

$$b[i][j] = h(a[i][j]) \qquad (1.5.3)$$

are elementary operations used in a variety of applications, e.g. gamma correction [FOL90], image sharpening [JAI89]. If the transformation function $h(x)$ is implemented by using a table $h[0\ldots255]$, transform (1.5.3) can be reduced to a look-up table operation. Special point-wise operations are *clipping*:

$$b[i][j] = \begin{cases} cmax & \text{if } a[i][j] > cmax \\ a[i][j] & \text{if } cmin \leq a[i][j] \leq cmax \\ cmin & \text{if } a[i][j] < cmin \end{cases} \qquad (1.5.4)$$

and *thresholding*:

$$b[i][j] = \begin{cases} a_1 & \text{if } a[i][j] < T \\ a_2 & \text{if } a[i][j] \geq T \end{cases} \qquad (1.5.5)$$

If $a_1 = 0$ and $a_2 = 1$, the thresholding operation transforms a grayscale image to a binary one. Both clipping and thresholding are used in elementary region segmentation operations.

Digital images are of type image (unsigned char). Therefore, elementary binary operations (and, or, xor) can be easily performed:

$$c[i][j] = a[i][j] \& b[i][j] \qquad (1.5.6)$$

$$c[i][j] = a[i][j] | b[i][j] \qquad (1.5.7)$$

$$c[i][j] = a[i][j] \wedge b[i][j] \qquad (1.5.8)$$

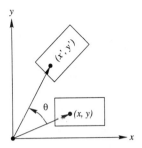

Fig. 1.5.2 Image rotation.

Binary operations have important applications in binary and morphological image processing [SER82], [PIT90]. The routines for the binary operations are very similar to the subroutine shown in Figure 1.5.1. In the case of binary images $a[i][j] = 0, 1$, the **not** operation can be described as follows:

$$b[i][j] =!a[i][j] \tag{1.5.9}$$

The **not** operation (1.5.9) must not be confused with the negative operation:

$$b[i][j] = 255 - a[i][j] \tag{1.5.10}$$

that produces the *negative* of an image. Operation (1.5.10) essentially produces 1's complement of an image byte: $b[i][j] =\sim a[i][j]$.

Another set of elementary digital image processing operations is related to geometric transforms. Image *translation* of the form:

$$b[i][j] = a[i+k][j+l] \tag{1.5.11}$$

can be easily implemented by a copy operation from image buffer a to buffer b. Image *rotation* is depicted in Figure 1.5.2. If the image point $a(x,y)$ is rotated by θ degrees, its new coordinates (x', y') are given by:

$$\begin{bmatrix} x' \\ y' \end{bmatrix} = \begin{bmatrix} \cos\theta & -\sin\theta \\ \sin\theta & \cos\theta \end{bmatrix} \cdot \begin{bmatrix} x \\ y \end{bmatrix} \tag{1.5.12}$$

Thus, the rotated version $b[x'][y']$ of the image $a[x][y]$ is given by:

$$b[x'][y'] = b[x\cos\theta - y\sin\theta][x\sin\theta + y\cos\theta] = a[x][y] \tag{1.5.13}$$

The subroutine which implements image rotation is shown in Figure 1.5.3. It performs rotation about an arbitrary center.

```
int rotim(a,b,theta,ic,jc,ND1,MD1,N1,M1,N2,M2)
image a,b;
float theta;
```

```
int ic,jc;
int ND1, MD1;
int N1,M1,N2,M2;
/* Subroutine rotim() rotates the image around
   (ic,jc) for rotation angle theta.
  a: input buffer
  b: output buffer
  theta: rotation angle (degrees)
  ic,jc: rotation center coordinates
  ND1,MD1: destination coordinates
  N1,M1: start coordinates
  N2,M2: end   coordinates          */

{
int i,j,ii,jj,k,l,NS; double rad,costh,sinth;
float N22,M22; double PI=4.0*atan((double)1.0);

/* Transform degrees to rad */
  rad=(double)theta*PI/180.0;
  costh=cos(rad); sinth=sin(rad);
/* Find enclosing rectangle in the rotated image plane */
  N22=(float)(N1-N2)/2.0; M22=(float)(M1-M2)/2.0;
  NS=(int)sqrt((double)(N22*N22+M22*M22))+1;

/* Start rotation loop     */
  for(i=-NS; i<NS; i++)
    for(j=-NS; j<NS; j++)
      {
        ii=(int)((float)(j)*sinth+(float)(i)*costh)+ic;
        jj=(int)((float)(j)*costh-(float)(i)*sinth)+jc;
        k=i+ND1; l=j+MD1;
        if ( ii>=N1 && ii<N2 && jj>=M1 && jj<M2 &&
             k>=0 && k<NMAX && l>=0 && l<NMAX)
                              b[k][l]=a[ii][jj];
      }
  return(0);
}
```

Fig. 1.5.3 Algorithm for image rotation.

This subroutine maps pixels of the destination buffer to the source image buffer. The opposite is possible and it is easier to implement. However, it must be noted that digital image coordinates (k, l) are integers, whereas the arithmetic used in (1.5.13) is floating point arithmetic. Therefore, truncation or rounding operations are needed. These operations may force two image pixels $a[i][j], a[i'][j']$ lying in adjacent positions to be mapped on the same pixel $b[k][l]$. This fact creates pattern spikes in the rotated image, if the original image $a[i][j]$ is scanned and rotated on a pixel-by-pixel basis. Scanning the destination buffer and mapping its pixels to the source buffer does not create such problems.

Scaling is another basic geometrical transformation. It is described by the equation:

$$\begin{bmatrix} x' \\ y' \end{bmatrix} = \begin{bmatrix} \alpha & 0 \\ 0 & \beta \end{bmatrix} \cdot \begin{bmatrix} x \\ y \end{bmatrix} \qquad (1.5.14)$$

Image scaling (or *zooming*) is given by:

$$b[k][l] = b[\alpha \cdot i][\beta \cdot j] = a[i][j] \qquad (1.5.15)$$

Image scaling requires image interpolation [JAI89], when $i/\alpha, j/\beta$ are not integers. The related algorithms are described in a subsequent chapter. A negative scale factor causes image reflection. Such a reflection operation is shown in Figure 1.5.4 for $a = 1, b = -1$. The reflected image $b[k][l]$ is given by:

$$b[k][l] = a[N - k][l] \qquad (1.5.16)$$

for an image of dimension $N \times M$. Similarly, the reflections in the y axis or

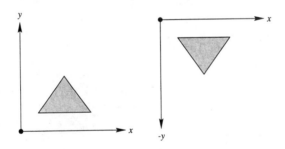

Fig. 1.5.4 Geometrical reflection in the x axis.

in the origin are described by:

$$b[k][l] = a[k][M - l] \qquad (1.5.17)$$

$$b[k][l] = a[N - k][M - l] \qquad (1.5.18)$$

Reflection in the line $i = j$ is essentially image matrix transposition:

$$b[k][l] = a[l][k] \qquad (1.5.19)$$

Finally, another reflection is given by the transformation:

$$b[k][l] = a[M - l][N - k] \qquad (1.5.20)$$

18 DIGITAL IMAGE PROCESSING FUNDAMENTALS

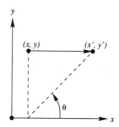

Fig. 1.5.5 Shear along the horizontal axis.

The last geometrical transform to be described here is shear, shown in Figure 1.5.5. The shear equations are given by [ANG90]:

$$\begin{bmatrix} x' \\ y' \end{bmatrix} = \begin{bmatrix} 1 & \cot\theta \\ 0 & 1 \end{bmatrix} \cdot \begin{bmatrix} x \\ y \end{bmatrix} \tag{1.5.21}$$

A similar equation can be written for a y-axis shear. The amount of shear is determined by the angle θ. The shear transform of an image is given by:

$$b[x'][y'] = b[x + y\cot\theta][y] = a[x][y] \tag{1.5.22}$$

The coordinate $x + y\cot\theta$ must be truncated or rounded to an integer.

1.6 DIGITAL IMAGE DISPLAY

Digital image display is straightforward if a color graphics card is available that is capable of producing 256 shades for each primary color RGB (Red, Green and Blue). If a procedure of the form:

```
setpixel (r,g,b,i,j)
unsigned char r; /* red channel value;   */
unsigned char g; /* green channel value; */
unsigned char b; /* blue channel value;  */
int i,j; /* display coordinates */
```

is available in the graphics package, which usually accompanies the graphics card, it can be used to display the digital image in the way described in Figure 1.6.1.

```
int imdisp(r,g,b, NS,MS, N1,M1,N2,M2)
image r,g,b;
int cn;
int NS,MS;
```

```
int N1,M1,N2,M2;

/* r,g,b: buffer
   NS,MS: screen starting point (integers)
   N1,M1: buffer display starting point
   N2,M2: buffer display end point                    */

{ int i,j;

   for(i=N1; i<N2; i++)
     for(j=M1; j<M2; j++)
       setpixel(r[i][j],g[i][j],b[i][j],j-M1+MS,i-N1+NS);
   return(0);
}
```

Fig. 1.6.1 Display routine for a digital image.

It is clearly seen that the pixel (i,j) is displayed at the position (j,i) due to the difference in the coordinate systems used in image representation and graphics display, shown in Figure 1.4.2. If the display of a black and white image is desired, the following change must be made in the previous program:

```
setpixel(a[i][j],a[i][j], a[i][j],j,i);
```

A slightly modified procedure, shown in Figure 1.6.2, can be used for the display of color images under MICROSOFT WINDOWS.

```
int win3imdisp(hWnd,r,g,b, NS,MS, N1,M1,N2,M2)
HWND hWnd;
image r;
image g;
image b;
int NS,MS;
int N1,M1,N2,M2;
/* Subroutine to display an image under Microsoft Windows.
    r,g,b: image buffers of the RGB components
    NS,MS: screen starting point
    N1,M1: buffer display starting point
    N2,M2: buffer display end point*/
{
register i,j;
float xval,yval;
RECT   rectClient;
HDC hDC;
```

```
GetClientRect (hWnd, &rectClient);
hDC = GetDC(hWnd);

for(i=N1;i<N2;i++)
 for (j=M1;j<M2;j++)
  SetPixel(hDC,j+MS,i+NS,
           RGB((int)r[i][j],(int)g[i][j],(int)b[i][j]));
ReleaseDC(hWnd, hDC);
return(0);
}
```

Fig. 1.6.2 Display of a digital color image under MICROSOFT WINDOWS.

This routine is relatively slow, because it performs a pixel-by-pixel access to video RAM. Much faster display routines can be written by sending entire image blocks or image buffers to the video RAM by using only one function call. Such a display routine is the `SetDIBitsToDevice` function, which is available in the native Win32 API (Windows 9.x and Windows NT) and in the Win32 extensions of Windows 3.x (Win32s library). This function sets the pixels in the specified rectangle on the device that is associated with the destination device context using color data from a `device-independent-bitmap` (DIB). The use of the DIB structure permits an image buffer to be mapped to a bitmap that can be displayed on screen by a single call of this function.

1.7 FUNDAMENTALS OF COLOR IMAGE PROCESSING

The visual perception of color images is much richer than that of achromatic (black and white) images. Color images or color image sequences (e.g. video) are the preferred visual communication medium. Thus digital color image acquisition, processing and display are very important for a variety of applications ranging from the printing industry to high-definition TV (HDTV). Color representation is based on the theory of T. Young (1802) [WYZ67] which states that any color can be produced by mixing three primary colors $\mathbf{C}_1, \mathbf{C}_2, \mathbf{C}_3$ at appropriate percentages:

$$C = a\mathbf{C}_1 + b\mathbf{C}_2 + c\mathbf{C}_3 \qquad (1.7.1)$$

This theory is consistent with the fact that the human eye has three different types of cones in the retina. They have peaks at the yellow–green, green and blue regions of the visible light spectrum, respectively. According to (1.7.1), each color image pixel can be represented as a vector $[a, b, c]^T$ in the three-dimensional space (C_1, C_2, C_3). In the following, we shall drop the coefficients a, b, c and we shall use the notations C_1, C_2, C_3 for color representation. The chromaticities of a color are defined by the ratios:

$$c_i = \frac{C_i}{C_1 + C_2 + C_3} \qquad i = 1, 2, 3 \qquad (1.7.2)$$

It is clear that only two chromaticity coordinates, for example, c_1, c_2, are independent because:

$$c_1 + c_2 + c_3 = 1 \qquad (1.7.3)$$

The chromaticity coordinates c_1, c_2 project the three-dimensional space (C_1, C_2, C_3) to a two-dimensional plane (c_1, c_2). The entire color space is represented by the triplet (c_1, c_2, Y), where Y is given by:

$$Y = C_1 + C_2 + C_3 \qquad (1.7.4)$$

and represents a chrominance plane.

Several color coordinate systems have been proposed in the past [WYZ67]. CIE (Commission Internationale de l'Eclairage) has proposed the spectral primary system RGB corresponding to monochromatic primary sources R_{CIE} (red 700 nm), G_{CIE} (green 546.1 nm) and B_{CIE} (blue 435.8 nm). Reference white color has $R_{CIE} = G_{CIE} = B_{CIE} = 1$. The CIE spectral primary system RGB cannot yield all possible reproducible colors. Therefore, CIE has proposed the XYZ primary system having hypothetical (physically unrealizable) coordinates X, Y, Z. The XYZ primaries are linearly related to the RGB primary system:

$$\begin{bmatrix} X \\ Y \\ Z \end{bmatrix} = \begin{bmatrix} 0.490 & 0.310 & 0.200 \\ 0.177 & 0.813 & 0.011 \\ 0.000 & 0.010 & 0.990 \end{bmatrix} \cdot \begin{bmatrix} R_{CIE} \\ G_{CIE} \\ B_{CIE} \end{bmatrix} \qquad (1.7.5)$$

The reference white is represented by $X = Y = Z = 1$. The chromaticity coordinates

$$x = \frac{X}{X + Y + Z} \qquad (1.7.6)$$

$$y = \frac{Y}{X + Y + Z} \qquad (1.7.7)$$

can be used to produce the chromaticity diagram shown in Figure 1.7.1. The ellipses shown in this diagram correspond to indistinguishable colors. The colors within each ellipse cannot be distinguished by the human eye. It is clearly seen that their orientation and size varies. Therefore, color differences cannot be defined in a uniform way on the (x, y) plane. A new uniform chromaticity scale (UCS) system has been proposed in order to alleviate the above-mentioned problem. Its coordinates u, v, Y are given by:

$$u = \frac{4X}{X + 15Y + 3Z} = \frac{4x}{-2x + 12y + 3} \qquad (1.7.8)$$

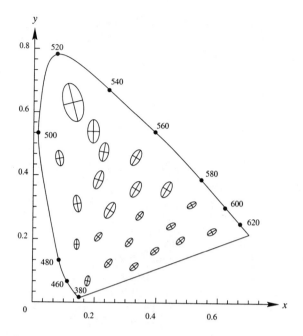

Fig. 1.7.1 Chromaticity diagram of the XYZ system.

$$v = \frac{6Y}{X + 15Y + 3Z} = \frac{6y}{-2x + 12y + 3} \qquad (1.7.9)$$

The component Y of the UCS system is the same as luminance Y of the XYZ system. The tristimulus coordinates U, V, W of this system, corresponding to the ratios $u, v, w = 1 - u - v$, are given by:

$$U = \frac{2X}{3} \qquad (1.7.10)$$
$$V = Y \qquad (1.7.11)$$
$$W = \frac{-X + 3Y + Z}{2} \qquad (1.7.12)$$

A modified UCS system, called the $U^*V^*W^*$ system, results from a shift of UCS origin to the reference white u_0, v_0. The resulting transform is described by the following equations:

$$W^* = 25(100Y)^{1/3} - 17 \quad 0.01 \leq Y \leq 1 \qquad (1.7.13)$$
$$U^* = 13W^*(u - u_0) \qquad (1.7.14)$$
$$V^* = 13W^*(v - v_0) \qquad (1.7.15)$$

This system can be used to measure color differences, because they are expressed by the Euclidean distance in the $U^*V^*W^*$ space. The difference of two colors C_a, C_b, denoted by ΔC, has the following form:

$$\Delta C^2 = (U_a^* - U_b^*)^2 + (V_a^* - V_b^*)^2 + (W_a^* - W_b^*)^2 = (\Delta U^*)^2 + (\Delta V^*)^2 + (\Delta W^*)^2 \quad (1.7.16)$$

The $L^*a^*b^*$ system:

$$L^* = 25(100Y/Y_0)^{1/3} - 16 \quad 1 \leq 100Y \leq 100 \quad (1.7.17)$$

$$a^* = 500[(X/X_0)^{1/3} - (Y/Y_0)^{1/3}] \quad (1.7.18)$$

$$b^* = 200[(Y/Y_0)^{1/3} - (Z/Z_0)^{1/3}] \quad (1.7.19)$$

is also used to measure color differences. The coordinates (X_0, Y_0, Z_0) in (1.7.17–1.7.19) correspond to the reference white light. L^* denotes lightness, whereas a^*, b^* describe the red–green and yellow–blue light contents respectively. The color difference ΔC in the L^*, a^*, b^* system is given by:

$$\Delta C^2 = (\Delta L^*)^2 + (\Delta a^*)^2 + (\Delta b^*)^2 \quad (1.7.20)$$

Finally, the $L^*u^*v^*$ system is used to measure color differences as well. The corresponding transform is given by:

$$L^* = 25(100Y/Y_0)^{1/3} - 16 \quad (1.7.21)$$
$$u^* = 13L^*(u - u_0) \quad (1.7.22)$$
$$v^* = 13L^*(1.5v - v_0) \quad (1.7.23)$$

The coordinates Y_0, u_0, v_0 correspond to the reference white. The color difference formula for the $L^*U^*V^*$ plane is given by:

$$\Delta C^2 = (\Delta L^*)^2 + (\Delta u^*)^2 + (\Delta v^*)^2 \quad (1.7.24)$$

All the above-mentioned transforms can be easily computed, as can be seen in the procedures of Figure 1.7.2. Their computation is slow because it requires relatively complex floating point computations. Furthermore, the transform results are floating point numbers. Thus, the storage requirements for the transformed color images are four times higher than for the original color image, because a floating point number occupies four bytes.

```
int cieRGB_to_XYZ(r,g,b,x,y,z,N1,M1,N2,M2)
image r,g,b,x,y,z;
int N1,M1,N2,M2;
/* Subroutine to perform cieRGB to XYZ transform
   r,g,b: input image buffers
   x,y,z: transform buffers
   N1,M1: upper left corner coordinates
   N2,M2: lower right corner coordinates */
{
```

```
int i,j; double R,G,B; double X,Y,Z;
for(i=N1;i<N2;i++)
 for(j=M1;j<M2;j++)
  { R=(double)r[i][j]; G=(double)g[i][j]; B=(double)b[i][j];
    X=0.490*R+0.310*G+0.200*B; Y=0.177*R+0.813*G+0.011*B;
    Z=0.010*G+0.990*B;
    if(X>255.0) x[i][j]=255; else x[i][j]=(unsigned  char)X;
    if(Y>255.0) y[i][j]=255; else y[i][j]=(unsigned char)Y;
    if(Z>255.0) z[i][j]=255; else z[i][j]=(unsigned char)Z;
  }
return(0);
}


XYZ_to_cieRGB(x,y,z,r,g,b,N1,M1,N2,M2)
image x,y,z,r,g,b;
int N1,M1;
int N2,M2;
/* Subroutine to perform XYZ to cieRGB transform
   x,y,z: transform buffers
   r,g,b: input image buffers
   N1,M1: upper left corner coordinates
   N2,M2: lower right corner coordinates */
{
int i,j; double R,G,B; double X,Y,Z;
for(i=N1;i<N2;i++)
 for(j=M1;j<M2;j++)
  { X=(double)x[i][j]; Y=(double)y[i][j]; Z=(double)z[i][j];
    R=2.365*X-0.896*Y-0.468*Z; G=-0.515*X+1.425*Y+0.088*Z;
    B=0.005*X-0.014*Y+1.009*Z;
    if(R>255.0) r[i][j]=255; else if (R<0.0) r[i][j]=0;
     else r[i][j]=(unsigned char)R;
    if(G>255.0) g[i][j]=255; else if (G<0.0) g[i][j]=0;
     else g[i][j]=(unsigned char)G;
    if(B>255.0) b[i][j]=255; else if (B<0.0) B=0.0;
     else b[i][j]=(unsigned char)B;
  }
return(0);
}


XYZ_to_cieuvY(x,y,z,u,v,Y,N1,M1,N2,M2)
image x,y,z;
matrix u,v,Y;
int N1,M1;
int N2,M2;
/* Subroutine to perform XYZ to CIE uvY transform
```

```
   x,y,z: XYZ transform buffers
   u,v,Y: uvY transform buffers
   N1,M1: upper left corner coordinates
   N2,M2: lower right corner coordinates */
{
int i,j; double X,Z,w;
for(i=N1;i<N2;i++)
 for(j=M1;j<M2;j++)
   { X=((double)x[i][j])/255.0; Y[i][j]=((double)y[i][j])/255.0;
     Z=((double)z[i][j])/255.0;
     w = X+15.0*Y[i][j]+3.0*Z; u[i][j]=(4.0*X/w);
     v[i][j] = (6.0*Y[i][j]/w);
    }
return(0);
}


cieuvY_to_XYZ(u,v,Yi,x,y,z,N1,M1,N2,M2)
matrix u,v,Yi;
image x,y,z;
int N1,M1;
int N2,M2;
/* Subroutine to perform CIE uvY to XYZ transform
   u,v,Yi: uvY transform buffers
   x,y,z: XYZ transform buffers
   N1,M1: upper left corner coordinates
   N2,M2: lower right corner coordinates */

{
int i,j; int X,Y,Z;
for(i=N1;i<N2;i++)
 for(j=M1;j<M2;j++)
   { Y = (int)(Yi[i][j]*255.0);
     Yi[i][j] /= v[i][j];
     X=(int)(((1.5*u[i][j])*Yi[i][j])*255.0);
     Z=(int)(((-0.5*u[i][j]-5.0*v[i][j]+2.0)*Yi[i][j])*255.0);
     if(X>255)X=255;if(Y>255)Y=255;if(Z>255)Z=255;
     if(X<0)X=0;if(Y<0)Y=0;if(Z<0)Z=0;
     x[i][j]=(unsigned char)X;y[i][j]=(unsigned char)Y;
     z[i][j]=(unsigned char)Z;
   }
return(0);
}

XYZ_to_cieUsVsWs(x,y,z,u,v,w,N1,M1,N2,M2)
image x,y,z;
```

```c
matrix u,v,w;
int N1,M1;
int N2,M2;
/* Subroutine to perform XYZ to CIE U* V* W* transform
   x,y,z: XYZ transform buffers
   u,v,w: U* V* W* transform buffers
   N1,M1: upper left corner coordinates
   N2,M2: lower right corner coordinates */
{
int i,j; double X,Y,U,V,Z,W,Ws;
double uo_reference=0.210; double vo_reference=0.316;
for(i=N1;i<N2;i++)
 for(j=M1;j<M2;j++)
   { X=((double)x[i][j])/255.0;
     Y=((double)y[i][j])/255.0;
     Z=((double)z[i][j])/255.0;
     if(Y<0.01)Y=0.01;
     W=X+15.0*Y+3.0*Z;
     Ws=25.0*pow((double)(100.0*Y),(double)(0.333333))-17.0;
     U=V=13.0*Ws;
     U*=((4.0*X/W)-uo_reference);
     V*=((6.0*Y/W)-vo_reference);
     u[i][j]=U;v[i][j]=V;w[i][j]=Ws;
   }
return(0);
}

cieUsVsWs_to_XYZ(u,v,w,x,y,z,N1,M1,N2,M2)
matrix u,v,w;
image x,y,z;
int N1,M1;
int N2,M2;
/* Subroutine to perform CIE U* V* W* to XYZ transform
   u,v,w: U* V* W* transform buffers
   x,y,z: XYZ transform buffers
   N1,M1: upper left corner coordinates
   N2,M2: lower right corner coordinates */
{
int i,j; int X,Y,Z; double w1,w2,U,V,Yi;
double uo_reference=0.210; double vo_reference=0.316;
for(i=N1;i<N2;i++)
 for(j=M1;j<M2;j++)
   { w1=((w[i][j]+17.0)/25.0);
     w2=13.0*w[i][j];
     U=u[i][j]/w2+uo_reference;
```

```
        V=v[i][j]/w2+vo_reference;
        Yi=0.01*w1*w1*w1;
        Y = (int)(Yi*255.0);
        Yi /= V;
        X=(int)(((1.5*U)*Yi)*255.0);
        Z=(int)(((-0.5*U-5.0*V+2.0)*Yi)*255.0);
        if(X>255)X=255;if(Y>255)Y=255;if(Z>255)Z=255;
        if(X<0)X=0;if(Y<0)Y=0;if(Z<0)Z=0;
        x[i][j]=(unsigned char)X;y[i][j]=(unsigned char)Y;
        z[i][j]=(unsigned char)Z;
    }
return(0);
}
```

Fig. 1.7.2 Color image transform algorithms.

Besides the above-mentioned color coordinate systems, color models have also been proposed for convenient image display on specific hardware platforms. Thus, an RGB color system has been proposed by the National Television Systems Committee (NTSC) for color image display on CRT monitors. The Cyan Magenta Yellow (CMY) model is extensively used in printing color images. Finally the NTSC transmission system (YIQ) is used for color image transmission that is compatible with monochrome image transmission. The NTSC RGB system is a three-dimensional color model as shown in Figure 1.7.3. It is different from the CIE RGB spectral primary system, although

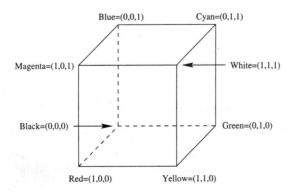

Fig. 1.7.3 RGB color cube.

they are linearly related to each other:

$$\begin{bmatrix} R_{CIE} \\ G_{CIE} \\ B_{CIE} \end{bmatrix} = \begin{bmatrix} 1.167 & -0.146 & -0.151 \\ 0.114 & 0.753 & 0.159 \\ -0.001 & 0.059 & 1.128 \end{bmatrix} \begin{bmatrix} R \\ G \\ B \end{bmatrix} \quad (1.7.25)$$

Thus, the XYZ system is also linearly related to the NTSC RGB system, according to equations (1.7.5) and (1.7.25):

$$\begin{bmatrix} X \\ Y \\ Z \end{bmatrix} = \begin{bmatrix} 0.607 & 0.174 & 0.201 \\ 0.299 & 0.587 & 0.114 \\ 0.000 & 0.066 & 1.117 \end{bmatrix} \begin{bmatrix} R \\ G \\ B \end{bmatrix} \quad (1.7.26)$$

As can be seen in Figure 1.7.3, cyan, magenta and yellow are complementary colors to red, green and blue respectively. They are called subtractive primaries, whereas red, green and blue are called additive primaries. In the CMY system, the colors are specified by what is subtracted from the white light. The CMY coordinates are easily produced from the RGB system:

$$C = 1 - R \quad (1.7.27)$$

$$M = 1 - G \quad (1.7.28)$$

$$Y = 1 - B \quad (1.7.29)$$

and vice versa. The CMY system can be used for color image printing but because of ink imperfections, this system cannot reproduce black correctly. Therefore the $CMYK$ system is used instead:

$$K = \min(C, M, Y) \quad (1.7.30)$$

$$C = C - K \quad (1.7.31)$$

$$M = M - K \quad (1.7.32)$$

$$Y = Y - K \quad (1.7.33)$$

The component K represents black as a fourth color.

The YIQ model is used in NTSC TV broadcasting. Its major advantage is that it guarantees downward compatibility with monochrome TV. The NTSC RGB to YIQ conversion is given by:

$$\begin{bmatrix} Y \\ I \\ Q \end{bmatrix} = \begin{bmatrix} 0.299 & 0.587 & 0.114 \\ 0.596 & -0.274 & -0.322 \\ 0.211 & -0.523 & 0.312 \end{bmatrix} \begin{bmatrix} R \\ G \\ B \end{bmatrix} \quad (1.7.34)$$

Luminance is represented by the Y component. The components I, Q encode image chrominance. All the above-mentioned transforms are easily programmed, as shown in Figure 1.7.4.

```
int ntscRGB_to_XYZ(r,g,b,x,y,z,N1,M1,N2,M2)
image r,g,b,x,y,z;
int N1,M1,N2,M2;
/* Subroutine to perform NTSC RGB to XYZ transform
   r,g,b: input image buffers
   x,y,z: transform buffers
   N1,M1: upper left corner coordinates
   N2,M2: lower right corner coordinates */
{
int i,j; double R,G,B; double X,Y,Z;
for(i=N1;i<N2;i++)
 for(j=M1;j<M2;j++)
  { R=(double)r[i][j]; G=(double)g[i][j]; B=(double)b[i][j];
    X=0.607*R+0.174*G+0.201*B; Y=0.299*R+0.587*G+0.114*B;
    Z=0.066*G+1.117*B;
    if(X>255.0) x[i][j]=255; else x[i][j]=(unsigned char)X;
    if(Y>255.0) y[i][j]=255; else y[i][j]=(unsigned char)Y;
    if(Z>255.0) z[i][j]=255; else z[i][j]=(unsigned char)Z;
  }
return(0);
}

XYZ_to_ntscRGB(x,y,z,r,g,b,N1,M1,N2,M2)
image x,y,z,r,g,b;
int N1,M1;
int N2,M2;
/* Subroutine to perform XYZ to NTSC RGB transform
   x,y,z: transform buffers
   r,g,b: input image buffers
   N1,M1: upper left corner coordinates
   N2,M2: lower right corner coordinates */
{
int i,j; double R,G,B; double X,Y,Z;
for(i=N1;i<N2;i++)
 for(j=M1;j<M2;j++)
  { X=(double)x[i][j]; Y=(double)y[i][j]; Z=(double)z[i][j];
    R=1.910364*X-0.533748*Y-0.289289*Z;
    G=-0.984377*X+1.998384*Y-0.026818*Z;
    B=0.058164*X-0.118078*Y+0.896840*Z;
    if(R>255.0) r[i][j]=255; else if (R<0.0) r[i][j]=0;
      else r[i][j]=(unsigned char)R;
    if(G>255.0) g[i][j]=255; else if (G<0.0) g[i][j]=0;
      else g[i][j]=(unsigned char)G;
    if(B>255.0) b[i][j]=255; else if (B<0.0) B=0.0;
      else b[i][j]=(unsigned char)B;
```

```
        }
return(0);
}


RGB_to_CMYK(r,g,b,c,m,y,k,N1,M1,N2,M2)
image r,g,b,c,m,y,k;
int N1,M1;
int N2,M2;
/* Subroutine to perform RGB to CMYK transform
   r,g,b: input image buffers
   c,m,y,k: transform buffers
   N1,M1: upper left corner coordinates
   N2,M2: lower right corner coordinates */
{
int i,j;
for(i=N1;i<N2;i++)
 for(j=M1;j<M2;j++)
  { c[i][j]=(unsigned char)255-r[i][j];
    m[i][j]=(unsigned char)255-g[i][j];
    y[i][j]=(unsigned char)255-b[i][j];
    k[i][j]=c[i][j];
    if (k[i][j]>m[i][j]) k[i][j]=m[i][j];
    if (k[i][j]>y[i][j]) k[i][j]=y[i][j];
    c[i][j]=c[i][j]-k[i][j];
    m[i][j]=m[i][j]-k[i][j];
    y[i][j]=y[i][j]-k[i][j];
  }
return(0);
}


CMYK_to_RGB(c,m,y,k,r,g,b,N1,M1,N2,M2)
image r,g,b,c,m,y,k;
int N1,M1;
int N2,M2;
/* Subroutine to perform CMYK to RGB transform
   c,m,y,k: transform buffers
   r,g,b: input image buffers
   N1,M1: upper left corner coordinates
   N2,M2: lower right corner coordinates */
{
int i,j;
for(i=N1;i<N2;i++)
 for(j=M1;j<M2;j++)
  { c[i][j]=c[i][j]+k[i][j]; m[i][j]=m[i][j]+k[i][j];
```

```
        y[i][j]=y[i][j]+k[i][j];
        r[i][j]=(unsigned char)255-c[i][j];
        g[i][j]=(unsigned char)255-m[i][j];
        b[i][j]=(unsigned char)255-y[i][j];
    }
return(0);
}

ntscRGB_to_YIQ(r,g,b,y,ii,q,N1,M1,N2,M2)
image r,g,b,y,ii,q;
int N1,M1;
int N2,M2;
/* Subroutine to perform NTSC RGB to YIQ transform
   r,g,b: input image buffers
   y,ii,q: transform buffers
   N1,M1: upper left corner coordinates
   N2,M2: lower right corner coordinates */
{
int i,j; double R,G,B; double Y,II,Q;
for(i=N1;i<N2;i++)
 for(j=M1;j<M2;j++)
  { R=(double)r[i][j]; G=(double)g[i][j]; B=(double)b[i][j];
    Y=0.299*R+0.587*G+0.114*B; II=0.596*R-0.274*G-0.322*B;
    Q=0.211*R-0.523*G+0.312*B;
    if(Y>255.0) y[i][j]=255; else if (Y<0.0) y[i][j]=0;
      else y[i][j]=(unsigned char)Y;
    if(II>255.0) ii[i][j]=255; else if (II<0.0) ii[i][j]=0;
      else ii[i][j]=(unsigned char)II;
    if(Q>255.0) q[i][j]=255; else if (Q<0.0) q[i][j]=0;
      else q[i][j]=(unsigned char)Q;
  }
return(0);
}
```

Fig. 1.7.4 Transformations of the NTSC RGB color space.

All the previous color models described are derived either from hardware considerations (RGB, $CMYK$, YIQ) or from colorimetry issues (XYZ, $L^*a^*b^*$, UCS, $U^*V^*W^*$, $L^*U^*V^*$). Such systems cannot match well human visual perception. Color perception usually involves three attributes known as *hue*, *saturation* and *brightness* (or *lightness*). Hue is used to distinguish colors (e.g. red, yellow, blue) and to determine the *redness* or *greenness* etc. of the light. If the light source is monochromatic, hue is an indicator of the light's wavelength. Saturation is the measure of the percentage of white light

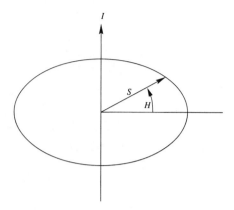

Fig. 1.7.5 Definitions of hue, saturation and brightness of a color.

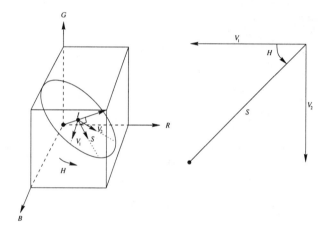

Fig. 1.7.6 HSI color space.

that is added to a pure color. For example, red is a highly saturated color, whereas pink is less saturated. Lightness or brightness refer to the perceived light intensity. The hue, saturation and brightness coordinates of a color are shown in Figure 1.7.5. The coordinates form a cylindrical system. Brightness varies from pure black to pure white and saturation varies from pure gray to highly saturated colors. Hue is measured as an angle between the actual color vector and a reference pure color vector (e.g. red). Several color models have been developed that represent the color's hue, saturation and brightness. Such a system is the HSI model (hue, saturation, intensity). It is a cylindrical coordinate system, whose axis is the line $R = G = B$ of the RGB space, as can be seen in Figure 1.7.6. The displayable colors of the cylindrical coordinate system HSI are the ones that are included in the RGB cube. Thus, the allowable saturation range is very small for large light intensities (close to white) and for small light intensities (close to black). The conversion from the RGB color space to the HSI coordinate system is done in two steps. As the first step, the RGB coordinates are rotated to form the coordinate system (I, V_1, V_2) whose axis is the line $R = G = B$. The rotation is described by the following linear transformation [HAR87]:

$$\begin{bmatrix} I \\ V_1 \\ V_2 \end{bmatrix} = \begin{bmatrix} \sqrt{3}/3 & \sqrt{3}/3 & \sqrt{3}/3 \\ 0 & 1/\sqrt{2} & -1/\sqrt{2} \\ 2/\sqrt{6} & -1/\sqrt{6} & -1/\sqrt{6} \end{bmatrix} \begin{bmatrix} R \\ G \\ B \end{bmatrix} \quad (1.7.35)$$

The rectangular coordinates (V_1, V_2) can be transformed to polar coordinates in the second step:

$$H = \tan^{-1}(V_2/V_1) \quad (1.7.36)$$

$$S = \sqrt{V_1^2 + V_2^2} \quad (1.7.37)$$

The hue coordinate range is $[0, \ldots, 2\pi]$, or, equivalently, $[0, \ldots, 360]$ degrees. The HSI to RGB transformation can be done by the inverse procedure:

$$V_1 = S \cos H \quad (1.7.38)$$

$$V_2 = S \sin H \quad (1.7.39)$$

$$\begin{bmatrix} R \\ G \\ B \end{bmatrix} = \begin{bmatrix} \sqrt{3}/3 & 0 & 2/\sqrt{6} \\ \sqrt{3}/3 & 1/\sqrt{2} & -1/\sqrt{6} \\ \sqrt{3}/3 & -1/\sqrt{2} & -1/\sqrt{6} \end{bmatrix} \begin{bmatrix} I \\ V_1 \\ V_2 \end{bmatrix} \quad (1.7.40)$$

Several variations of this system exist [NIB86], [PRA91]. However, all of them are based on the same philosophy [HAR87]. An algorithm for RGB to HSI conversion and vice versa is shown in Figure 1.7.7.

```
RGB_to_HSI(r,g,b,h,s,I,N1,M1,N2,M2)
image r,g,b;
matrix h,s,I;
int N1,M1;
int N2,M2;
/* Subroutine to perform RGB to HSI transform
   r,g,b: input buffers
   h,s,I: HSI transform buffers
   N1,M1: upper left corner coordinates
   N2,M2: lower right corner coordinates */
{
int i,j; double I1,V1,V2;

for(i=N1;i<N2;i++)
  for(j=M1;j<M2;j++)
   { I1=0.57735*((double)r[i][j]+
```

```
          (double)g[i][j]+(double)b[i][j]);
      V1=0.7071*((double)g[i][j]-(double)b[i][j]);
      V2=0.816496*(double)r[i][j]
          -0.40824*((double)g[i][j]+(double)b[i][j]);
      I[i][j]=I1; s[i][j]=sqrt(V1*V1+V2*V2);
      /* hue is in degrees */
      h[i][j]=atan3(V1,V2);
  }
return(0);
}

HSI_to_RGB(h,s,I,r,g,b,N1,M1,N2,M2)
matrix h,s,I;
image r,g,b;
int N1,M1;
int N2,M2;
/* Subroutine to perform HSI to RGB transform
   h,s,I: HSI transform buffers
   r,g,b: output buffers
   N1,M1: upper left corner coordinates
   N2,M2: lower right corner coordinates */
{
int i,j; double I1,V1,V2,H; int R,G,B;
for(i=N1;i<N2;i++)
 for(j=M1;j<M2;j++)
   { H=h[i][j];
     V1=s[i][j]*cos(H);
     V2=s[i][j]*sin(H);
     I1=I[i][j];
     R=(int)(0.816496*V2+0.57735*I1);
     G=(int)(0.7071*V1-0.40824*V2+0.57735*I1);
     B=(int)(-0.7071*V1-0.40824*V2+0.57735*I1);

     if(R>255) r[i][j]=255;
       else if (R<0) r[i][j]=0; else r[i][j]=(unsigned char)R;
     if(G>255) g[i][j]=255;
       else if (G<0) g[i][j]=0; else g[i][j]=(unsigned char)G;
     if(B>255) b[i][j]=255;
       else if (B<0) B=0.0; else b[i][j]=(unsigned char)B;
   }
return(0);
}
```

Fig. 1.7.7 RGB to HSI conversion.

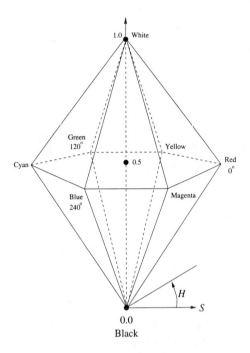

Fig. 1.7.8 Definition of the HLS color space.

Another color coordinate system based on perceptual color attributes is the HLS model (hue, lightness, saturation). It is defined in a double hexcone subset of a cylindrical coordinate system. The HLS color model is shown in Figure 1.7.8. Hue is represented by an angle whose value is zero for the red color. When the color space boundary is traversed in a counterclockwise manner the colors appear in the following order: red, yellow, green, cyan, blue and magenta. The gray light has saturation $S = 0$. The maximal saturation appears at colors having $S = 1$ and $L = 0.5$. The RGB to HLS transformation and vice versa are shown in Figures 1.7.9 and 1.7.10. They are essentially C-like versions of the algorithms reported in [FOL90].

```
RGB_to_HLS(r,g,b,h,l,s,N1,M1,N2,M2)
image r,g,b;
matrix h,l,s;
int N1,M1;
int N2,M2;
/* Subroutine to perform RGB to HLS transform
   r,g,b: input buffers
   h,l,s: HLS transform buffers
   N1,M1: upper left corner coordinates
   N2,M2: lower right corner coordinates */
```

```
{
int i,j;
double R,G,B; double H,L,S;
double max,min,delta;
double UNDEFINED= -1.0;

for(i=N1;i<N2;i++)
 for(j=M1;j<M2;j++)
  { R=((double)r[i][j])/255.0; G=((double)g[i][j])/255.0;
    B=((double)b[i][j])/255.0;
    max=R;
    if(max<G)max=G; if(max<B)max=B; min=R;
    if(min>G)min=G; if(min>B)min=B;
    L=(max+min)/2.0;
    if(max==min){ S=0; H=UNDEFINED; }
     else{
          delta=max-min;
          if(L<0.5) S=delta/(max+min);
           else S=delta/(2.0-max-min);
          if(R==max) H=(G-B)/delta;
           else{ if(G==max) H=2.0+(B-R)/delta;
                 else {if(B==max) H=4+(R-G)/delta; }
                }
          H*=60.0; if(H<0.0)H+=360.0;
         }
    h[i][j]=H; l[i][j]=L; s[i][j]=S;
  }
return(0);
}
```

Fig. 1.7.9 RGB to HLS transform.


```
HLS_to_RGB(h,l,s,r,g,b,N1,M1,N2,M2)
matrix h,l,s;
image r,g,b;
int N1,M1;
int N2,M2;
/* Subroutine to perform HLS to RGB transform
   h,l,s: HLS transform buffers
   r,g,b: RGB buffers
   N1,M1: upper left corner coordinates
   N2,M2: lower right corner coordinates */
{
```

```
int i,j; int R1,G1,B1;
double R,G,B; double H,L,S; double m1,m2;
double UNDEFINED= -1.0;

for(i=N1;i<N2;i++)
 for(j=M1;j<M2;j++)
  { H=h[i][j]; L=l[i][j]; S=s[i][j];
    if(L<=0.5) m2=L*(1.0+S); else m2=L+S-L*S;
    m1=2.0*L-m2;
    if(S==0.0){ if(H==UNDEFINED) R=G=B=L;
                else R=G=B=0.0; /* Error */
              }
     else{
          R=value(m1,m2,H+120.0); G=value(m1,m2,H);
          B=value(m1,m2,H-120.0);
         }
    R1=(int)(R*255.0);G1=(int)(G*255.0);B1=(int)(B*255.0);
    if(R1>255)R1=255;if(G1>255)G1=255;if(B1>255)B1=255;
    if(R1<0)R1=0;if(G1<0)G1=0;if(B1<0)B1=0;
    r[i][j]=(unsigned char)R1; g[i][j]=(unsigned char)G1;
    b[i][j]=(unsigned char)B1;
  }
return(0);
}

static double value(n1,n2,hue)
double n1,n2,hue;
{
 if(hue>360.0)hue-=360.0; if(hue<0.0)hue+=360.0;
 if(hue<60.0) return(n1+(n2-n1)*hue/60.0);
 if(hue<180.0) return(n2);
 if(hue<240.0) return(n1+(n2-n1)*(240.0-hue)/60.0);
 return(n1);
}
```

Fig. 1.7.10 Algorithm for HLS to RGB transform.

Correct color information acquisition is very important for graphical arts, notably for color image scanning and printing. Therefore, color scanner calibration and color correction have been very active research topics recently [KAN97]. Several techniques have been used for calibration, notably regression methods that minimize the mean square error (MMSE) and neural networks. The same techniques can be used for an entirely different application, the color correction of digitized old paintings that are covered with dirt or suffer from varnish oxidation. Digital color 'cleaning' can be an important aid to the restorer before the actual chemical cleaning. If small cleaned patches exist, their information can be used for training the color correction technique with very promising results [PAP98].

Another important problem in color image processing is color image quantization [HEC82], [ORC90]. Sometimes, true color image display is not feasible due to hardware constraints. In this case the color image must be quantized and displayed using color palettes. Color quantization provides color palettes that optimize certain criteria, for example, minimize the approximation error. Essentially color quantization is vector quantization that is performed in the three-dimensional color space. Therefore, adaptations of vector quantization techniques coming from image compression (e.g. the LBG algorithm) can be used for color quantization as well. Alternatively, Learning Vector Quantizers (LVQs) [KOH97] and their modifications [PIT96TIP] can be used for this purpose.

1.8 NOISE GENERATORS FOR DIGITAL IMAGE PROCESSING

Digital images are corrupted by noise either during image acquisition or during image transmission. The image acquisition noise is photoelectronic noise (for photoelectronic sensors) or film-grain noise (in the case of photography) [AND77], [PIT90]. It can be proven that in both cases the noise is signal-dependent. Let $f(x,y)$ be the original image that is recorded on a film slide. Let us denote by $g(x,y)$ the image that is observed if the film is used as a transparency. The observed image is a nonlinear transform of the original image, corrupted by multiplicative noise [PIT90]:

$$g(x,y) = c(f(x,y))^{-\gamma} n(x,y) \qquad (1.8.1)$$

The noise process $n(x,y)$ has a log-normal distribution. In the case of photoelectronic detectors, the photoelectronic noise $n(x,y)$ is signal-dependent noise of the form:

$$g(x,y) = c_2(b(x,y))^{\gamma} + (c_2 b(x,y)^{\gamma})^{1/2} n(x,y) \qquad (1.8.2)$$

where $b(x,y)$ is the input image to the photoelectronic sensor and $g(x,y)$ is the observed image. Photoelectronic sensors produce thermal noise $n_t(x,y)$ as well, which is additive zero-mean white Gaussian noise. Thus, the noisy image acquisition can be described by the equation:

$$g(x,y) = c_2(h(x,y) \ast\ast f(x,y))^{\gamma} + (c_2(h(x,y) \ast\ast f(x,y))^{\gamma})^{1/2} n(x,y) + n_t(x,y) \qquad (1.8.3)$$

where $f(x,y)$ is the original image and $h(x,y)$ is the transfer function of the optical subsystem. The symbol $\ast\ast$ denotes two-dimensional convolution. The complete image formation model is shown in Figure 1.8.1.

Another type of noise, which usually appears during image transmission, is the *salt–pepper* noise. It appears as black and/or white impulses on the

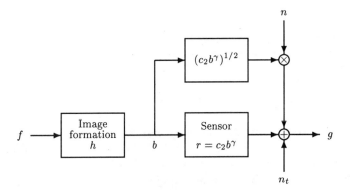

Fig. 1.8.1 Image formation model.

image. It is impulsive noise and its source is either atmospheric or man-made (e.g. car engines). It can be modelled as follows:

$$g(i,j) = \begin{cases} z(i,j) & \text{with probability } p \\ f(i,j) & \text{with probability } 1-p \end{cases} \quad (1.8.4)$$

The noise impulses are denoted by $z(i,j)$ and appear with probability p. The impulses may have fixed values (e.g. 0 or 255) or they may have a long-tailed probability distribution. In many cases, the Laplacian distribution is used as a long-tailed probability distribution model [PAP65].

Digital image noise is removed by using noise filtering techniques to be described in a subsequent chapter. In many cases, digital images must be corrupted artificially in order to assess the performance of various filtering techniques. Furthermore, artificial noise generation is needed for certain artistic applications (e.g. for the simulation of film-grain noise). Such noise generators that can produce additive noise or multiplicative noise:

$$g(i,j) = f(i,j) + n(i,j) \quad (1.8.5)$$

$$g(i,j) = f(i,j)n(i,j) \quad (1.8.6)$$

or impulsive noise of the form (1.8.4) must be implemented.

All digital image noise generators to be described in this section will be based on the uniform random number generator. This generator will be used to produce random numbers having uniform, Gaussian and Laplacian distributions [PAP65]. Most C libraries have routines to initialize and then generate pseudorandom numbers having a uniform distribution. A typical example of such a set is the following [MIC87]:

```
#include<stdlib.h>
void srand(seed):
unsigned seed;   /* Seed for random number generator */
int rand(void);
```

The random number generator is initialized by calling srand() with a specific seed. The rand() routine is executed subsequently in order to produce a sequence of random numbers which are integers in the range $[0, MAX)$. A typical value for the upper bound MAX is 32,767. The pseudorandom number sequence has an approximately uniform distribution in the range $[0, MAX)$. Thus, a uniformly distributed pseudorandom sequence in the range $[0, 1]$ can be obtained by appropriate scaling:

$$x = \frac{rand()}{MAX + 1} \tag{1.8.7}$$

Most implementations of the rand() subroutine are based on the recursive relation:

$$rand()_i = (\alpha \cdot rand()_{i-1} + b) \mod MAX \tag{1.8.8}$$

where $rand()_i$ denotes the output of rand() at the i-th iteration. This recurrence relation repeats itself after at most MAX iterations. This fact may have disastrous effects in the randomness of the generated number sequence, especially when MAX is small. Furthermore, the pseudorandom sequence possesses local correlation on successive random numbers and thus is unacceptable in most applications. A variation of the rand() subroutine can be used that improves the randomness of the output sequence. It uses a vector **v** of integers that is full of random numbers. On each call, an element of vector **v**, chosen in a random way, is presented to the subroutine output. The subroutine rand() is used to refill the element of **v** that has been selected. Such a randomized choice from a vector of random numbers decreases output correlation considerably. This subroutine (called rand0()) is described in [PRE88]. The random number generator rand0() produces output in the range [0,1).

The rand0() routine is used in the algorithm described in Figure 1.8.2 for the generation of impulsive noise on a digital image.

```
int noise_imp(a,b,p,imin,imax,N1,M1,N2,M2)
image a,b;
float p;
int imin,imax;
int N1,M1,N2,M2;
/* Subroutine to add impulsive noise to an image.
   a,b: buffers
```

```
    p: noise probability
    imin: minimum impulse value
    imax: maximum impulse value
    N1,M1: start coordinates
    N2,M2: end   coordinates          */
{ int i,j,turn,k,l;
  unsigned long np,inp;
  float rand0();
  int idum=-1;
  np=(unsigned long)(p*((float)(N2-N1)*(float)(M2-M1)));
  turn=0;

  if(b!=a)
    { for(i=N1; i<N2; i++)
        for(j=M1; j<M2; j++)
          b[i][j]=a[i][j];
    }

  for(inp=0; inp<np; inp++)
    { k=N1+(int)(((float)(N2-N1-1))*(rand0(&idum)));
      l=M1+(int)(((float)(M2-M1-1))*(rand0(&idum)));
      if(turn==0) { turn=1; b[k][l]=imax; }
         else     { turn=0; b[k][l]=imin; }
    }
  return(0);
}
```

Fig. 1.8.2 Generation of impulsive noise.

If the image size is $N \times M$ and the probability of occurrence is p, the random number generator is called pNM times to produce random pixel coordinates that will be corrupted by impulses. The corrupted pixels take values $IMAX$ or $IMIN$ by using a **turn** variable. Thus, half of the impulses are positive and half are negative. The algorithm works relatively well if pNM is less than MAX.

The uniformly distributed random numbers can be easily transformed to random numbers having other distributions by using nonlinear transformations. Let x be a random variable having probability density function (pdf) $f_x(x)$. If $y = g(x)$ is a pointwise transformation, it can be easily proven that the distribution of y is given by [PAP65]:

$$f_y(y) = \frac{f_x(x_1)}{\left|\frac{dg(x_1)}{dx}\right|} + \ldots + \frac{f_x(x_n)}{\left|\frac{dg(x_n)}{dx}\right|} + \ldots \qquad (1.8.9)$$

where $x_1, \ldots, x_n, \ldots$ are the real roots of the equation $y = g(x)$. In many cases, this equation has only one root x_1. If $f_x(x)$ is known and $f_y(y)$ is the target distribution, (1.8.9) can be used to find the desired transform function $g(x)$. If $f_x(x)$ is a uniform distribution in the range [0,1] and $f_y(y)$ is a unit Gaussian pdf having zero mean and unit variance:

$$f_y(y) = \frac{1}{\sqrt{2\pi}} e^{-x^2} \qquad (1.8.10)$$

the resulting transformation function is given by:

$$y = \sqrt{-2\ln x} \cos(2\phi) \qquad (1.8.11)$$

or

$$y = \sqrt{-2\ln x} \sin(2\phi) \qquad (1.8.12)$$

where ϕ is uniformly distributed in the interval $(-\pi, \pi]$. This transformation is used to produce a white multiplicative Gaussian noise generator in the algorithm of Figure 1.8.3.

```
int noise_mult_gauss(a,b,v,N1,M1,N2,M2)
image a,b;
float v;
int N1,M1,N2,M2;

/* Subroutine to add multiplicative Gaussian noise to an image
   a,b: buffers
   v: noise range [-v/2, v/2]
   N1,M1: start coordinates
   N2,M2: end   coordinates       */

{ int i,j, ime;
  float rand0(),uniform,rayleigh,gauss;
  double PI=4.0*atan(1.0);
  float EPSILON=1.0e-5;
  int INFINITY=255;
  int idum=-1;

  for(i=N1; i<N2; i++)
  for(j=M1; j<M2; j++)
     { uniform=rand0(&idum);
       if (uniform<=EPSILON) rayleigh=(float)INFINITY;
       else rayleigh=sqrt(-2.0*log(uniform));
       uniform=rand0(&idum);
       gauss=rayleigh*cos(2.0*PI*uniform);
       ime=0.5+a[i][j]+a[i][j]*v*gauss;
       if (ime<0) b[i][j]=0;
```

```
        else if (ime>255) b[i][j]=255;
        else b[i][j]=ime;
     }
  return(0);
}
```

Fig. 1.8.3 Multiplicative Gaussian noise generator.

The following transformation can be used to transform a uniform distribution in [0,1] to a Laplacian distribution:

$$y = \begin{cases} \ln(2x) & 0 \leq x \leq 1/2 \\ -\ln(2-2x) & 1/2 \leq x < 1 \end{cases} \quad (1.8.13)$$

The Laplacian distribution produced by (1.8.13) is described by:

$$f_y(y) = \frac{1}{2}e^{-|y|} \quad (1.8.14)$$

The transformation (1.8.13) is used in the algorithm of Figure 1.8.4 to produce a white additive Laplacian noise generator.

```
int noise_add_laplace(a,b,v,N1,M1,N2,M2)
image a,b;
float v;
int N1,M1,N2,M2;

/* Subroutine to add Laplacian noise to an image
   a,b: buffers
   v: Laplace coefficient
   N1,M1: start coordinates
   N2,M2: end coordinates      */

{ int i,j, ime;
  float rand0(),s,u,u1;
  float EPSILON=1.0e-5;
  int INFINITY=255;
  int idum=-1;

  for(i=N1; i<N2; i++)
  for(j=M1; j<M2; j++)
     { u=rand0(&idum);
       if(u<=0.5)
          { if (u<=EPSILON) ime=a[i][j]-INFINITY;
            else ime=0.5+a[i][j]+v*log(2.0*u);
```

```
      }
      else
      { if((u1=1.0-u)<=0.5*EPSILON) ime=a[i][j]+INFINITY;
          else ime=0.5+a[i][j]-v*log(2.0*u1);
      }
      if (ime<0) b[i][j]=0;
        else if (ime>255) b[i][j]=255; else b[i][j]=ime;
    }
  return(0);
}
```

Fig. 1.8.4 Additive Laplacian noise generator.

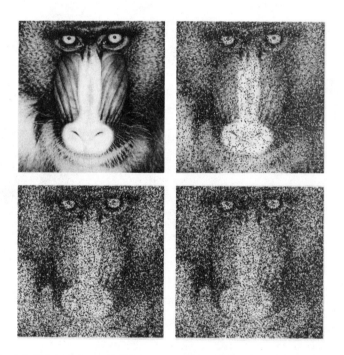

Fig. 1.8.5 (a) Original image; (b) image corrupted by impulsive noise; (c) image corrupted by multiplicative Gaussian noise; (d) image corrupted by additive Laplacian noise.

The above-mentioned noise generators have been used to corrupt the digital image shown in Figure 1.8.5a. The result of the impulsive noise generator having noise probability 10% is shown in Figure 1.8.5b. The output of the multiplicative Gaussian noise generator is shown in Figure 1.8.5c. It is clearly seen that the noise is stronger in the image regions having high intensity values. Finally, the result of the additive Laplacian noise generator is shown in Figure 1.8.5d.

References

AND77. H.C. Andrews, B.R. Hunt, *Digital image restoration*, Prentice-Hall, 1977.

ANG90. E. Angel, *Computer graphics*, Addison-Wesley, 1990.

BAL82. D.H. Ballard, C.M. Brown, *Computer vision*, Prentice-Hall, 1982.

BAX94. G.A. Baxes, *Digital image processing*, Wiley, 1994.

BIJ98. A. Bijaoui, F. Murtagh, J.L. Starck, *Image processing and data analysis*, Cambridge University, 1998.

BRO95. C.W. Brown, B.J. Shepherd, *Graphics file format*, Manning Publications, 1995.

CHA94. N. Chaddha, W.-C. Tan, and T. Meng, "Color quantization of images based on human vision perception," in *Proc. IEEE Int. Conf. on Acoustics, Speech and Signal Processing*, vol. V, pp. 89-92, Adelaide, Australia 1994.

CRA97. R. Crane, *A simplified approach to image processing*, Prentice-Hall, 1997.

FOL90. J.D. Foley, A. van Dam, S.K. Feiner, J.F. Hughes, *Computer graphics: Principles and practice*, Addison-Wesley, 1990.

GON87. R.C. Gonzalez, P. Wintz, *Digital image processing*, Addison-Wesley, 1987.

HAR87. S. Harrington, *Computer graphics: A programming approach*, McGraw-Hill, 1987.

HEC82. P. Heckbert, "Color image quantization for frame buffer display," *Comput. Graph.*, vol. 16, no. 3, pp. 297-307, July 1982.

HOR98. R.E.N. Horne, S.J. Sangwine, *The color image processing handbook*, Chapman-Hall, 1998.

JAI89. A.K. Jain, *Fundamentals of digital image processing*, Prentice-Hall, 1989.

KAN97. H.R. Kang, *Color technology for electronic imaging devices*, SPIE, 1997.

KLE96. R. Klette, P. Zamperoni, *Handbook of image processing operators*, Wiley, 1996.

KOH97. T. Kohonen, *Self-organizing maps*, 3rd edition, New York, Springer Verlag, 1997.

LEV85. M.D. Levine, *Vision in man and machine*, McGraw-Hill, 1985.

LIN91. C.A. Lindley, *Practical image processing in C*, Wiley, 1991.

LIN94. T.-S. Lin, and L.-W. Chang, "Greedy tree growing for color image quantization," in *Proc. IEEE Int. Conf. on Acoustics, Speech and Signal Processing*, vol. V, pp. 97-100, Adelaide, Australia, 1994.

LOH98. G. Lohnmann, *Volumetric image analysis*, Wiley, Teubner, 1998.

MIC87. *Microsoft C: Runtime library reference*, Microsoft Press, 1987.

NIB86. W. Niblack, *Digital image processing*, Prentice-Hall, 1986.

OPP89. A. Oppenheim, R.W. Schafer, *Discrete-time signal processing*, Prentice-Hall, 1989.

ORC90. M. Orchard, and C. Bouman, "Color quantization of images," *IEEE Trans. Signal Processing*, vol. 39, no. 12, pp. 2677-2690, December 1990.

PAP65. A. Papoulis, *Probability, random variables and stochastic processes*, McGraw-Hill, 1965.

PAP98. M. Pappas, I. Pitas, "Old painting digital color restoration," in *Proc. NMBIA*, pp. 188-192, Glasgow, UK, July 1998.

PIT90. I. Pitas, A.N. Venetsanopoulos, *Nonlinear digital filters: Principles and applications*, Kluwer Academic, 1990.

PIT93. I. Pitas, *Parallel algorithms for digital image processing*, Wiley, 1993.

PIT96TIP. I. Pitas, C. Kotropoulos, N. Nikolaidis, R. Yang, and M. Gabbouj, "Order Statistics Learning Vector Quantizer," *IEEE Trans. Image Processing*, vol. 5, no. 6, pp. 1048-1053, 1996.

PRA91. W.K. Pratt, *Digital image processing*, Wiley, 1991.

PRE88. W.H. Press, B.P. Flannery, S.A. Teukolsky, W.T. Vetterling, *Numerical recipes in C*, Cambridge University Press, 1988.

RIM90. S. Rimmer, *Bit-mapped graphics*, Windcrest, 1990.

ROS82. A. Rosenfeld, A.C. Kak, *Digital picture processing*, Academic Press, 1982.

RUS92. J.C. Russ, *The image processing handbook*, CRC Press, 1992.

RUS99. J.C. Russ, *The image processing handbook*, Springer-Verlag, 1999.

SCH89. R.J. Schalkof, *Digital image processing and computer vision*, Wiley, 1989.

SER82. J. Serra, *Image analysis and mathematical morphology*, Academic Press, 1982.

WIL87. R. Wilto, *Programmer's guide to PC and PS/2 video systems*, Microsoft Press, 1987.

WYZ67. G.W. Wyzecki, W.S. Stiles, *Color science*, Wiley, 1967.

2
Digital Image Transform Algorithms

2.1 INTRODUCTION

Image transforms play an important role in digital image processing as a theoretical and implementational tool in numerous tasks, notably in digital image filtering, restoration, encoding and analysis. Image transforms are often linear ones. They are represented by the transform matrices **A**:

$$\mathbf{X} = \mathbf{A}\mathbf{x} \tag{2.1.1}$$

where $\mathbf{x}, \mathbf{X}$ are the original and the transformed image, respectively. In most cases, the transform matrices are *unitary*:

$$\mathbf{A}^{-1} = \mathbf{A}^{*T} \tag{2.1.2}$$

The columns of $\mathbf{A}^{*T}$ are the *basis vectors* of the transform. In the case of two-dimensional transforms, the basis vectors correspond to *basis images*. Thus, a transform decomposes a digital image to a weighted sum of basis images. Two transforms will be treated in detail in this chapter: the discrete Fourier transform (DFT) and the discrete cosine transform (DCT). The DFT possesses very interesting theoretical properties and is extensively used in digital filter implementations and in power spectrum estimation. The DCT is frequently used in transform image coding schemes, because it is an excellent tool for digital image compression. Besides these two transforms, a multitude of image transforms has been proposed in the related literature [AHM75], [RAO90], [JAI89], for example, Walsh transform, Haar transform, slant transform, discrete sine transform, mainly for digital image coding applications [CLA85].

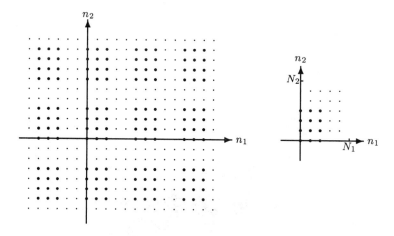

Fig. 2.2.1 (a) Rectangularly periodic sequence; (b) fundamental period.

Their description is beyond the scope of this book. This chapter will focus only on the DFT and the DCT, because they are by far the most frequently used. Fast algorithms for DFT and DCT will be described in detail.

2.2 TWO-DIMENSIONAL DISCRETE FOURIER TRANSFORM

Let $\tilde{x}(n_1, n_2)$ denote a two-dimensional rectangularly periodic sequence, whose periodicity is defined as follows:

$$\tilde{x}(n_1, n_2) = \tilde{x}(n_1 + N_1, n_2) = \tilde{x}(n_1, n_2 + N_2) \qquad (2.2.1)$$

The integers N_1, N_2 denote the horizontal and vertical periods respectively. Such a periodic sequence is depicted in Figure 2.2.1. The fundamental period of the periodic sequence $\tilde{x}(n_1, n_2)$ is the rectangle $R_{N_1 N_2}$ having size $N_1 \times N_2$:

$$R_{N_1 N_2} \triangleq \{(n_1, n_2) : 0 \leq n_1 < N_1, 0 \leq n_2 < N_2\} \qquad (2.2.2)$$

It is well known that a periodic sequence can be represented by a Fourier series. The two-dimensional discrete Fourier series can be used for the representation of a two-dimensional rectangularly periodic signal $\tilde{x}(n_1, n_2)$:

$$\tilde{x}(n_1, n_2) = \frac{1}{N_1 N_2} \sum_{k_1=0}^{N_1-1} \sum_{k_2=0}^{N_2-1} \tilde{X}(k_1, k_2) \exp\left(i\frac{2\pi n_1 k_1}{N_1} + i\frac{2\pi n_2 k_2}{N_2}\right) \qquad (2.2.3)$$

The Fourier series coefficients are given by:

$$\tilde{X}(k_1, k_2) = \sum_{n_1=0}^{N_1-1} \sum_{n_2=0}^{N_2-1} \tilde{x}(n_1, n_2) \exp\left(-i\frac{2\pi n_1 k_1}{N_1} - i\frac{2\pi n_2 k_2}{N_2}\right) \qquad (2.2.4)$$

Let us suppose that a two-dimensional sequence $x(n_1, n_2)$ is defined on the rectangle $R_{N_1 N_2}$. It can be easily extended to form a periodic sequence $\tilde{x}(n_1, n_2)$:

$$\tilde{x}(n_1, n_2) = \sum_{r_1=-\infty}^{\infty} \sum_{r_2=-\infty}^{\infty} x(n_1 - r_1 N_1, n_2 - r_2 N_2) \qquad (2.2.5)$$

The original sequence $x(n_1, n_2)$ can be easily reconstructed from the periodic sequence $\tilde{x}(n_1, n_2)$:

$$x(n_1, n_2) = \begin{cases} \tilde{x}(n_1, n_2) & (n_1, n_2) \in R_{N_1 N_2} \\ 0 & \text{elsewhere} \end{cases} \qquad (2.2.6)$$

Similarly, the Fourier series coefficients $\tilde{X}(k_1, k_2)$ of the periodic sequence $\tilde{x}(n_1, n_2)$ can be considered to be the periodic extension of the sequence $X(k_1, k_2)$ given by:

$$X(k_1, k_2) = \begin{cases} \tilde{X}(k_1, k_2) & (k_1, k_2) \in R_{N_1 N_2} \\ 0 & \text{elsewhere} \end{cases} \qquad (2.2.7)$$

$$\tilde{X}(k_1, k_2) = \sum_{r_1} \sum_{r_2} X(k_1 - r_1 N_1, k_2 - r_2 N_2) \qquad (2.2.8)$$

By combining (2.2.3–2.2.8), it can be easily seen that a discrete two-dimensional space-limited signal $x(n_1, n_2)$ can be completely represented by the sequence $X(k_1, k_2)$:

$$X(k_1, k_2) = \sum_{n_1=0}^{N_1-1} \sum_{n_2=0}^{N_2-1} x(n_1, n_2) \exp\left(-i\frac{2\pi n_1 k_1}{N_1} - i\frac{2\pi n_2 k_2}{N_2}\right) \qquad (2.2.9)$$

$$x(n_1, n_2) = \frac{1}{N_1 N_2} \sum_{k_1=0}^{N_1-1} \sum_{k_2=0}^{N_2-1} X(k_1, k_2) \exp\left(i\frac{2\pi n_1 k_1}{N_1} + i\frac{2\pi n_2 k_2}{N_2}\right) \qquad (2.2.10)$$

Equations (2.2.9–2.2.10) define the two-dimensional discrete Fourier transform (2-d DFT). The discrete Fourier transform is a very important tool in the processing and analysis of two-dimensional discrete signals, because it lends itself to numerical computation. Its computation can be done by fast algorithms, as will be seen subsequently in this chapter. Before proceeding to a short description of the properties of the two-dimensional DFT, we give the DFT pair (2.2.9-2.2.10) in a different form that is commonly used in the literature:

$$X(k_1, k_2) = \sum_{n_1=0}^{N_1-1} \sum_{n_2=0}^{N_2-1} x(n_1, n_2) W_{N_1}^{n_1 k_1} W_{N_2}^{n_2 k_2} \qquad (2.2.11)$$

$$x(n_1, n_2) = \frac{1}{N_1 N_2} \sum_{k_1=0}^{N_1-1} \sum_{k_2=0}^{N_2-1} X(k_1, k_2) W_{N_1}^{-n_1 k_1} W_{N_2}^{-n_2 k_2} \qquad (2.2.12)$$

where:
$$W_{N_j} = \exp\left(-i\frac{2\pi}{N_j}\right) \quad j = 1, 2 \quad (2.2.13)$$

The discrete Fourier transform is directly related to the Fourier transform $X(\omega_1, \omega_2)$ of a discrete two-dimensional signal defined as follows [DUD84, LIM90]:

$$X(\omega_1, \omega_2) = \sum_{n_1=-\infty}^{\infty} \sum_{n_2=-\infty}^{\infty} x(n_1, n_2) \exp(-i\omega_1 n_1 - i\omega_2 n_2) \quad (2.2.14)$$

$$x(n_1, n_2) = \frac{1}{4\pi^2} \int_{-\pi}^{\pi} \int_{-\pi}^{\pi} X(\omega_1, \omega_2) \exp(i\omega_1 n_1 + i\omega_2 n_2) d\omega_1 d\omega_2 \quad (2.2.15)$$

The discrete frequencies ω_1, ω_2 are related to the analog frequencies Ω_1, Ω_2 of the continuous signal $X(t_1, t_2)$:

$$\omega_i = \Omega_i T_i \quad i = 1, 2 \quad (2.2.16)$$

where T_1, T_2 are the sampling periods along the horizontal and vertical dimensions. It can be easily seen by comparing (2.2.9) and (2.2.14) that the DFT coefficients $X(k_1, k_2)$ are a sampled version of the two-dimensional Fourier transform of a discrete sequence:

$$X(k_1, k_2) = X(\omega_1, \omega_2)\big|_{\omega_1 = \frac{2\pi k_1}{N_1}, \omega_2 = \frac{2\pi k_2}{N_2}} \quad (2.2.17)$$

over the unit bicircle $z_1 = \exp(i\omega_1)$, $z_2 = \exp(i\omega_2)$.

The definition of the *circular shift* of a two-dimensional discrete sequence $x(n_1, n_2)$:

$$y(n_1, n_2) = x(((n_1 + m_1))_{N_1}, ((n_2 + m_2))_{N_2}) \quad (n_1, n_2) \in R_{N_1 N_2} \quad (2.2.18)$$

$$((n))_N \triangleq n \bmod N \quad (2.2.19)$$

is needed for the definition of certain properties of the two-dimensional DFT. Such a circular shift is illustrated in Figure 2.2.2. The circular shift can be used in the definition of the two-dimensional circular convolution:

$$y(n_1, n_2) \triangleq x(n_1, n_2) \circledast\circledast h(n_1, n_2)$$
$$= \sum_{m_1=0}^{N_1-1} \sum_{m_2=0}^{N_2-1} x(m_1, m_2) h(((n_1 - m_1))_{N_1}, ((n_2 - m_2))_{N_2}) \quad (2.2.20)$$

of two sequences $x(n_1, n_2)$, $h(n_1, n_2)$ defined over $R_{N_1 N_2}$. One of the most important properties of the 2-d DFT is the support of the circular convolution:

$$y(n_1, n_2) = x(n_1, n_2) \circledast\circledast h(n_1, n_2) \leftrightarrow Y(k_1, k_2) = X(k_1, k_2) H(k_1, k_2) \quad (2.2.21)$$

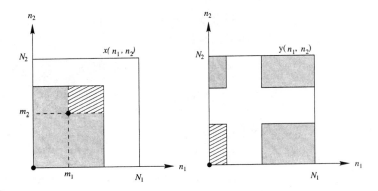

Fig. 2.2.2 Definition of the circular shift of a two-dimensional sequence $x(n_1, n_2)$.

Thus, the following scheme can be used for the fast computation of the circular convolution:

$$y(n_1, n_2) = IDFT[DFT[x(n_1, n_2)]DFT[h(n_1, n_2)]] \quad (2.2.22)$$

provided that fast algorithms for the calculation of the 2-d DFT exist. However, in most applications the fast computation of the *linear convolution* is needed:

$$y(n_1, n_2) = \sum_{m_1=0}^{Q_1-1} \sum_{m_2=0}^{Q_2-1} h(m_1, m_2) x(n_1 - m_1, n_2 - m_2) \quad (2.2.23)$$

If the sequences $x(n_1, n_2), h(n_1, n_2)$ have regions of support $R_{P_1 P_2} = [0, P_1) \times [0, P_2)$ and $R_{Q_1 Q_2} = [0, Q_1) \times [0, Q_2)$ respectively, the linear convolution output has the following region of support:

$$R_{L_1 L_2} = [0, L_1) \times [0, L_2) \quad L_i = P_i + Q_i - 1 \quad i = 1, 2 \quad (2.2.24)$$

The calculation of the linear convolution (2.2.23) can be cast in a circular convolution calculation as follows. A region of support $R_{N_1 N_2}$ is chosen such that $N_1 \geq L_1$, $N_2 \geq L_2$. The sequences $x(n_1, n_2), h(n_1, n_2)$ are padded with zeros to form the sequences $x_p(n_1, n_2), h_p(n_1, n_2)$ that are defined over $R_{N_1 N_2}$:

$$x_p(n_1, n_2) = \begin{cases} x(n_1, n_2) & (n_1, n_2) \in R_{P_1 P_2} \\ 0 & (n_1, n_2) \in R_{N_1 N_2} - R_{P_1 P_2} \end{cases} \quad (2.2.25)$$

$$h_p(n_1, n_2) = \begin{cases} h(n_1, n_2) & (n_1, n_2) \in R_{Q_1 Q_2} \\ 0 & (n_1, n_2) \in R_{N_1 N_2} - R_{Q_1 Q_2} \end{cases} \quad (2.2.26)$$

It can be easily proven that the cyclic convolution:

$$y_p(n_1, n_2) = x_p(n_1, n_2) \circledast\circledast h_p(n_1, n_2) \quad (2.2.27)$$

gives exactly the same output with the linear convolution (2.2.23). The linear convolution output is given by:

$$y(n_1, n_2) = y_p(n_1, n_2) \quad (n_1, n_2) \in R_{L_1 L_2} \qquad (2.2.28)$$

Thus, the linear convolution output can be calculated by using the following algorithm:

1. Choose $N_1 N_2$ such that $N_i \geq P_i + Q_i - 1$, $i = 1, 2$.

2. Pad the sequences $x(n_1, n_2), h(n_1, n_2)$ with zeros, according to (2.2.25–2.2.26).

3. Calculate the DFTs of $x_p(n_1, n_2), h_p(n_1, n_2)$.

4. Calculate the DFT $Y_p(k_1, k_2)$, as the product of $X_p(k_1, k_2)$ and $H_p(k_1, k_2)$.

5. Calculate $y_p(n_1, n_2)$ by using the inverse DFT. The linear convolution output is given by (2.2.28).

This algorithm is illustrated by a subroutine in Chapter 3 of this book. The 2-d discrete Fourier transform also supports the 2-d *correlation*:

$$R_{xy}(m_1, m_2) = \sum_{n_1=0}^{N_1-1} \sum_{n_2=0}^{N_2-1} x(n_1, n_2) y(n_1 + m_1, n_2 + m_2) \qquad (2.2.29)$$

The images x, y are assumed to have regions of support $R_{P_1 P_2}, R_{Q_1 Q_2}$ respectively. They are appended with zeros to form images having a region of support $R_{N_1 N_2}$, $N_i \geq P_i + Q_i - 1, i = 1, 2$, that are used in (2.2.29). Correlation (2.2.29) is essentially a convolution of $y(n_1, n_2)$ by $x(-n_1, -n_2)$. If both sequences are real, it can be easily proven that the correlation is equivalent to the following relation in the frequency domain:

$$P_{xy}(k_1, k_2) = X^*(k_1, k_2) Y(k_1, k_2) \qquad (2.2.30)$$

where $P_{xy}(k_1, k_2)$ is the DFT of the correlation $R_{xy}(m_1, m_2)$. Therefore, the following method can be used for the fast calculation of 2-d correlation:

$$R_{xy}(n_1, n_2) = IDFT[\, DFT[x(n_1, n_2)]^* \, DFT[y(n_1, n_2)] \,] \qquad (2.2.31)$$

The resulting algorithm is shown in Figure 2.2.3. Subroutine correlation() can be used for the calculation of both auto- and cross-correlation. The temporary floating point matrices needed are created by using subroutine matf2(). Its structure is similar to that of matuc2() that was presented in Chapter 1. A subroutine fft2d() is used for the calculation of the two-dimensional DFT and IDFT. Such subroutines will be described in subsequent sections. The output of this subroutine is given by (2.2.29) divided by $P_1 P_2$, so that it is

TWO-DIMENSIONAL DISCRETE FOURIER TRANSFORM

consistent with the definitions used in power spectrum estimation that will be given in a subsequent section. The same algorithm can be used for the calculation of the autocorrelation of an image $x(n_1, n_2)$:

$$R_{xx}(m_1, m_2) = \sum_{n_1=0}^{N_1-1} \sum_{n_2=0}^{N_2-1} x(n_1, n_2) x(n_1 + m_1, n_2 + m_2) \qquad (2.2.32)$$

```
int correlation(mats1,mats2, matd, ND1,MD1, N1,M1,N2,M2)
matrix mats1,mats2,matd;
int ND1,MD1;
int N1,M1,N2,M2;

/* Subroutine to compute the correlation of two signals
   mats1,mats2 : source matrices
   matd : destination matrix (correlation)
   ND1,MD1 : destination coordinates
   N1,M1 : upper left  corner of source matrices
   N2,M2 : lower right corner of source matrices
   NOTE! size of destination matrix : 2*(N2-N1) x 2*(M2-M1) */

{ int nr,nc,rows2,cols2;
  int ret;
  float a,b,c,d;
  matrix matmere1,matmeim1,matmere2,matmeim2;

  ret=0; rows2=2*(N2-N1); cols2=2*(M2-M1);

  /* Create temporary matrices having double the ROI size */
  matmere1=matf2(rows2,cols2);
  matmeim1=matf2(rows2,cols2);

  if(mats1 != mats2)    /* Cross-correlation */
    { matmere2=matf2(rows2,cols2);
      matmeim2=matf2(rows2,cols2);
      for(nr=0; nr<rows2; nr++)
      for(nc=0; nc<cols2; nc++)
        { matmere1[nr][nc]=0.0;
          matmeim1[nr][nc]=0.0;
          matmere2[nr][nc]=0.0;
          matmeim2[nr][nc]=0.0;
        }
      for(nr=N1; nr<N2; nr++)
      for(nc=M1; nc<M2; nc++)
        { matmere1[nr-N1][nc-M1]=mats1[nr][nc];
          matmere2[nr-N1][nc-M1]=mats2[nr][nc];
```

```
      }
  /* Perform FFT transforms */
  ret=fft2d(matmere1,matmeim1,
            matmere1,matmeim1, 0, 0,0,0,0,rows2,cols2);
  if(ret) goto R22;
  ret=fft2d(matmere2,matmeim2,
            matmere2,matmeim2, 0, 0,0,0,0,rows2,cols2);
  if(ret) goto R22;
  /* Perform multiplication in the transform domain. The
     conjugate of the second signal (mats2) is used !! */
  for(nr=0; nr<rows2; nr++)
  for(nc=0; nc<cols2; nc++)
   { a=matmere1[nr][nc];
     c=matmere2[nr][nc];
     b=matmeim1[nr][nc];
     d=matmeim2[nr][nc];
     matmere1[nr][nc]=a*c+b*d;
     matmeim1[nr][nc]=b*c-a*d;
   }
  /* Perform inverse FFT */
  ret=fft2d(matmere1,matmeim1,
            matmere1,matmeim1, 1, 0,0,0,0,rows2,cols2);
  if(ret) goto R22;
 }

else    /* Auto-correlation */
 { for(nr=0; nr<rows2; nr++)
   for(nc=0; nc<cols2; nc++)
    { matmere1[nr][nc]=0.0;
      matmeim1[nr][nc]=0.0;
    }
   for(nr=N1; nr<N2; nr++)
   for(nc=M1; nc<M2; nc++)
    matmere1[nr-N1][nc-M1]=mats1[nr][nc];
   /* Perform FFT transform */
   ret=fft2d(matmere1,matmeim1,
             matmere1,matmeim1, 0, 0,0,0,0,rows2,cols2);
   if(ret) goto R12;
   /* Perform multiplication in the transform domain */
   for(nr=0; nr<rows2; nr++)
   for(nc=0; nc<cols2; nc++)
    { a=matmere1[nr][nc];
      b=matmeim1[nr][nc];
      matmere1[nr][nc]=a*a+b*b;
      matmeim1[nr][nc]=0.0;
```

```
        }
        /* Perform inverse FFT */
        ret=fft2d(matmere1,matmeim1,
                matmere1,matmeim1, 1, 0,0,0,0,rows2,cols2);
        if(ret) goto R12;
    }

    /* Divide by (N2-N1)*(M2-M1) */
    a=(float)(N2-N1)*(float)(M2-M1);
    for(nr=0; nr<rows2; nr++)
    for(nc=0; nc<cols2; nc++)
        matd[ND1+nr][MD1+nc] = matmere1[nr][nc] / a;

    if(mats1 == mats2) goto R12;

R22: mfreef2(matmeim2,rows2);
R21: mfreef2(matmere2,rows2);
R12: mfreef2(matmeim1,rows2);
R11: mfreef2(matmere1,rows2);
RET: return(ret);
}
```

Fig. 2.2.3 Algorithm for the calculation of digital correlation.

Another interesting property of the 2-d DFT is Parseval's theorem:

$$\sum_{n_1=0}^{N_1-1}\sum_{n_2=0}^{N_2-1} \mid x(n_1,n_2) \mid^2 = \frac{1}{N_1 N_2} \sum_{k_1=0}^{N_1-1}\sum_{k_2=0}^{N_2-1} \mid X(k_1,k_2) \mid^2 \qquad (2.2.33)$$

Thus, the energy in the space domain is equal to the energy in the transform domain. The 2-d DFT possesses several other interesting properties. Since their description is outside the scope of this book, they are summarized in Table 2.2.1, without proof. Having described the basic properties of the 2-d DFT, we proceed to the description of 2-d fast Fourier transform algorithms. Such algorithms are very important in a variety of applications including 2-d FIR filter implementation and power spectrum estimation.

2.3 ROW–COLUMN FFT ALGORITHM

The calculation of the 2-d DFT involves very large computational complexity. Its direct computation by employing the definitions (2.2.11–2.2.12) requires $N_1^2 N_2^2$ complex multiplications and $N_1 N_2 (N_1 N_2 - 1)$ complex additions.

Table 2.2.1 Properties of the 2-d discrete Fourier transform.

---

1. Separable sequence.
   $x(n_1, n_2) = x_1(n_1) x_2(n_2) \longleftrightarrow X(k_1, k_2) = X_1(k_1) X_2(k_2)$

2. Linearity.
   $x(n_1, n_2) = a \cdot v(n_1, n_2) + b \cdot w(n_1, n_2) \longleftrightarrow$
   $X(k_1, k_2) = a \cdot V(k_1, k_2) + b \cdot W(k_1, k_2)$

3. Circular shift.
   $y(n_1, n_2) = x(((n_1 - m_1))_{N_1}, ((n_2 - m_2))_{N_2}) \longleftrightarrow$
   $Y(k_1, k_2) = W_{N_1}^{m_1 k_1} W_{N_2}^{m_2 k_2} X(k_1, k_2)$

4. Modulation.
   $y(n_1, n_2) = W_{N_1}^{-n_1 l_1} W_{N_2}^{-n_2 l_2} x(n_1, n_2) \longleftrightarrow$
   $Y(k_1, k_2) = X(((k_1 - l_1))_{N_1}, ((k_2 - l_2))_{N_2})$

5. Complex conjugate property.
   $x^*(((N_1 - n_1))_{N_1}, ((N_2 - n_2))_{N_2}) \longleftrightarrow X^*(k_1, k_2)$

   If $x(n_1, n_2)$ is a real-valued signal:
   $X^*(k_1, k_2) = X(((N_1 - k_1))_{N_1}, ((N_2 - k_2))_{N_2})$

6. Reflection.
   $x(n_2, n_1) \longleftrightarrow X(k_2, k_1)$

7. Initial value and DC value.
   $x(0,0) = \frac{1}{N_1 N_2} \sum_{k_1=0}^{N_1-1} \sum_{k_2=0}^{N_2-1} X(k_1, k_2)$

   $X(0,0) = \sum_{n_1=0}^{N_1-1} \sum_{n_2=0}^{N_2-1} x(n_1, n_2)$

8. Parseval's theorem.
   $\sum_{n_1=0}^{N_1-1} \sum_{n_2=0}^{N_2-1} x(n_1, n_2) y^*(n_1, n_2) = \frac{1}{N_1 N_2} \sum_{k_1=0}^{N_1-1} \sum_{k_2=0}^{N_2-1} X(k_1, k_2) Y^*(k_1, k$

   $\sum_{n_1=0}^{N_1-1} \sum_{n_2=0}^{N_2-1} |x(n_1, n_2)|^2 = \frac{1}{N_1 N_2} \sum_{k_1=0}^{N_1-1} \sum_{k_2=0}^{N_2-1} |X(k_1, k_2)|^2$

9. Circular convolution and multiplication.
   $x(n_1, n_2) \circledast \circledast \; h(n_1, n_2) \longleftrightarrow X(k_1, k_2) \cdot H(k_1, k_2)$

   $x(n_1, n_2) h(n_1, n_2) \longleftrightarrow \frac{1}{N_1 N_2} X(k_1, k_2) \circledast \circledast \; H(k_1, k_2)$.

Therefore, the computational complexity is of the order $O(kN^4)$ which is very high for most practical applications. Several algorithms have been proposed for the calculation of the 2-d DFTs. The simplest of them is the so-called row–column FFT (RCFFT) algorithm that employs one-dimensional FFTs along rows and columns, respectively. The 2-d DFT (2.2.11) can be decomposed in N_1 DFTs along rows and in N_2 DFTs along columns as follows:

$$X'(n_1, k_2) = \sum_{n_2=0}^{N_2-1} x(n_1, n_2) W_{N_2}^{n_2 k_2} \qquad (2.3.1)$$

$$X(k_1, k_2) = \sum_{n_1=0}^{N_1-1} X'(n_1, k_2) W_{N_1}^{n_1 k_1} \qquad (2.3.2)$$

Therefore, the computation of 2-d DFTs can be split into the computation of 1-d DFTs along rows and columns, respectively. The resulting algorithm is shown in Figure 2.3.1.

```
#define SWAP(a,b) FTEMP=(a);(a)=(b);(b)=FTEMP
float FTEMP;

int rcfft(matsre,matsim,matdre,matdim,mode,ND1,MD1,N1,M1,N2,M2)
matrix matsre,matsim,matdre,matdim;
int mode;
int ND1,MD1;
int N1,M1,N2,M2;

/* Subroutine to compute the 2-dimensional Discrete Fourier
   Transform by using the row-column decomposition method in
   conjunction with the 1-dimensional FFT.
   matsre,matsim: source matrices(real and imaginary components)
   matdre,matdim: destination matrices(real and imag components)
   mode : 0 forward / 1 inverse transform
   ND1,MD1 : destination coordinates
   N1,M1 : upper left  corner of source matrices
   N2,M2 : lower right corner of source matrices      */

{ int i,j,m,ii,jj,nr,nc,nr1,nc1,rows,cols,sign,intme,max;
  float *RE, *IM;
  float *WR, *WI;
  double theta;
  float FTEMP,fme;

  rows=N2-N1; cols=M2-M1;
  max=rows>cols?rows:cols;
```

```
/* 1-d temporary vectors for the real and
   imaginary parts of rows (columns) */
RE=(float *)malloc((unsigned int)max*sizeof(float));
IM=(float *)malloc((unsigned int)max*sizeof(float));

/* 1-d temporary vectors for the real and imaginary
   parts of sin and cos  coefficients */
WR=(float *)malloc((unsigned int)(max/2)*sizeof(float));
WI=(float *)malloc((unsigned int)(max/2)*sizeof(float));

/* FFT is performed always on the destination buffers */
if(matsre!=matdre || ND1!=N1 || MD1!=M1)
   for(nr=0; nr<rows; nr++)
   for(nc=0; nc<cols; nc++)
   matdre[ND1+nr][MD1+nc]=matsre[N1+nr][M1+nc];

if(matsim!=matdim || ND1!=N1 || MD1!=M1)
   for(nr=0; nr<rows; nr++)
   for(nc=0; nc<cols; nc++)
   matdim[ND1+nr][MD1+nc]=matsim[N1+nr][M1+nc];

/* Perform bit reversion along rows */
j=0;
for(i=0; i<cols; i++)
   { if(j>i)
      { ii=MD1+i; jj=MD1+j;
        for(nr=0; nr<rows; nr++)
         { nr1=ND1+nr;
           SWAP(matdre[nr1][jj],matdre[nr1][ii]);
           SWAP(matdim[nr1][jj],matdim[nr1][ii]);
         }
      }
     m=cols>>1;
     while(m>0 && j>=m) { j -= m; m >>= 1; }
     j += m;
   }

/* Perform bit reversion along columns*/
j=0;
for(i=0; i<rows; i++)
   { if(j > i)
      { ii=ND1+i; jj=ND1+j;
        for(nc=0; nc<cols; nc++)
         { nc1=MD1+nc;
```

```
                    SWAP(matdre[jj][nc1],matdre[ii][nc1]);
                    SWAP(matdim[jj][nc1],matdim[ii][nc1]);
                }
            }
        m=rows>>1;
        while(m>0 && j>=m) { j -= m; m >>= 1; }
        j += m;
    }

/* Calculate the sin and cos coefficients */
sign=mode?1:-1;

theta=sign*6.28318530717959/cols;
for(i=0; i<cols/2; i++)
    { WR[i]=cos(i*theta);
      WI[i]=sin(i*theta);
    }

/* Perform row transform */
for(nr=0; nr<rows; nr++)
    { for(nc=0; nc<cols; nc++)
        { RE[nc]=matdre[ND1+nr][MD1+nc];
          IM[nc]=matdim[ND1+nr][MD1+nc];
        }
      fft1d(RE,IM,cols,WR,WI);
      for(nc=0; nc<cols; nc++)
        { matdre[ND1+nr][MD1+nc]=RE[nc];
          matdim[ND1+nr][MD1+nc]=IM[nc];
        }
    }

/* If the column number is different from row number,
   calculate the sin and cos coefficients */
if(rows!=cols)
    { theta=sign*6.28318530717959/rows;
      for(i=0; i<rows/2; i++)
        { WR[i]=cos(i*theta);
          WI[i]=sin(i*theta);
        }
    }

/* Perform column transform */
for(nc=0; nc<cols; nc++)
    { for(nr=0; nr<rows; nr++)
        { RE[nr]=matdre[ND1+nr][MD1+nc];
```

```
            IM[nr]=matdim[ND1+nr][MD1+nc];
          }
        fft1d(RE,IM,rows,WR,WI);
        for(nr=0; nr<rows; nr++)
          { matdre[ND1+nr][MD1+nc]=RE[nr];
            matdim[ND1+nr][MD1+nc]=IM[nr];
          }
      }

  free(RE); free(IM); free(WR); free(WI);

  /* Divide by (N2-N1)*(M2-M1) in the case of inverse FFT */
  if(mode)
    { fme=(float)rows*(float)cols;
      for(nr=0; nr<rows; nr++)
      for(nc=0; nc<cols; nc++)
        { matdre[ND1+nr][MD1+nc] /= fme ;
          matdim[ND1+nr][MD1+nc] /= fme ;
        }
    }

  return(0);
}
```

Fig. 2.3.1 Row–column FFT algorithm.

The subroutine rcfft() employs a 1-d FFT routine fft1d(). This can be any one-dimensional FFT algorithm [OPP89], [BRI74], [NUS81], [BUR85]. The sine and cosine functions are calculated externally and are passed to fft1d() in order to speed up execution, since their calculation must be performed only once for row transform and once for column transform. Furthermore, bit reversion along rows and columns is calculated externally before the execution of row and column transforms. rcfft() can perform direct or inverse transform according to the sign of the sine function that is passed to the subroutine fft1d(). The number of complex multiplications needed for a radix-2 FFT algorithm of length N is $\frac{N}{2} \log_2 N$. The output of the row–column FFT algorithm is two floating point matrices containing the real and the imaginary parts $X_R(k_1, k_2), X_I(k_1, k_2)$ of the image transform, respectively. In many cases, the transform magnitude and phase are required:

$$X_M(k_1, k_2) = \sqrt{X_R(k_1, k_2)^2 + X_I(k_1, k_2)^2} \qquad (2.3.3)$$

$$\phi(k_1, k_2) = \tan^{-1}(X_I(k_1, k_2)/X_R(k_1, k_2)) \qquad (2.3.4)$$

A routine for the transformation from rectangular to polar coordinates and vice versa is shown in Figure 2.3.2.

```c
int reim2magnphase(matre,matim, matmagn,matphase,
                   ND1,MD1, N1,M1,N2,M2)
matrix matre,matim,matmagn,matphase;
int ND1,MD1;
int N1,M1,N2,M2;

/* Subroutine to convert the real-imaginary representation
   of a signal to the magnitude-phase one
   matre,matim: source matrices(real and imaginary components)
   matmagn,matphase: destination matrices
                          (magnitude and phase components)
   ND1,MD1 : destination coordinates
   N1,M1 : upper left  corner of source matrices
   N2,M2 : lower right corner of source matrices
   NOTE!  -3.14 < phase angle <= 3.14   */

{ int nc,nr,NDD,MDD;
  float fmere,fmeim;

  NDD=ND1-N1;
  MDD=MD1-M1;

  for(nr=N1; nr<N2; nr++)
  for(nc=M1; nc<M2; nc++)
     { fmere=matre[nr][nc];
       fmeim=matim[nr][nc];
       matmagn[NDD+nr][MDD+nc]=
                sqrt((double)(fmere*fmere+fmeim*fmeim));
       if(fabs((double)fmere) > 1.0e-37)
         matphase[NDD+nr][MDD+nc]=
                atan2((double)fmeim,(double)fmere);
       else if(fmeim >= 0.0)
           matphase[NDD+nr][MDD+nc]= 3.14159/2.0;
             else matphase[NDD+nr][MDD+nc]= -3.14159/2.0;
     }

  return(0);
}

int magnphase2reim(matmagn,matphase, matre,matim,
                   ND1,MD1, N1,M1,N2,M2)
matrix matmagn,matphase,matre,matim;
int ND1,MD1;
int N1,M1,N2,M2;
```

```
/* Subroutine to convert the magnitude-phase representation
   of a signal to the real-imaginary one
   matre,matim: source matrices(magnitude and phase components)
   matmagn,matphase : destination matrices
                (real and imaginary components)
   ND1,MD1 : destination coordinates
   N1,M1 : upper left  corner of source matrices
   N2,M2 : lower right corner of source matrices        */

{ int nc,nr,NDD,MDD;
  float fmemagn,fmephase;

  NDD=ND1-N1;
  MDD=MD1-M1;

  for(nr=N1; nr<N2; nr++)
  for(nc=M1; nc<M2; nc++)
     { fmemagn=matmagn[nr][nc];
       fmephase=matphase[nr][nc];
       matre[NDD+nr][MDD+nc]=fmemagn*cos((double)fmephase);
       matim[NDD+nr][MDD+nc]=fmemagn*sin((double)fmephase);
     }

  return(0);
}
```

Fig. 2.3.2 Routines for transformation from polar to rectangular coordinates and vice versa.

The display of the magnitude of the transform can give useful information about the frequency content of an image. The magnitude of the transform is stored in a **matrix**. It must be scaled and transformed to an **image** (array of unsigned characters), before it can be displayed. This transformation is shown in the algorithm of Figure 2.3.3. A logarithmic display is used in order to facilitate the display of a relatively large dynamic range. Finally, the DFT output is shifted circularly so that the DC term $X(0,0)$ appears at the center of the display window.

```
int fft2image(matre,matim, immagn, ND1,MD1, N1,M1,N2,M2)
matrix matre,matim;
image immagn;
int ND1,MD1;
int N1,M1,N2,M2;
```

```
/* Subroutine to compute the image of the magnitude of DFT
   matre,matim: source matrices
                (real and imaginary components of DFT)
   immagn : destination image
   ND1,MD1 : destination coordinates
   N1,M1 : upper left  corner of source matrices
   N2,M2 : lower right corner of source matrices      */

{ int nc,nr,cols,rows,cols2,rows2;
  float temp,max,coef;
  float fmere,fmeim;
  matrix matmagn;

  rows=N2-N1;
  cols=M2-M1;
  rows2=rows/2;
  cols2=cols/2;

  matmagn=matf2(rows,cols);

  /* Find the maximal element of the FFT magnitude */
  max=0.0;
  for(nr=N1; nr<N2; nr++)
  for(nc=M1; nc<M2; nc++)
     { fmere=matre[nr][nc];
       fmeim=matim[nr][nc];
       temp=sqrt((double)(fmere*fmere+fmeim*fmeim));
       matmagn[nr-N1][nc-M1]=temp;
       if(temp>max) max=temp;
     }

  /* Perform scaling so that the FFT magnitude can be
     displayed in the [0,..,255] range
     Perform shifts so that the transform point (0,0)
     appears at the center of the image */
  coef=255.0/log((double)(1.0+max));

  for(nr=0; nr<rows2; nr++)
     { for(nc=0; nc<cols2; nc++)
          immagn[ND1+nr][MD1+nc]=(unsigned char)(0.5+log((double)
              (1.0+matmagn[nr+rows2][nc+cols2]))*coef);
       for(nc=cols2; nc<cols; nc++)
          immagn[ND1+nr][MD1+nc]=(unsigned char)(0.5+log((double)
              (1.0+matmagn[nr+rows2][nc-cols2]))*coef);
     }
```

```
    for(nr=rows2; nr<rows; nr++)
    { for(nc=0; nc<cols2; nc++)
        immagn[ND1+nr][MD1+nc]=(unsigned char)(0.5+log((double)
           (1.0+matmagn[nr-rows2][nc+cols2]))*coef);
      for(nc=cols2; nc<cols; nc++)
        immagn[ND1+nr][MD1+nc]=(unsigned char)(0.5+log((double)
           (1.0+matmagn[nr-rows2][nc-cols2]))*coef);
    }

    mfreef2(matmagn,rows);
    return(0);
}
```

Fig. 2.3.3 Routine for the transformation of the DFT magnitude to an image that can be displayed.

The number of complex multiplications needed for the RCFFT is:

$$C = N_1 \frac{N_2}{2} \log_2 N_2 + N_2 \frac{N_1}{2} \log_2 N_1 = \frac{N_1 N_2}{2} \log_2(N_1 N_2) \qquad (2.3.5)$$

The radix-2 FFT algorithm of length N requires $N \log_2 N$ complex additions. Thus, the number of complex additions needed by the RCFFT is:

$$A = N_1 N_2 \log_2(N_1 N_2) \qquad (2.3.6)$$

The RCFFT reduces the order of the computational complexity from $O(kN^4)$ to $O(kN^2 \log_2 N)$. The computational savings over the direct computation of the 2-d DFT are very large. The direct 1024×1024 DFT computation requires 2^{40} complex multiplications, whereas the RCFFT requires $10 \cdot 2^{20}$ multiplications according to (2.3.5). Thus a reduction by a factor of approximately 500 is achieved. The speed-up obtained by the RCFFT and the use of well-known 1-d FFT routines explain the popularity of this algorithm. Other algorithms described in this chapter may be even faster. However, they are less popular because their programming requires extensive knowledge of digital signal processing techniques for fast algorithm development.

2.4 MEMORY PROBLEMS IN 2-D DFT CALCULATIONS

The calculation of the 2-d DFT by using the RCFFT algorithm requires the storage of the entire image array $X(n_1, n_2)$, $0 \leq n_1 \leq N_1 - 1$, $0 \leq n_2 \leq N_2 - 1$ in the RAM of the computer. Since complex arithmetic is required, the memory problems become severe for image sizes larger than 256×256. For example, if 8 bytes are needed for the storage of a complex number, an

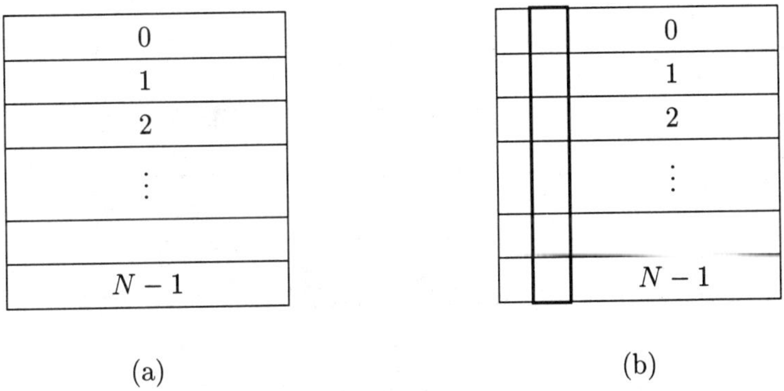

Fig. 2.4.1 (a) Storage of an image in a row-wise manner. Each image row occupies one data record. (b) Access of image data in a column-wise manner.

8 MB RAM is needed for the calculation of the 2-d DFT of an image having size 1024×1024. Such memory requirements cannot be met easily.

A solution to this problem is the storage of the image on a hard disk in a row-wise manner as shown in Figure 2.4.1. Each image row occupies one record. For reasons of simplicity, we assume that the image dimensions are $N \times N$. Let us suppose that the computer's RAM can store only one row (or column). The row FFTs of the RCFFT algorithm can be calculated fairly easily by fetching one row at a time to the RAM. Thus $2N$ I/O operations (read and write) are required for the computation of the row transforms. The computation of the column transforms requires much more I/O operations. All image records must be sequentially fetched to the RAM in order to form a complete image column, as can be seen in Figure 2.4.1b. This column must be transformed and the transform results must be redistributed in the image records. The storage of the results is performed by reading the image records, by writing the corresponding transform coefficient $X(k_1, k_2)$ and by storing the record back to the hard disk. This step can be combined with the read operations for the next column to be transformed. Thus $3N$ I/O operations are needed for the transform of the first column and $2N$ operations for the transform of the subsequent $N-1$ columns. The total number of I/O operations for the entire transform is:

$$N_{IO} = 2N + 3N + 2N(N-1) = 2N^2 + 3N \qquad (2.4.1)$$

This implementation of the RCFFT is extremely slow, because disk access is a slow operation. The situation is better if K signal rows or columns ($K < N$) can be stored in RAM simultaneously. In this case, K columns can be fetched and transformed in one step. The total number of I/O operations required in

70 DIGITAL IMAGE TRANSFORM ALGORITHMS

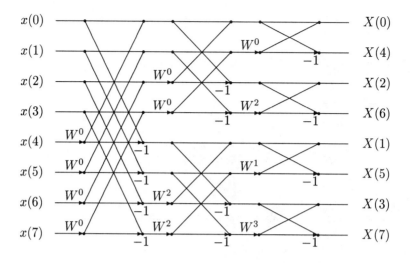

Fig. 2.4.2 Radix-2 FFT algorithm for $N=8$.

this case is given by [LIM90]:

$$N_{IO} = \frac{2N^2}{K} + 3N \quad (2.4.2)$$

Although the speed-up obtained is linearly related to K, many I/O operations are still needed.

A further speed-up can be obtained by combining the row transform computation with the computation of a part of the column transform [PIT87]. Let us suppose that the image dimension N is a power of 2 ($N = 2^n$) and that K is a power of 2 ($K = 2^k$) as well. If K appropriately chosen image rows are stored on RAM, their row transform can be calculated without extra I/O operations. Furthermore, the first k steps of the column transform can be computed and the results stored back on the hard disk. This stage requires only $2K$ I/O operations. An example of this procedure can be given for an 8×8 transform. The radix-2 FFT for length $N=8$ is given in Figure 2.4.2. Let us suppose that we can store only two image rows on RAM ($K = 2$). Only the first step of the column transform can be calculated in the first stage. This procedure can be repeated N/K times until the entire row transform and the first $k = \log_2 K$ steps of all column transforms are computed. At the second stage, K rows (appropriately chosen) are fetched to the RAM and the next k steps of the column transforms are computed. The second stage is repeated N/K times, until all column transforms are advanced by k steps. At the $\lfloor n/k \rfloor$-th stage, the column transforms are completed. One more stage is needed to perform column unscrambling. The algorithm is shown in C-like format in Figure 2.4.3. Subroutine fft_IO_mc() can perform both forward

MEMORY PROBLEMS IN 2-D DFT CALCULATIONS 71

and inverse 2-d FFTs. It uses a subroutine `fft()` to perform row transforms. This can be any one-dimensional FFT algorithm [OPP89], [BRI74], [NUS81]. The computation of sine and cosine functions is performed externally and the resulting tables are passed to `fft()`. The last step of the forward FFT is output unscrambling. In the case of inverse FFT, subroutine `fft_IO_mc()` performs both output unscrambling and division by N (for the inverse column transform). In this subroutine it is supposed that k divides n.

```
int fft_IO_mc(fr,fi,N1,N2,k,mode)
FILE *fr,*fi;
int N1,N2,k,mode;
/* Subroutine to perform Row-Column FFT for
    images stored on disk files.
    fr,fi: stream files containing the real and imaginary parts
        of 2-d input signal respectively. FFT output
        is stored in these files at exit.
        !!! Important !!! THE FILES MUST BE OPEN
    N1,N2: image sizes. They MUST be powers of 2.
    NOTE:  The data is stored in the files in a row-wise manner:
        x(0,0) x(0,1) ... x(0,N2-1), x(1,0) .. etc.
    k    : number of lines to be read as a block in memory.
        It MUST be a power of 2.
    mode : 0 for forward, 1 for inverse DFT.
*/
{
  matrix xr,xi;
  int i,j,BL,s,BS,k0,N0,l,ak,BB,d,b,s0,n,r,m,p,q,nm,numread;
  register ii;
  float wr,wi,pi=4.0*atan(1.0);
  float xpr,xpi,xqr,xqi;
  float *BRE, *BIM , *WR, *WI;
  int sign;
  sign = mode ? -1 : 1;

  /* Local buffers to store the real and the imaginary
     parts of the image.
     Each local buffer has size k x N2 */
  xr = matf2(k,N2);
  xi = matf2(k,N2);

  /* Local vectors to be used in 1-d FFT calculation */
  BRE = (float *)malloc(N2*sizeof(float));
  BIM = (float *)malloc(N2*sizeof(float));
```

72 DIGITAL IMAGE TRANSFORM ALGORITHMS

```
/* These buffers hold the sine and cosine
   coefficients for fft1d */
WR = (float *)malloc((N2/2)*sizeof(float));
WI = (float *)malloc((N2/2)*sizeof(float));
wr = -sign*2.0*pi/(double)N2;
for(i=0;i<N2/2;i++) { WR[i] = cos(i*wr); WI[i] = sin(i*wr); }

/* Each step s in the FFT computation is divided in BL blocks
   for FFT computation. Each block has size BS/2 butterflies. */
BL = 1; s = 0; BS = N1;

/* Compute log_2(k), log_2(N1) */
k0 = 0; i=k;  while(i>1) { k0++; i>>=1; }
N0 = 0; i=N1; while(i>1) { N0++; i>>=1; }

/* Main loop */
while(s<N0)
{
  BB=BS/k; if (BB==0) BB++;
  for(i=0;i<BL;i++)
    for(j=0;j<BB;j++)
    {
     /* Read at most BS image lines */
     ak = min(k,BS);
     for(l=0;l<ak;l++)
       {
        fseek(fr,(long)((long)
              (i*BS+l*BB+j)*N2*sizeof(float)),SEEK_SET);
        fread(BRE,sizeof(float),N2,fr);
        fseek(fi,(long)((long)
              (i*BS+l*BB+j)*N2*sizeof(float)),SEEK_SET);
        fread(BIM,sizeof(float),N2,fi);

        /* If we are at the first stage of 2-d FFT
           calculation, perform first row transforms and
           store the output in the local buffers xr, xi. */
        if (s==0) fft(BRE,BIM,N2,WR,WI,mode);
        for(ii=0;ii<N2;ii++) xr[l][ii]=BRE[ii];
        for(ii=0;ii<N2;ii++) xi[l][ii]=BIM[ii];
       }

     /* Perform part of the column transform. */
     s0=0;
     for(d=ak/2;d>0;d>>=1)
     {
```

```
            for( b=0 ; b<ak ; b+=(d<<1) )
             for(n=0;n<d;n++)
                {
                   p=b+n;
                   q=b+n+d;
                   r=((i*BS+j+p*BB)%(BS>>s0))<<(s+s0);
                   wr=cos(2*pi*r/N1);
                   wi=-(float)sign*sin(2*pi*r/N1);
                   for(m=0;m<N2;m++)
                    {
                      xpr = xr[p][m] + xr[q][m];
                      xpi = xi[p][m] + xi[q][m];
                      xqr = xr[p][m] - xr[q][m];
                      xqi = xi[p][m] - xi[q][m];
                      xr[p][m] = xpr; xi[p][m] = xpi;
                      xr[q][m] = xqr*wr - xqi*wi;
                      xi[q][m] = xqi*wr + xqr*wi;
                     }
                }
             s0++;
          }

          /* Store intermediate results in the disk file. */
          for(l=0;l<ak;l++)
           {
            for(ii=0;ii<N2;ii++) BRE[ii]=xr[l][ii];
            for(ii=0;ii<N2;ii++) BIM[ii]=xi[l][ii];
             fseek(fr,(long)((long)
                    (i*BS+l*BB+j)*N2*sizeof(float)),SEEK_SET);
             fwrite(BRE,sizeof(float),N2,fr);
             fseek(fi,(long)((long)
                    (i*BS+l*BB+j)*N2*sizeof(float)),SEEK_SET);
             fwrite(BIM,sizeof(float),N2,fi);
           }

       }

    s+=s0;
    BL<<=s0;
    BS>>=s0;
 }

 /* Perform output unscrambling */
 j=0;
 for (i=1;i<N1-1;i++)
```

```c
{ l=N1;
  while (j+l>=N1) l >>= 1;
  j = j%l +l;
  if (j>i)
   {
    fseek(fr,((long)(i*N2)*sizeof(float)),SEEK_SET);
    fread(BRE,sizeof(float),N2,fr);
    fseek(fr,((long)(j*N2)*sizeof(float)),SEEK_SET);
    fread(BIM,sizeof(float),N2,fr);

    /* If an inverse 2-d FFT is calculated, divide
       output elements by N1 */
    if (sign<0)
     for(m=0;m<N2;m++)
       {
        BRE[m]/=(double)N1;
        BIM[m]/=(double)N1;
       }

    fseek(fr,((long)(i*N2)*sizeof(float)),SEEK_SET);
    fwrite(BIM,sizeof(float),N2,fr);
    fseek(fr,((long)(j*N2)*sizeof(float)),SEEK_SET);
    fwrite(BRE,sizeof(float),N2,fr);

    fseek(fi,((long)(i*N2)*sizeof(float)),SEEK_SET);
    fread(BRE,sizeof(float),N2,fi);
    fseek(fi,((long)(j*N2)*sizeof(float)),SEEK_SET);
    fread(BIM,sizeof(float),N2,fi);


    /* If an inverse 2-d FFT is calculated, divide
       output elements by N1 */
    if (sign<0)
     for(m=0;m<N2;m++)
       {
        BRE[m]/=(double)N1;
        BIM[m]/=(double)N1;
       }

    fseek(fi,((long)(i*N2)*sizeof(float)),SEEK_SET);
    fwrite(BIM,sizeof(float),N2,fi);
    fseek(fi,((long)(j*N2)*sizeof(float)),SEEK_SET);
    fwrite(BRE,sizeof(float),N2,fi);
   }
}
```

```
/* If an inverse 2-d FFT is calculated, divide
   output elements by N1 */
if(sign<0)
 for(i=0;i<N1;i++)
  {
    l=0;
    n=i; m=i;
    for(j=0;j<(N0/2);j++)
     { l=l<<1; n=n&0x1; l=l^n;
       n=m>>1; m=m>>1;
     }
    l=l<<((N0+1)/2); n=1<<((N0+1)/2);
    n=n-1; m=i&n;
    if ((l+m)==i)
     {
        fseek(fr,((long)(i*N2)*sizeof(float)),SEEK_SET);
        fread(BRE,sizeof(float),N2,fr);
        fseek(fi,((long)(i*N2)*sizeof(float)),SEEK_SET);
        fread(BIM,sizeof(float),N2,fi);
        for(m=0;m<N2;m++)
          {
            BRE[m]/=(double)N1;
            BIM[m]/=(double)N1;
          }
        fseek(fr,((long)(i*N2)*sizeof(float)),SEEK_SET);
        fwrite(BRE,sizeof(float),N2,fr);
        fseek(fi,(long)(j*N2)*sizeof(float),SEEK_SET);
        fwrite(BIM,sizeof(float),N2,fi);
     }
  }

 free(BRE); free(BIM); free(WR); free(WI);
 mfreef2(xr,k); mfreef2(xi,k);
 return(0);
}
```

Fig. 2.4.3 Modification of the RCFFT algorithm to reduce I/O operations.

This algorithm has $\lfloor n/k \rfloor$ stages. At each stage $2K\frac{N}{K}$ I/O operations are needed. Therefore, the number of the I/O operations required is:

$$N_{IO} = 2N\lfloor n/k \rfloor \tag{2.4.3}$$

Another approach to obtain a reduction of the I/O operations is to use fast matrix transposition algorithms [EKL72]. Let us suppose that a matrix X has size $N \times N$ where $N = 2^n$. It can be split into four submatrices $X_{ij}, i,j = 1,2$, of size $(N/2) \times (N/2)$ each:

$$X = \begin{bmatrix} X_{11} X_{12} \\ X_{21} X_{22} \end{bmatrix} \qquad (2.4.4)$$

The transposed matrix X^T is given by:

$$X^T = \begin{bmatrix} X_{11}^T X_{21}^T \\ X_{12}^T X_{22}^T \end{bmatrix} \qquad (2.4.5)$$

Thus the problem of transposing a matrix of size $N \times N$ is split into four matrix transpositions for matrix sizes $(N/2) \times (N/2)$. Each of the submatrices $X_{ij}, i,j = 1,2$, can be split further into four submatrices and a transposition formula similar to (2.4.5) can be used. This procedure can be repeated until submatrices of size 2×2 are reached. This fast transposition algorithm has $n = \log_2 N$ steps and the transposition can be performed in place. This algorithm is illustrated in Figure 2.4.4 for an 8×8 matrix.

Let us suppose that matrix X is stored in a row-wise manner on the hard disk. Its 2-d DFT can be computed as follows. The first stage computes the row transforms and the first step of the matrix transposition. If the computer's memory has the capacity to store two image rows, two rows are fetched at a time. The rows are transformed and they interchange data, as illustrated in Figure 2.4.4a. This process is repeated until all rows are transformed and the first step of the transposition is completed. The number of I/O operations needed is $2N$. The next $\log_2 N - 1$ stages complete the matrix transposition. At each stage, two rows are fetched at a time and swap part of their data, as shown in Figure 2.4.4b,c. At the last stage, matrix transposition is completed. Two columns are available on RAM at a time and column transforms can be calculated before these columns are stored back on the hard disk. Thus, the last step of the transposition can be combined with the column transforms. Since $2N$ I/O operations are needed at each stage, the total number of I/O operations is given by:

$$N_{IO} = 2N \log_2 N \qquad (2.4.6)$$

If two rows are stored in memory ($K = 2, k = 1$), the same number of I/O operations is needed both for the algorithm fft_IO_mc() and for the fast transposition RCFFT algorithm. The resulting subroutine fft_IO_trans() is shown in Figure 2.4.5. It uses a subroutine fft() to perform 1-d transforms. The fast transposition RCFFT algorithm can be easily modified for the case when $K > 2$ rows can be stored on RAM.

```
int fft_IO_trans(fr,fi,N,mode)
```

x_{00}	x_{01}	x_{02}	x_{03}	x_{04}	x_{05}	x_{06}	x_{07}
x_{10}	x_{11}	x_{12}	x_{13}	x_{14}	x_{15}	x_{16}	x_{17}
x_{20}	x_{21}	x_{22}	x_{23}	x_{24}	x_{25}	x_{26}	x_{27}
x_{30}	x_{31}	x_{32}	x_{33}	x_{34}	x_{35}	x_{36}	x_{37}
x_{40}	x_{41}	x_{42}	x_{43}	x_{44}	x_{45}	x_{46}	x_{47}
x_{50}	x_{51}	x_{52}	x_{53}	x_{54}	x_{55}	x_{56}	x_{57}
x_{60}	x_{61}	x_{62}	x_{63}	x_{64}	x_{65}	x_{66}	x_{67}
x_{70}	x_{71}	x_{72}	x_{73}	x_{74}	x_{75}	x_{76}	x_{77}

(a)

x_{00}	x_{01}	x_{02}	x_{03}	x_{40}	x_{41}	x_{42}	x_{43}
x_{10}	x_{11}	x_{12}	x_{13}	x_{50}	x_{51}	x_{52}	x_{53}
x_{20}	x_{21}	x_{22}	x_{23}	x_{60}	x_{61}	x_{62}	x_{63}
x_{30}	x_{31}	x_{32}	x_{33}	x_{70}	x_{71}	x_{72}	x_{73}
x_{04}	x_{05}	x_{06}	x_{07}	x_{44}	x_{45}	x_{46}	x_{47}
x_{14}	x_{15}	x_{16}	x_{17}	x_{54}	x_{55}	x_{56}	x_{57}
x_{24}	x_{25}	x_{26}	x_{27}	x_{64}	x_{65}	x_{66}	x_{67}
x_{34}	x_{35}	x_{36}	x_{37}	x_{74}	x_{75}	x_{76}	x_{77}

(b)

x_{00}	x_{01}	x_{20}	x_{21}	x_{40}	x_{41}	x_{60}	x_{61}
x_{10}	x_{11}	x_{30}	x_{31}	x_{50}	x_{51}	x_{70}	x_{71}
x_{02}	x_{03}	x_{22}	x_{23}	x_{42}	x_{43}	x_{62}	x_{63}
x_{12}	x_{13}	x_{32}	x_{33}	x_{52}	x_{53}	x_{72}	x_{73}
x_{04}	x_{05}	x_{24}	x_{25}	x_{44}	x_{45}	x_{64}	x_{65}
x_{14}	x_{15}	x_{34}	x_{35}	x_{54}	x_{55}	x_{74}	x_{75}
x_{06}	x_{07}	x_{26}	x_{27}	x_{46}	x_{47}	x_{66}	x_{67}
x_{16}	x_{17}	x_{36}	x_{37}	x_{56}	x_{57}	x_{76}	x_{77}

(c)

x_{00}	x_{10}	x_{20}	x_{30}	x_{40}	x_{50}	x_{60}	x_{70}
x_{01}	x_{11}	x_{21}	x_{31}	x_{41}	x_{51}	x_{61}	x_{71}
x_{02}	x_{12}	x_{22}	x_{32}	x_{42}	x_{52}	x_{62}	x_{72}
x_{03}	x_{13}	x_{23}	x_{33}	x_{43}	x_{53}	x_{63}	x_{73}
x_{04}	x_{14}	x_{24}	x_{34}	x_{44}	x_{54}	x_{64}	x_{74}
x_{05}	x_{15}	x_{25}	x_{35}	x_{45}	x_{55}	x_{65}	x_{75}
x_{06}	x_{16}	x_{26}	x_{36}	x_{46}	x_{56}	x_{66}	x_{76}
x_{07}	x_{17}	x_{27}	x_{37}	x_{47}	x_{57}	x_{67}	x_{77}

(d)

Fig. 2.4.4 Three stages in the transposition of an 8×8 matrix.

```
FILE *fr,*fi;
int N,mode;
/* Subroutine to perform Row-Column FFT of signal stored on
   disk files. Fast matrix transposition is used to calculate
   the column FFTs. Input matrices MUST be square!!!
  fr,fi: stream files containing the real and imaginary parts of
         2-d input signal respectively. FFT output is stored in
         these files at exit. Output is !!! TRANSPOSED !!! with
         respect to the input signal.
         !!! Important !!! THE FILES MUST BE OPEN
   N :input signal size(square matrix).It MUST be a power of 2.
   mode  : 0 for forward, 1 for inverse DFT. */
{
  int pass,se;
```

78 DIGITAL IMAGE TRANSFORM ALGORITHMS

```
register int i,j,k,l;
float *x1r,*x1i,*x2r,*x2i;
float t;
int sign;
float *WR,*WI;

sign = mode ? -1 : 1;
/* Create four 1-d buffers of size N each. */
x1r = (float *)malloc(4*N*sizeof(float));
if (x1r==NULL) return -2010;
x1i = x1r+N;
x2r = x1i+N;
x2i = x2r+N;

/* Create buffers that hold the sine and cosine
   coefficients for fft1d */
WR = (float *)malloc(N*sizeof(float));
WI = WR + N/2;
t = -sign*8.0*atan(1.0)/(double)N;
for(i=0;i<N/2;i++) { WR[i] = cos(i*t); WI[i] = sin(i*t); }

pass=1;
se=N/2;

while ( se>0 )
{
  for(i=0;i<pass;i++)
  for(j=0;j<se;j++)
  {
    /* Read input rows */
    fseek(fr,(long)(2*i*se+j)*N*sizeof(float),SEEK_SET);
    fread(x1r,sizeof(float),N,fr);
    fseek(fi,(long)(2*i*se+j)*N*sizeof(float),SEEK_SET);
    fread(x1i,sizeof(float),N,fi);
    fseek(fr,(long)(2*i*se+se+j)*N*sizeof(float),SEEK_SET);
    fread(x2r,sizeof(float),N,fr);
    fseek(fi,(long)(2*i*se+se+j)*N*sizeof(float),SEEK_SET);
    fread(x2i,sizeof(float),N,fi);

    /* Calculate row FFTs in the first pass */
    if(pass==1)
    {
     fft(x1r,x1i,N,WR,WI,mode);
     fft(x2r,x2i,N,WR,WI,mode);
    }
```

```
      /* Perform transpositions */
      for(k=0;k<pass;k++)
      for(l=0;l<se;l++)
      {
        t=x1r[2*k*se+se+l];
        x1r[2*k*se+se+l]=x2r[2*k*se+l]; x2r[2*k*se+l]=t;
        t=x1i[2*k*se+se+l];
        x1i[2*k*se+se+l]=x2i[2*k*se+l]; x2i[2*k*se+l]=t;
      }

      /* After last transposition, perform column FFTs
         before storing the output to disk */
      if(se==1)
      {
        fft(x1r,x1i,N,WR,WI,mode);
        fft(x2r,x2i,N,WR,WI,mode);
      }

      /* Store intermediate or output data on disk */
      fseek(fr,(long)(2*i*se+j)*N*sizeof(float),SEEK_SET);
      fwrite(x1r,sizeof(float),N,fr);
      fseek(fi,(long)(2*i*se+j)*N*sizeof(float),SEEK_SET);
      fwrite(x1i,sizeof(float),N,fi);
      fseek(fr,(long)(2*i*se+se+j)*N*sizeof(float),SEEK_SET);
      fwrite(x2r,sizeof(float),N,fr);
      fseek(fi,(long)(2*i*se+se+j)*N*sizeof(float),SEEK_SET);
      fwrite(x2i,sizeof(float),N,fi);
    }
    pass<<=1; se>>=1;
  }

  free(x1r); free(WR);
  return(0);
}
```

Fig. 2.4.5 Fast matrix transposition RCFFT algorithm.

If the image $x(n_1, n_2)$ is real, the complex conjugate property:

$$X(k_1, k_2) = X^*(((N_1 - k_1))_{N_1}, ((N_2 - k_2))_{N_2}) \qquad (2.4.7)$$

can be used to reduce the 2-d DFT memory requirements considerably [PIT86]. Let us suppose that both N_1, N_2 are powers of 2 and that the image is stored

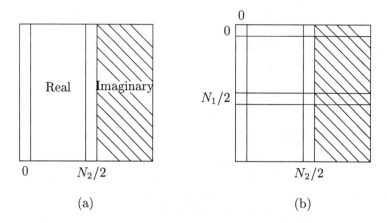

Fig. 2.4.6 In-place storage of the 2-d DFT of a real-valued signal. (a) Storage after row transform; (b) storage after column transform. The hatched areas denote the storage places for the imaginary part of the transform.

on a matrix A of size $N_1 \times N_2$. The first stage of the RCFFT:

$$X'(n_1, k_2) = \sum_{n_2=0}^{N_2-1} x(n_1, n_2) W_{N_2}^{n_2 k_2} \qquad (2.4.8)$$

is a row transform that processes the following conjugate symmetry:

$$X'(n_1, k_2) = X'^*(n_1, ((N_2 - k_2))_{N_2}) \qquad (2.4.9)$$

The columns $X'(n_1, 0), X'(n_1, N_2/2)$ are real numbers and can be stored in place at the matrix positions $A(n_1, 0), A(n_1, N_2/2), 0 \le n_1 \le N_1 - 1$. The property (2.4.9) can be used to store the real and imaginary parts of $X'(n_1, k_2)$ in place, as follows:

$$Re[X'(n_1, k_2)] \to A(n_1, k_2) \quad 1 \le k_2 < N_2/2, \ 0 \le n_1 \le N_1 - 1 \quad (2.4.10)$$
$$Im[X'(n_1, k_2)] \to A(n_1, N_2 - k_2)$$
$$1 \le k_2 < N_2/2, \ 0 \le n_1 \le N_1 - 1 \qquad (2.4.11)$$

This storage scheme requires only 50% of the memory required for the storage of the complete complex signal $X'(n_1, k_2)$. Another advantage is that it is performed in place. The storage scheme is shown in Figure 2.4.6a. The result of the column transform:

$$X(k_1, k_2) = \sum_{n_1=0}^{N_1-1} X'(n_1, k_2) W_{N_1}^{n_1 k_1} \qquad (2.4.12)$$

satisfies the conjugate property (2.4.7). The samples $X(0,0)$, $X(N_1/2,0)$, $X(0,N_2/2)$, $X(N_1/2,N_2/2)$ are real and can be stored on the corresponding positions of matrix A. The samples $X(k_1,0), X(k_1,N_2/2)$ can be stored as follows:

$$Re[X(k_1,0)] \to A(k_1,0) \quad 1 \le k_1 < N_1/2 \qquad (2.4.13)$$
$$Im[X(k_1,0)] \to A(N_1 - k_1, 0) \quad 1 \le k_1 < N_1/2$$
$$Re[X(k_1,N_2/2)] \to A(k_1,N_2/2) \quad 1 \le k_1 < N_1/2 \qquad (2.4.14)$$
$$Im[X(k_1,N_2/2)] \to A(N_1 - k_1, N_2/2) \quad 1 \le k_1 < N_1/2$$

The real and imaginary parts of the samples $X(k_1,k_2)$, $1 \le k_1 \le N_1 - 1$, $1 \le k_2 < N_2/2$, can be stored as follows:

$$Re[X(k_1,k_2)] \to A(k_1,k_2) \quad 1 \le k_1 \le N_1-1, 1 \le k_2 < N_2/2 \qquad (2.4.15)$$
$$Im[X(k_1,k_2)] \to A(((N_1-k_1))_{N_1}, ((N_2-k_2))_{N_2})$$
$$1 \le k_1 \le N_1-1, 1 \le k_2 < N_2/2 \qquad (2.4.16)$$

```
int real_2d_fft(x,N1,N2)
matrix x;
int N1,N2;
{
/* Subroutine to perform Row-Column FFT of a
   real two-dimensional signal.
   x : input and output image matrix.
   N1,N2 : image size. They MUST be powers of 2. */


   register int i,j;
   int N22,N222,jr,ji;
   float *Xr,*Xi,*WR,*WI;
   float wr,wi,wnr,wni,wr1,pi2=8.0*atan(1.0);

   /* Create local vectors to be used in the 1-d FFT */
   N22=N2/2; N222=N22+1;
   i = max(N22+1,N1);
   Xr = (float *) malloc(i*sizeof(float));
   Xi = (float *) malloc(i*sizeof(float));

   /* Create and load vectors for cosine and sine functions */
   WR = (float *) malloc(max(N22,N1)*sizeof(float));
   WI = WR + max(N22,N1)/2;

   wr = -pi2/(double)N22;
   for(j=0;j<N22/2;j++) { WR[j] = cos(j*wr); WI[j] = sin(j*wr);}
```

```
wnr=cos(pi2/N2); wni=-sin(pi2/N2);
for(i=0;i<N1;i++)
{
 /* First calculate row FFTs. N2/2-point complex DFTs
    are used, because two k-point REAL DFTs can be
    calculated by a k-point complex DFT. In our case, k=N2/2
    and the two real inputs are taken to be x(i,2*j), and
    x(i,2*j+1). The odd-numbered and even-numbered elements
    are transformed separately by using one N2/2-point FFT
    and their output is combined to produce the N2-point
    FFT output.*/

  /*  Store x(i,2*j) and x(i,2*j+1) into Xr,Xi respectively.*/
  for(j=0;j<N22;j++)
       { Xr[j]=x[i][j<<1]; Xi[j]=x[i][(j<<1)+1]; }

  /*  Calculate DFT */
  fft(Xr,Xi,N22,WR,WI,0);  Xr[N22]=Xr[0]; Xi[N22]=Xi[0];

  /* Produce N2-point FFT output. The complex output
     is stored in place. */
  wr=1.0; wi=0.0;
  for(j=0;j<N22;j++)
  {
    /* Real part of result */
    x[i][j]=0.5*(Xr[j]+Xr[N22-j]+ wr*(Xi[j]+Xi[N22-j])
           +wi*(Xr[j]-Xr[N22-j]));
    /* Imaginary part of result */
    if(j!=0)
      x[i][N2-j]=0.5*(Xi[j]-Xi[N22-j]-wr* (Xr[j]-Xr[N22-j])
              +wi*(Xi[j]+Xi[N22-j]));
    wr1 = wr*wnr - wi*wni;
    wi = wr*wni + wi*wnr;
    wr=wr1;
  }
  x[i][N22] = Xr[0] - Xi[0] ;
}

/* Perform column FFTs having length N1. */
wr = -pi2/(double)N1;
for(i=0;i<N1/2;i++) {WR[i]= cos(i*wr); WI[i]= sin(i*wr);}

for(j=0;j<N222;j++)
{ for(i=0;i<N1;i++)
  {
```

```
        Xr[i]=x[i][j];
          /* Initialize vector Xi for the real-
             valued rows 0, N1/2 */
          if ((j==0) || (j==N22)) Xi[i] = 0.;
            else Xi[i]=x[i][N2-j];
        }
     fft(Xr,Xi,N1,WR,WI,0);

     /* Store the results in place. */
     if ((j==0) || (j==N22))
       {
        for(i=0;i<N1/2;i++)
          {
            x[i][j] = Xr[i];
            if(i!=0) x[N1-i][j] = Xi[i];
          }
         x[N1/2][j] = Xr[N1/2];
       }
       else
        for(i=0;i<N1;i++)
          {
           x[i][j] = Xr[i];
           if(i==0) x[0][N2-j]=Xi[i];
             else if(i==N1/2) x[N1/2][N2-j]=Xi[i];
                 else x[N1-i][N2-j]=Xi[i];
          }
   }

  free(WR); free(Xr); free(Xi);
  return(0);
}

int reffft2images(mat1,mat2,N1,N2,)
matrix mat1,mat2;
int N1,N2;
/* Subroutine to unscramble the output of real_2d_fft().
   mat1: input matrix.
         It contains the real 2-d DFT part at exit.
   mat2: output matrix containing the imaginary 2-d DFT part.
   N1,N2: image size */
{
  int i,j;

   /* The imaginary part of points (0,0), (0,N2/2),
      (N1/2,0), (N1/2,N2/2) */
```

```
mat2[0][0]=0; mat2[0][N2/2]=0;
mat2[N1/2][0]=0; mat2[N1/2][N2/2]=0;

/* Unscramble row 0 */
for(j=1; j<N2/2; j++)
 {
  mat2[0][j]=mat1[0][N2-j];
  mat2[0][N2-j]=-mat2[0][j];
 }

for(j=1; j<N2/2; j++)
  mat1[0][N2-j]=mat1[0][j];

/* Unscramble row N1/2 */
for(j=1; j<N2/2; j++)
 {
  mat2[N1/2][j]=mat1[N1/2][N2-j];
  mat2[N1/2][N2-j]=-mat2[N1/2][j];
 }

for(j=1; j<N2/2; j++)
  mat1[N1/2][N2-j]=mat1[N1/2][j];

for(i=N1/2+1; i<N1; i++)
 for(j=0; j<N2; j++)
  mat2[i-N1/2][j]=mat1[i][j];

/* Unscramle column 0, N2/2 */
for(i=1; i<N1/2;i++)
 { mat2[i][0]=mat1[N1-i][0];
   mat2[N1-i][0]=-mat2[i][0];
   mat1[N1-i][0]=mat1[i][0];
   mat2[i][N2/2]=mat1[N1-i][N2/2];
   mat2[N1-i][N2/2]=-mat2[i][N2/2];
   mat1[N1-i][N2/2]=mat1[i][N2/2];
 }

/* Unscramble rest elements (two square blocks) */
for(i=1; i<N1/2;i++)
 for(j=1;j<N2/2;j++)
   { mat2[i][j]=mat1[N1-i][N2-j];
     mat2[N1-i][N2-j]=-mat2[i][j];
     mat1[N1-i][N2-j] = mat1[i][j];
   }
```

```
  for(i=N1/2+1; i<N1;i++)
   for(j=1;j<N2/2;j++)
    { mat2[i][j]=mat1[N1-i][N2-j];
      mat2[N1-i][N2-j]=-mat2[i][j];
      mat1[N1-i][N2-j] = mat1[i][j];
    }

 return(0);
}
```

Fig. 2.4.7 Algorithm for the in-place calculation of the 2-d DFT of real-valued signals.

This algorithm saves both memory and computational effort, because only the $N_1/2$ column DFTs must be computed, owing to the conjugate symmetry (2.4.7). Subroutine `real_2d_fft()` illustrating this algorithm is shown in Figure 2.4.7. It is accompanied by subroutine `refft2images()`, which performs output unscrambling. Subroutine `real_2d_fft()` uses a subroutine `fft()` to perform one-dimensional transforms. The above-mentioned algorithm can be easily extended to images having odd dimensions N_1, N_2 and to multidimensional images as well [PIT86].

2.5 VECTOR-RADIX FAST FOURIER TRANSFORM ALGORITHM

The row–column approach decomposes the 2-d DFT into row and column transforms. However, several other decompositions are possible. For notational simplicity, we shall assume that the image to be transformed is square having size $N_1 = N_2 = N$. Its DFT can be split it into four 2-d DFTs of size $(N/2) \times (N/2)$ each, by following a *decimation-in-time* approach [OPP89]:

$$X(k_1, k_2) = \sum_{n_1=0}^{N-1} \sum_{n_2=0}^{N-1} x(n_1, n_2) W_N^{n_1 k_1} W_N^{n_2 k_2} = G_{ee}(k_1, k_2)$$
$$+ G_{eo}(k_1, k_2) W_N^{k_2} + G_{oe}(k_1, k_2) W_N^{k_1} + G_{oo}(k_1, k_2) W_N^{k_1+k_2} \quad (2.5.1)$$

The terms $G_{ee}(k_1, k_2), G_{eo}(k_1, k_2), G_{oe}(k_1, k_2), G_{ee}(k_1, k_2)$ denote sums over the samples $x(n_1, n_2)$ for even/odd indices n_1, n_2:

$$G_{ee}(k_1, k_2) = \sum_{l_1=0}^{N/2-1} \sum_{l_2=0}^{N/2-1} x(2l_1, 2l_2) W_N^{2l_1 k_1} W_N^{2l_2 k_2} \quad (2.5.2)$$

$$G_{eo}(k_1, k_2) = \sum_{l_1=0}^{N/2-1} \sum_{l_2=0}^{N/2-1} x(2l_1, 2l_2 + 1) W_N^{2l_1 k_1} W_N^{2l_2 k_2} \quad (2.5.3)$$

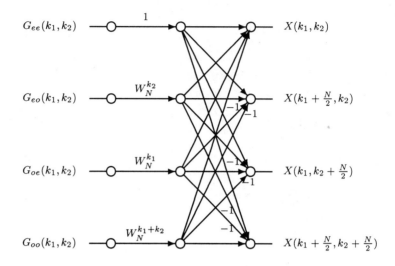

Fig. 2.5.1 Radix-2 × 2 butterfly.

$$G_{oe}(k_1, k_2) = \sum_{l_1=0}^{N/2-1} \sum_{l_2=0}^{N/2-1} x(2l_1+1, 2l_2) W_N^{2l_1 k_1} W_N^{2l_2 k_2} \quad (2.5.4)$$

$$G_{oo}(k_1, k_2) = \sum_{l_1=0}^{N/2-1} \sum_{l_2=0}^{N/2-1} x(2l_1+1, 2l_2+1) W_N^{2l_1 k_1} W_N^{2l_2 k_2} \quad (2.5.5)$$

Each sum is a 2-d DFT of size $(N/2) \times (N/2)$. It can be easily shown that the output samples $X(k_1 + N/2, k_2)$, $X(k_1, k_2 + N/2)$, $X(k_1 + N/2, k_2 + N/2)$ can be expressed in terms of the sums (2.5.2–2.5.5):

$$X(k_1 + N/2, k_2) = G_{ee}(k_1, k_2) + W_N^{k_2} G_{eo}(k_1, k_2)$$
$$- W_N^{k_1} G_{oe}(k_1, k_2) - W_N^{k_1+k_2} G_{oo}(k_1, k_2) \quad (2.5.6)$$

$$X(k_1, k_2 + N/2) = G_{ee}(k_1, k_2) - W_N^{k_2} G_{eo}(k_1, k_2)$$
$$+ W_N^{k_1} G_{oe}(k_1, k_2) - W_N^{k_1+k_2} G_{oo}(k_1, k_2) \quad (2.5.7)$$

$$X(k_1 + N/2, k_2 + N/2) = G_{ee}(k_1, k_2) - W_N^{k_2} G_{eo}(k_1, k_2)$$
$$- W_N^{k_1} G_{oe}(k_1, k_2) + W_N^{k_1+k_2} G_{oo}(k_1, k_2) \quad (2.5.8)$$

The relations (2.5.1), (2.5.6–2.5.8) form a 2 × 2 butterfly shown in Figure 2.5.1. The $(N/2) \times (N/2)$ DFTs (2.5.2–2.5.5) can be split into DFTs of size $(N/4) \times (N/4)$ and this procedure can continue until DFTs of length 2 × 2 are reached. The resulting 2-d DFT algorithm is called the vector-radix FFT (VRFFT) [HAR77], [RIV77]. The flow diagram of a 4 × 4 vector-radix FFT is shown in Figure 2.5.2.

VECTOR-RADIX FAST FOURIER TRANSFORM ALGORITHM

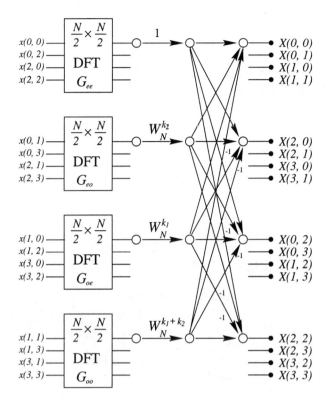

Fig. 2.5.2 Flow diagram of a 4 × 4 vector-radix FFT.

```
int vrfft(matsre,matsim,matdre,matdim,mode,ND1,MD1,N1,M1,N2,M2)
matrix matsre,matsim,matdre,matdim;
int mode;
int ND1,MD1;
int N1,M1,N2,M2;

/* Subroutine to compute the 2-dimensional Discrete Fourier
   Transform by using the vector-radix 2 x 2 method.
   matsre,matsim : source matrices
                   (real and imaginary components)
   matdre,matdim : destination matrices
                   (real and imaginary components)
   mode : 0 forward / 1 inverse transform
   ND1,MD1 : destination coordinates
   N1,M1 : upper left  corner of source matrices
   N2,M2 : lower right corner of source matrices   */

{ int i,j,ii,jj,nr,nc,nr1,nc1,sign,intme;
```

88   DIGITAL IMAGE TRANSFORM ALGORITHMS

```
      double theta;
      float FTEMP,fme;
      float *WR, *WI;
      float WRk1,WIk1,WRk2,WIk2,WRk3,WIk3;
      int N,NN,Ni,k1,k2,k3,k33,yi,yi1,yi2,xj,xj1,xj2,m,mmax,istep;
      float S0r,S0i,S1r,S1i,S2r,S2i,S3r,S3i;
      float sum1r,sum1i,sum2r,sum2i,dif1r,dif1i,dif2r,dif2i;
      matrix RE,IM;

N=N2-N1;
NN=N/2;

/* 1-d temporary vectors for the real
   and imaginary parts of sin and cos coefficients */
WR=(float *)malloc((unsigned int)(NN+1)*sizeof(float));
WI=(float *)malloc((unsigned int)(NN+1)*sizeof(float));

/* FFT is performed always on the destination buffers */
if(matsre!=matdre || ND1!=N1 || MD1!=M1)
   for(nr=0; nr<N; nr++)
   for(nc=0; nc<N; nc++)
    matdre[ND1+nr][MD1+nc]=matsre[N1+nr][M1+nc];
if(matsim!=matdim || ND1!=N1 || MD1!=M1)
   for(nr=0; nr<N; nr++)
   for(nc=0; nc<N; nc++)
    matdim[ND1+nr][MD1+nc]=matsim[N1+nr][M1+nc];

/* Perform bit reversion along rows */
j=0;
for(i=0; i<N; i++)
   { if(j > i)
      { ii=MD1+i; jj=MD1+j;
        for(nr=0; nr<N; nr++)
         { nr1=ND1+nr;
           SWAP(matdre[nr1][jj],matdre[nr1][ii]);
           SWAP(matdim[nr1][jj],matdim[nr1][ii]);
         }
      }
     m=N>>1;
     while(m>0 && j>=m) { j -= m; m >>= 1; }
     j += m;
   }

/* Perform bit reversion along columns */
j=0;
```

VECTOR-RADIX FAST FOURIER TRANSFORM ALGORITHM   89

```
for(i=0; i<N; i++)
   { if(j > i)
       { ii=ND1+i; jj=ND1+j;
         for(nc=0; nc<N; nc++)
            { nc1=MD1+nc;
              SWAP(matdre[jj][nc1],matdre[ii][nc1]);
              SWAP(matdim[jj][nc1],matdim[ii][nc1]);
            }
       }
     m=N>>1;
     while(m>0 && j>=m) { j -= m; m >>= 1; }
     j += m;
   }

RE=matdre; IM=matdim;

/* Calculate the sin and cos coefficients */
sign=mode?1:-1;
theta=sign*6.28318530717959/N;
for(i=0; i<=NN; i++)
   { WR[i]=cos(i*theta); WI[i]=sin(i*theta); }

/* Decimation-in-time FFT */
/* Perform the first stage of the transform */
istep=2;

for(yi=0; yi<N; yi+=2)
 { yi1=yi+ND1;
   yi2=yi1+1;
   for(xj=0; xj<N; xj+=2)
     { xj1=xj+MD1;
       xj2=xj1+1;

       S0r=RE[yi1][xj1]; S0i=IM[yi1][xj1]; S1r=RE[yi1][xj2];
       S1i=IM[yi1][xj2]; S2r=RE[yi2][xj1]; S2i=IM[yi2][xj1];
       S3r=RE[yi2][xj2]; S3i=IM[yi2][xj2];

       sum1r=S0r+S1r; sum1i=S0i+S1i;
       sum2r=S2r+S3r; sum2i=S2i+S3i;
       dif1r=S0r-S1r; dif1i=S0i-S1i;
       dif2r=S2r-S3r; dif2i=S2i-S3i;

       RE[yi1][xj1]=sum1r+sum2r; IM[yi1][xj1]=sum1i+sum2i;
       RE[yi2][xj1]=sum1r-sum2r; IM[yi2][xj1]=sum1i-sum2i;
       RE[yi1][xj2]=dif1r+dif2r; IM[yi1][xj2]=dif1i+dif2i;
```

90 DIGITAL IMAGE TRANSFORM ALGORITHMS

```
            RE[yi2][xj2]=dif1r-dif2r; IM[yi2][xj2]=dif1i-dif2i;

     }
 }

/* Perform the remaining stages of the transform */
mmax=2;
while(N > mmax)
 { istep=2*mmax;
   Ni=N/istep;

   for(k1=0; k1<mmax; k1++)
    {
      WRk1=WR[k1*Ni]; WIk1=WI[k1*Ni];
      for(k2=0; k2<mmax; k2++)
       { k3=k1+k2;
         WRk2=WR[k2*Ni]; WIk2=WI[k2*Ni];
         k33=k3>mmax?istep-k3:k3;
         WRk3=WR[k33*Ni]; WIk3=WI[k33*Ni];
         if(k3>mmax) WIk3=-WIk3;

         for(yi=k1; yi<N; yi+=istep)
           { yi1=yi+ND1; yi2=yi1+mmax;
             for(xj=k2; xj<N; xj+=istep)
               { xj1=xj+MD1; xj2=xj1+mmax;

                 S0r=RE[yi1][xj1]; S0i=IM[yi1][xj1];
                 S1r=WRk2*RE[yi1][xj2]-WIk2*IM[yi1][xj2];
                 S1i=WRk2*IM[yi1][xj2]+WIk2*RE[yi1][xj2];
                 S2r=WRk1*RE[yi2][xj1]-WIk1*IM[yi2][xj1];
                 S2i=WRk1*IM[yi2][xj1]+WIk1*RE[yi2][xj1];
                 S3r=WRk3*RE[yi2][xj2]-WIk3*IM[yi2][xj2];
                 S3i=WRk3*IM[yi2][xj2]+WIk3*RE[yi2][xj2];

                 sum1r=S0r+S1r; sum1i=S0i+S1i;
                 sum2r=S2r+S3r; sum2i=S2i+S3i;
                 dif1r=S0r-S1r; dif1i=S0i-S1i;
                 dif2r=S2r-S3r; dif2i=S2i-S3i;

                 RE[yi1][xj1]=sum1r+sum2r;
                 IM[yi1][xj1]=sum1i+sum2i;
                 RE[yi2][xj1]=sum1r-sum2r;
                 IM[yi2][xj1]=sum1i-sum2i;
                 RE[yi1][xj2]=dif1r+dif2r;
                 IM[yi1][xj2]=dif1i+dif2i;
```

```
                RE[yi2][xj2]=dif1r-dif2r;
                IM[yi2][xj2]=dif1i-dif2i;

            }  /* x */
          }  /* y */
        }  /* k2 */
      }  /* k1 */
    mmax=istep;
  }  /* while */

  free(WR); free(WI);

  if(mode)
    { fme=(float)N*(float)N;
      for(nr=0; nr<N; nr++)
      for(nc=0; nc<N; nc++)
        { matdre[ND1+nr][MD1+nc] /= fme ;
          matdim[ND1+nr][MD1+nc] /= fme ;
        }
    }

  return(0);
}
```

Fig. 2.5.3 Algorithm of the vector-radix FFT.

The vector-radix FFT algorithm is shown in Figure 2.5.3.

VRFFT possesses $\log_2 N$ stages. Each stage contains $N^2/4$ butterflies and each butterfly requires three complex multiplications, as can be seen in Figure 2.5.1. Thus the total number of complex multiplications needed is:

$$C = \frac{3N^2}{4} \log_2 N \qquad (2.5.9)$$

The vector-radix FFT requires 25% less multiplications than the RCFFT, as can be seen by comparing (2.5.9) and (2.3.5). The vector-radix FFT needs the same number of complex additions as the RCFFT:

$$A = 2N^2 \log_2 N \qquad (2.5.10)$$

All computations of the vector-radix FFT can be performed in place. The algorithm can be easily generalized to higher dimensions and to arbitrary radices $r_1 \times r_2$.

2.6 POLYNOMIAL TRANSFORM FFT

Polynomial transforms are a very important theoretical tool for the construction of fast algorithms for 2-d convolutions and Fourier transforms [NUS81]. Let $X_m(z)$ be a polynomial of degree N:

$$X_m(z) = \sum_{n=0}^{N-1} x(m,n) z^n \qquad (2.6.1)$$

The polynomials $X_m(z)$, $m = 0, 1, \ldots, N-1$, can represent an image of size $N \times N$. Let $P(z)$ be an irreducible polynomial and $G(z)$ be a polynomial N-th root of unity defined as follows:

$$G^N(z) = 1 \bmod P(z) \qquad (2.6.2)$$

If N, $G(z)$ have inverses modulo $P(z)$ and:

$$S = \sum_{k=0}^{N-1} [G(z)]^{qk} \bmod P(z) = \begin{cases} 0 & \text{for } q \neq 0 \bmod N \\ N & \text{for } q = 0 \bmod N \end{cases} \qquad (2.6.3)$$

a polynomial transform can be defined, having the form:

$$\hat{X}_k(z) = \sum_{m=0}^{N-1} X_m(z) [G(z)]^{mk} \bmod P(z) \qquad (2.6.4)$$

$$X_m(z) = \frac{1}{N} \sum_{k=0}^{N-1} \hat{X}_k(z) [G(z)]^{-mk} \bmod P(z) \qquad (2.6.5)$$

The polynomial transforms (2.6.4), (2.6.5) have been extensively used in the construction of fast algorithms that can be used in the computation of 2-d convolution and 2-d FFT transforms. In this section, we shall describe a 2-d FFT algorithm based on polynomial transforms [NUS81]. The following polynomial transform will be employed:

$$\hat{X}_k(z) = \sum_{m=0}^{N-1} X_m(z) (z^2)^{mk} \bmod (z^N + 1) \qquad (2.6.6)$$

In this polynomial transform, the irreducible polynomial is $P(z) = z^N + 1$ and the root of unity is $G(z) = z^2$. No multiplications are needed for the calculation of (2.6.6). Multiplications by powers of z and reductions modulo $(z^N + 1)$ result only in shifts of polynomial coefficients and in additions/subtractions. If N is a power of 2, the polynomial transform can be computed by an algorithm similar to a radix-2 FFT, as can be seen in Figure 2.6.1.

POLYNOMIAL TRANSFORM FFT

```
#define SWAPPNT(a,b) PTRTEMP=(a);(a)=(b);(b)=PTRTEMP
float *PTRTEMP;

int pt(RE,IM,rs,cs)
matrix RE;
matrix IM;
int rs,cs;
/* Subroutine to perform polynomial transform
   RE: real matrix
   IM: imaginary matrix
   rs,cs: number of rows and columns */

{ int i,j,m,mmax,istep,x,sh,shift;
  float *fre; float *fim;
  float tre,tim;
  /* Create temporary 1-d vectors */
  fre=(float *)malloc((unsigned int)cs*sizeof(float));
  fim=(float *)malloc((unsigned int)cs*sizeof(float));
  /* Perform bit reversion along columns */
  j=0;
  for(i=0; i<rs; i++)
     { if(j > i)
         { SWAPPNT(RE[j],RE[i]);
           SWAPPNT(IM[j],IM[i]);
         }
       m=rs>>1;
       while(m>0 && j>=m) { j -= m; m >>= 1; }
       j += m;
     }
  mmax=1;
  while(rs > mmax)
     { istep=2*mmax;
       sh=rs/mmax;
       /* Perform the first stage of the transform */
       for(i=0; i<rs; i+=istep)
         { j=i+mmax;
           for(x=0; x<cs; x++)
             { tre=RE[j][x]; tim=IM[j][x];
               RE[j][x]=RE[i][x]-tre; IM[j][x]=IM[i][x]-tim;
               RE[i][x] += tre; IM[i][x] += tim;
             }
         }
       /* Perform the remaining stages of the transform */
       for(m=1; m<mmax; m++)
         { shift=m*sh;
```

```
        for(i=m; i<rs; i+=istep)
        { j=i+mmax;
          for(x=0; x<cs; x++)
          { fre[x]=RE[j][x]; fim[x]=IM[j][x]; }
          for(x=0; x<cs; x++)
          { if(x-shift >= 0)
              { tre=fre[x-shift];tim=fim[x-shift]; }
            else {tre=-fre[x-shift+cs];
                  tim=-fim[x-shift+cs];}
            RE[j][x]=RE[i][x]-tre; IM[j][x]=IM[i][x]-tim;
            RE[i][x] += tre; IM[i][x] += tim;
          }
        }
      }
      mmax=istep;
   }
   free(fre); free(fim);
   return(0);
}
```

Fig. 2.6.1 Polynomial transform algorithm.

The multiplications by powers of W_N are replaced by multiplications of polynomials by powers of z^2. Thus only $2N^2 \log_2 N$ real additions are needed for the calculation of the polynomial transform (2.6.6). This transform will be used in the computation of the 2-d DFT as follows.

Let us use a slightly modified definition of the 2-d DFT:

$$X(k_1, k_2) = \sum_{n_1=0}^{N-1} \sum_{n_2=0}^{N-1} x(n_1, n_2) W^{2n_1 k_1} W^{2n_2 k_2} \qquad (2.6.7)$$

where $W = \exp(-i\pi/N)$. It can be proven that the 2-d DFT can be computed by using the following steps [NUS81]:

$$X_{n_1}(z) = \sum_{n_2=0}^{N-1} x(n_1, n_2) W^{-n_2} z^{n_2} \qquad (2.6.8)$$

$$\hat{X}_{(2k_2+1)k_1}(z) = \sum_{n_1=0}^{N-1} X_{n_1}(z) z^{2n_1 k_1} \bmod (z^N + 1) \qquad (2.6.9)$$

$$X((2k_2+1)k_1, k_2) = \sum_{l=0}^{N-1} y(k_1, l) W^l W^{2lk_2} \qquad (2.6.10)$$

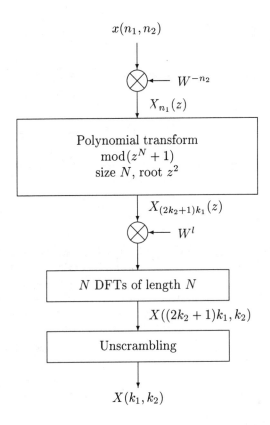

Fig. 2.6.2 Polynomial transform FFT.

where $y(k_1, l)$ are the coefficients of the polynomial $\hat{X}_{(2k_2+1)k_1}(z)$:

$$\hat{X}_{(2k_2+1)k_1}(z) = \sum_{l=0}^{N-1} y(k_1, l) z^l \qquad (2.6.11)$$

The derivation of (2.6.8–2.6.11) is rather complicated and thus omitted. The relation (2.6.8) depicts just a multiplication of the image samples $x(n_1, n_2)$ by a coefficient W^{-n_2} along the rows. The resulting polynomial sequence $X_{n_1}(z)$ is transformed by a polynomial transform similar to (2.6.6), as can be seen in (2.6.9). The transformed polynomial is denoted by $\hat{X}_{(2k_2+1)k_1}(z)$. Its coefficients $y(k_1, l)$ are described by (2.6.11). These coefficients are multiplied by W^l along the rows and transformed by a one-dimensional DFT in a row-wise manner, as can be seen in (2.6.10). The resulting algorithm, called the *polynomial transform FFT* (PTFFT), is illustrated in Figure 2.6.2.

The transformed signal $X((2k_2+1)k_1, k_2)$, which is produced by (2.6.10), is scrambled. Therefore, a final unscrambling step can be used to produce the correct 2-d DFT sequence $X(k_1, k_2)$. The 1-d DFTs that are used in (2.6.10) can be implemented by using a radix-2 FFT that requires $2N \log_2 N$ real multiplications and $3N \log_2 N$ real additions. The number of real multiplications needed for the calculation of the complex multiplications by W^{-n_2}, W^l is $2 \cdot 4N^2$. Thus, the total number of real multiplications is

$$C = 2N^2(4 + \log_2 N) \tag{2.6.12}$$

The polynomial transform requires $2N^2 \log_2 N$ real additions. The complex multiplications by W^{-n_2}, W^l require $2 \cdot 2N^2$ real additions. Thus, the following number of real additions is needed for the calculation of the polynomial transform FFT:

$$A = N^2(4 + 5\log_2 N) \tag{2.6.13}$$

additions. If the computational complexity of the polynomial transform FFT is compared with that of the RCFFT, it can be easily found that the PTFFT is better than the RCFFT for $N > 16$. For large transform lengths, the PTFFT has 50% fewer multiplications than the RCFFT.

2.7 TWO-DIMENSIONAL POWER SPECTRUM ESTIMATION

The estimation of the power spectrum of digital images is a very important and difficult problem. Power spectrum estimates are extensively used in the design of restoration filters (e.g. Wiener filters). The power spectrum carries important information about the 2-d signal content and therefore it is used in the description of image texture [LEV85]. Two-dimensional spectral estimation is an important topic that cannot be covered in this book. The interested reader is referred to the literature (e.g. [LIM90], [DUD84], [MAR87]). In the following, we shall restrict ourselves to the description of algorithmic techniques used in the computation of the power spectrum by classical methods (e.g. periodogram, Blackman–Tukey estimation, Bartlett estimation). The squared magnitude $\hat{P}_{xx}(k_1, k_2)$ of the 2-d DFT:

$$\hat{P}_{xx}(k_1, k_2) = \frac{1}{N_1 N_2} \mid X(k_1, k_2) \mid^2$$

$$= \frac{1}{N_1 N_2} \mid \sum_{n_1=0}^{N_1-1} \sum_{n_2=0}^{N_2-1} x(n_1, n_2) W_{N_1}^{n_1 k_1} W_{N_2}^{n_2 k_2} \mid^2 \tag{2.7.1}$$

is the 2-d *periodogram* of the discrete signal $x(n_1, n_2)$ and can be used as an estimator of its two-dimensional power spectrum. It can be calculated by using the 2-d FFT routines described in the previous sections, followed by a calculation of the squared DFT magnitude. This method has been used to calculate the periodogram of the image shown in Figure 2.7.1a. The resulting

Fig. 2.7.1 (a) Test image Baboon; (b) periodogram of Baboon.

power spectrum estimation is depicted in Figure 2.7.1b. The periodogram $\hat{P}_{xx}(\omega_1, \omega_2)$ is the Fourier transform of the estimator $\hat{R}_{xx}(n_1, n_2)$ of the autocorrelation function:

$$\hat{R}_{xx}(n_1, n_2) = \frac{1}{(2N_1+1)(2N_2+1)} \sum_{k_1=-N_1}^{N_1} \sum_{k_2=-N_2}^{N_2} x^*(k_1, k_2) x(k_1+n_1, k_2+n_2) \tag{2.7.2}$$

The properties of the 2-d periodogram (2.7.1) are similar to those of the 1-d periodogram [MAR87], [PRO89]. The periodogram is a smoothed version of the actual 2-d power spectrum $P_{xx}(\omega_1, \omega_2)$. If $N_1 \times N_2$ is the size of the image $x(n_1, n_2)$, the periodogram resolution is of the order $2\pi/N_1, 2\pi/N_2$ for the frequencies ω_1, ω_2 respectively [LIM90]. Therefore, it is very poor for small image sizes. Furthermore, the variance of the 2-d periodogram is large, because it is linearly related to the power spectrum magnitude. Thus, the periodogram is a noisy power spectrum estimator. The noise is evident in the periodogram shown in Figure 2.7.1b and can be reduced, at the expense of the resolution, by using a two-dimensional version of the *Bartlett* estimator [MAR87], [PRO89]. The signal $x(n_1, n_2)$ of size $N_1 \times N_2$ is split into $K_1 \times K_2$ non-overlapping sections having size $M_1 \times M_2$ each (where $M_i = N_i/k_i$, $i = 1, 2$):

$$x_{ij}(n_1, n_2) = x(n_1 + iM_1, n_2 + jM_2) \quad i = 0, \ldots, K_1 - 1, j = 0, \ldots, K_2 - 1 \tag{2.7.3}$$

The periodogram of each section can be calculated:

$$\hat{P}_{xx}^{(ij)}(k_1, k_2) = \frac{1}{M_1 M_2} \mid \sum_{n_1=0}^{M_1-1} \sum_{n_2=0}^{M_2-1} x_{ij}(n_1, n_2) W_{M_1}^{n_1 k_1} W_{M_2}^{n_2 k_2} \mid^2 \tag{2.7.4}$$

The new spectral estimator is the average of the periodograms of all sections:

$$\hat{P}_{xx}^B(K_1, K_2) = \frac{1}{K_1 K_2} \sum_{i=0}^{K_1-1} \sum_{j=0}^{K_2-1} \hat{P}_{xx}^{(ij)}(k_1, k_2) \tag{2.7.5}$$

The variance of the estimator $\hat{P}_{xx}^B(k_1, k_2)$ is $K_1 K_2$ times less than that of the periodogram. Its resolution is reduced by a factor K_i, $i = 1, 2$, along the coordinates ω_1, ω_2 respectively. If the original image has large size and can be split into a large number of reasonably sized image sections, this method performs relatively well.

Another 2-d power spectrum estimation technique is the extension of the Blackman–Tukey method to the two-dimensional case. An estimator of the two-dimensional autocorrelation function is used:

$$R_{xx}(m_1, m_2) = \frac{1}{N_1 N_2} \sum_{n_1=0}^{2N_1-2} \sum_{n_2=0}^{2N_2-2} x^*(n_1, n_2) x(n_1 + m_1, n_2 + m_2) \quad (2.7.6)$$

The image $x(n_1, n_2)$ has size $N_1 \times N_2$ and it is appended with zeros to the size $(2N_1 - 1) \times (2N_2 - 1)$ before it is used in (2.7.6). Periodic extension of the appended sequence is assumed. This estimator can be windowed and used subsequently for the calculation of the Blackman–Tukey estimator of the power spectrum:

$$P_{xx}^{BT}(k_1, k_2) = \sum_{m_1=0}^{2N_1-2} \sum_{m_2=0}^{2N_2-2} R_{xx}(m_1, m_2) w(m_1, m_2) W_{2N_1-1}^{m_1 k_1} W_{2N_2-1}^{m_2 k_2} \quad (2.7.7)$$

The size of the autocorrelation function is $(2N_1 - 1) \times (2N_2 - 1)$. However, the size $M_1 \times M_2$ of the window $w(n_1, n_2)$ may be much smaller than $(2N_1 - 1) \times (2N_2 - 1)$. The windowing operation is equivalent to convolution in the frequency domain. Therefore, Blackman–Tukey estimators are relatively smooth. The smaller the size of the window, the smoother is the Blackman–Tukey estimator (2.7.7). The periodogram (2.7.1) is directly related to the Blackman–Tukey estimator for $w(n_1, n_2) = 1$, $0 \leq n_1 \leq 2N_1 - 2$, $0 \leq n_2 \leq 2N_2 - 2$: it is a subsampled version of (2.7.7). A separable two-dimensional window can be used:

$$w(m_1, m_2) = w_1(m_1) w_2(m_2) \quad (2.7.8)$$

The triangular one-dimensional window:

$$w_i(m_i) = \begin{cases} 1 - \frac{|m_i|}{M_i} & \text{for } |m_i| \leq M_i \\ 0 & \text{for } M_i \leq |m_i| \leq N_i - 1 \end{cases} \quad i = 1, 2 \quad (2.7.9)$$

can be employed in (2.7.8). Alternatively, the 1-d Hamming window can be employed [MAR87]. For relatively large image sizes, the 2-d DFT of size $(2N_1 - 1) \times (2N_2 - 1)$ can be used for the calculation of $R_{xx}(m_1, m_2)$:

$$R_{xx}(m_1, m_2) = IDFT\{|DFT[x(n_1, n_2)]|^2\} \quad (2.7.10)$$

as has already been described in Section 2.2. The windowing operation can be performed in the spatial domain and the Blackman–Tukey estimator can

be obtained by using the $M_1 \times M_2$ 2-d DFT of the windowed autocorrelation function:
$$P_{xx}^{BT}(k_1,k_2) = DFT[w(m_1,m_2)R_{xx}(m_1,m_2)] \qquad (2.7.11)$$
The resulting algorithm is shown in Figure 2.7.2.

```
int blackman_tukey_psd(ima,psd,WN,WM,N,M)
image ima;
matrix psd;
int WN,WM,N,M;
{
/* Subroutine for Blackman-Tukey method
   Power Spectral Density (PSD) estimation.
   ima: input image buffer
   psd: output PSD matrix
   WN,WM: attenuation coefficients, the points in the
          N and M axes at which the window becomes zero
   N,M: size of the image (end coordinates)
   !!! PSD MATRIX SIZE  IS   DOUBLE  THE SIZE OF THE IMAGE !!!
*/
   int x,y,ret;
   matrix matcor,mathlp;

   matcor=matf2(2*N,2*M);
   mathlp=matf2(2*N,2*M);
   for(y=0; y<N; y++)
    for(x=0; x<M; x++)
       mathlp[y][x]=ima[y][x];
   ret=correlation(mathlp,mathlp, matcor, 0,0, 0,0,N,M);
   if(ret) goto RET_BLACKMANTUKEYPSD;
   ret=windowcorrel(matcor,matcor,WN,WM,2*N,2*M);
   if(ret) goto RET_BLACKMANTUKEYPSD;
   for(y=0; y<2*N; y++)
    for(x=0; x<2*M; x++)
       mathlp[y][x]=0.0;

   ret=fft2d(matcor,mathlp, psd,mathlp, 0, 0,0, 0,0,2*N,2*M);

RET_BLACKMANTUKEYPSD:
   mfreef2(matcor,2*N); mfreef2(mathlp,2*N);
   return(ret);
}
```

```
int windowcorrel(matin,matout,WN,WM,N,M)
matrix matin,matout;
int WN,WM,N,M;
{
/* Subroutine to apply a 2-d window (pyramid)
   on a correlation matrix
   matin: input correlation matrix
   matout: output windowed correlation matrix
   WN,WM: attenuation coefficients, the points in the
          N and M axes at which the window becomes zero
   N,M: size of the correlation matrix
   (!!! DOUBLE THE SIZE OF THE IMAGE !!!)
*/

  int x,y;
  int N2,M2;
  float coey,coef;

  N2=N/2; M2=M/2;
  for(y=0; y<N2; y++)
   { coey=1.0-(float)(y)/WN;
     if(coey<0.0) coey=0.0;
     for(x=0; x<M2; x++)
      { coef=(1.0-(float)(x)/WM)*coey;
        if(coef<=0.0)  matout[y][x]=0.0;
         else  matout[y][x] = matin[y][x]*coef;
      }
     for(x=M2; x<M; x++)
      { coef=(1.0-(float)(M-x)/WM)*coey;
        if(coef<=0.0)  matout[y][x]=0.0;
         else  matout[y][x] = matin[y][x]*coef;
      }
   }

  for(y=N2; y<N; y++)
   { coey=1.0-(float)(N-y)/WN;
     if(coey<0.0) coey=0.0;
     for(x=0; x<M2; x++)
      { coef=(1.0-(float)(x)/WM)*coey;
        if(coef<=0.0)  matout[y][x]=0.0;
         else  matout[y][x] = matin[y][x]*coef;
      }
     for(x=M2; x<M; x++)
      { coef=(1.0-(float)(M-x)/WM)*coey;
        if(coef<=0.0)  matout[y][x]=0.0;
```

```
        else matout[y][x] = matin[y][x]*coef;
      }
    }
    return(0);
}
```

Fig. 2.7.2 Algorithm for the calculation of the 2-d Blackman–Tukey estimator.

Beyond the classical techniques that have been described in this section, several other techniques have been proposed for two-dimensional power spectrum estimation. Some of them are based on 2-d AR modelling of images [JAI89]. Two-dimensional AR models [JAI89], [LIM90] of the form:

$$x(n_1, n_2) = \sum\sum_{(i,j)\in A} a(i,j) x(n_1 - i, n_2 - j) + w(n_1, n_2) \qquad (2.7.12)$$

are usually employed for this purpose. A is the *prediction window* and $a(i,j)$ are the *predictor coefficients*. The window can be quarter plane or non-symmetric half plane, as detailed in Chapter 4. The model is driven by a white noise process $w(n_1, n_2)$ having variance σ_w^2. Similar models are employed in digital image coding. Thus, a further discussion on AR image models and on the optimal choice of the prediction coefficients and of the noise variance is included in Chapter 3. If these coefficients are known, the power spectrum estimation can be done by the following formula:

$$P_{xx}^{AR}(k_1, k_2) = \frac{\sigma_w^2}{|\sum_{(m_1,m_2)\in A} a(m_1,m_2) W_{N_1}^{m_1 k_1} W_{N_2}^{m_2 k_2}|^2} \qquad (2.7.13)$$

The corresponding algorithm is shown in Figure 2.7.3.

```
int arpsd(ima,psd,N,M,K,LL,LR)
image ima;
matrix psd;
int N,M,K,LL,LR;
{
/* Subroutine for 2-d AR modelling
   Power Spectral Density (PSD) estimation.
   ima: input image buffer
   psd: output PSD matrix
   N,M: size of the image (end coordinates)
   K,LL,LR: size of the support region of the AR model */

   int x,y,i,k,l;
   matrix re,im;
   float *A;
```

```
    float Pu;
    int ret;

    /* Create temporary matrices and a 1-d vector */
    re=matf2(N,M);
    im=matf2(N,M);
    A=malloc(((K+1)*(LR+1+LL)-LR)*sizeof(float));

    /* Find 2-d AR NSHP coefficients */
    ret=arcoefficients(ima,N,M,K,LL,LR,A,&Pu);
    if(ret) goto RET_ARPSD;

    /* Put AR coefficients on a temporary matrix */
    for(y=0; y<N; y++)
    for(x=0; x<M; x++)
       { re[y][x]=0.0; im[y][x]=0.0; }

    i=-1;
    for(k=0; k<=K; k++)
     { if(k!=0)
        for(l=-LR; l<0; l++)   { i++; re[k][M+l]=A[i]; }
        for(l=0; l<=LL; l++)   { i++; re[k][l]=A[i]; }
     }

    /* Perform 2-d FFT */
    ret=fft2d(re,im, re,im, 0, 0,0, 0,0,N,M);
    if(ret) goto RET_ARPSD;

    /* Calculate power spectrum */
    for(y=0; y<N; y++)
    for(x=0; x<M; x++)
        psd[y][x]=Pu/(re[y][x]*re[y][x]+im[y][x]*im[y][x]);

RET_ARPSD:

    mfreef2(re,N); mfreef2(im,N); free(A);
    return(ret);
}
```

Fig. 2.7.3 Algorithm for the calculation of the power spectrum from the 2-d AR coefficients.

The performance of the algorithm depends greatly on the shape and the size of the prediction window A. Small windows produce a smoothed version of the power spectrum. Large windows tend to have better resolution and to produce spiky power spectra.

2.8 DISCRETE COSINE TRANSFORM

The discrete cosine transform (DCT) was introduced by Ahmed and his colleagues in 1974 [AHM74], [AHM75], [RAO90]. Since then it has been used extensively in various digital image coding schemes, because of its nearly optimal performance in typical images having high correlation in adjacent image pixels. In most practical cases, the transform schemes based on the DCT outperform any other orthogonal transform in terms of compression ratio. The DCT is used in the emerging JPEG standard for still image compression and in the moving image compression standard MPEG [KON95].

The transform kernel of the 1-d DCT is the following set of cosinusoidal functions:

$$g(n,0) = \frac{1}{\sqrt{N}} \qquad (2.8.1)$$

$$g(n,k) = \sqrt{\frac{2}{N}} \cos \frac{(2n+1)k\pi}{2N} \qquad (2.8.2)$$

for transform length $N: 0 \leq n, k \leq N-1$. Thus, the forward transform is given by:

$$C(0) = \frac{1}{\sqrt{N}} \sum_{n=0}^{N-1} x(n) \qquad (2.8.3)$$

$$C(k) = \sqrt{\frac{2}{N}} \sum_{n=0}^{N-1} x(n) \cos \frac{(2n+1)k\pi}{2N} \qquad (2.8.4)$$

where $C(k)$, $k = 0, \ldots, N-1$, are the DCT coefficients. It can be easily proven that the inverse DCT is given by the relation:

$$x(n) = \frac{1}{\sqrt{N}} C(0) + \sqrt{\frac{2}{N}} \sum_{k=1}^{N-1} C(k) \cos \frac{(2n+1)k\pi}{2N} \qquad (2.8.5)$$

Thus, both the forward and inverse DCT employ the same transform kernels (2.8.1, 2.8.2). The DCT is closely related to the DFT. Let us suppose that the sequence $x(n)$, $0 \leq n \leq N-1$, is extended to form an even sequence $f(n)$ of length $2N$:

$$f(n) = \begin{cases} x(n) & 0 \leq n \leq N-1 \\ x(2N-1-n) & N \leq n \leq 2N-1 \end{cases} \qquad (2.8.6)$$

as shown in Figure 2.8.1. The sequence is symmetric with respect to the point

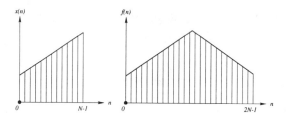

Fig. 2.8.1 Creation of an even sequence.

$n = N - 1/2$. Furthermore, its periodic extension does not have discontinuities. The DFT of $f(n)$ having length $2N$ is given by:

$$F(k) = \sum_{n=0}^{2N-1} f(n)W_{2N}^{nk} = \sum_{n=0}^{N-1} x(n)W_{2N}^{nk} + \sum_{n=N}^{2N-1} x(2N-1-n)W_{2N}^{nk}$$

$$= 2W_{2N}^{-k/2} \sum_{n=0}^{N-1} x(n) \cos \frac{(2n+1)k\pi}{2N} \qquad (2.8.7)$$

The DCT coefficient $C(k), k \neq 0$, is given by:

$$C(k) = \frac{W_{2N}^{k/2}}{\sqrt{2N}} F(k) \qquad (2.8.8)$$

Thus, there is a close relation between the DCT coefficient $C(k)$ and the DFT coefficients $F(k)$ of the even sequence $f(n)$. This fact explains partially the good compression properties of the DCT. The periodic extension of $f(n)$ does not have discontinuities and its DFT does not contain spurious high frequencies. On the contrary, the periodic extension of $x(n)$ has discontinuities and its DFT contains relatively strong high frequencies. The relation (2.8.8) can be used for the fast calculation of the DCT as follows:

1. The sequence $f(n)$ is formed by using (2.8.6).

2. The DFT of length $2N$ is calculated by using a 1-d FFT of length $2N$.

3. The DCT coefficients $C(k)$ are calculated by using (2.8.8) for $0 \leq k \leq N-1$.

An alternative method for the fast calculation of the 1-d DCT is based on the following expression for $C(k)$:

$$C(k) = \sqrt{\frac{2}{N}} \sum_{n=0}^{N-1} x(n) Re \left\{ \exp\left(-i\frac{(2n+1)k\pi}{2N}\right) \right\} \qquad (2.8.9)$$

where $Re\{\cdot\}$ denotes the real part of a complex number. If $x(n)$ is a real sequence, $C(k)$ is given by:

$$C(k) = \sqrt{\frac{2}{N}} Re\left\{\exp\left(-i\frac{k\pi}{2N}\right) \sum_{n=0}^{N-1} x(n) \exp\left(-i\frac{nk\pi}{N}\right)\right\}$$

$$= \sqrt{\frac{2}{N}} Re\left\{\exp\left(-i\frac{k\pi}{2N}\right) \sum_{n=0}^{2N-1} y(n) \exp\left(-i\frac{nk\pi}{N}\right)\right\} \quad (2.8.10)$$

where $y(n)$ is the sequence $x(n)$ appended with zeros:

$$y(n) = \begin{cases} x(n) & 0 \le n \le N-1 \\ 0 & N \le n \le 2N-1 \end{cases} \quad (2.8.11)$$

If $Y(k)$ is the DFT of the sequence $y(n)$:

$$Y(k) = \sum_{n=0}^{2N-1} y(n) \exp\left(-i\frac{2nk\pi}{2N}\right) \quad (2.8.12)$$

the DCT coefficient $C(k)$ is given by:

$$C(k) = \sqrt{\frac{2}{N}} Re\left\{\exp\left(-i\frac{k\pi}{2N}\right) Y(k)\right\} \quad (2.8.13)$$

Thus, the DCT coefficients can be calculated by employing a DFT of length $2N$ and by using (2.8.13).

An alternative definition of the DCT pair is the following [MAK80], [LIM90]:

$$C(k) = \sum_{n=0}^{N-1} 2x(n) \cos\frac{(2n+1)k\pi}{2N} \quad (2.8.14)$$

$$x(n) = \frac{1}{N} \sum_{k=0}^{N-1} w(k) C(k) \cos\frac{(2n+1)k\pi}{2N} \quad (2.8.15)$$

where the weight function $w(k)$ is given by:

$$w(k) = \begin{cases} 1/2 & k = 0 \\ 1 & 1 \le k \le N-1 \end{cases} \quad (2.8.16)$$

If the definition (2.8.14, 2.8.15) is used, the following formula relates the DCT coefficients to the DFT coefficients $F(k)$ given in (2.8.7):

$$C(k) = W_{2N}^{k/2} F(k) \quad (2.8.17)$$

In the following a fast DCT and inverse DCT algorithm will be described [MAK80]. The sequence $f(n)$ of length $2N$ can be divided into two subsequences of length N as follows:

$$v(n) = f(2n) \qquad 0 \le n \le N-1 \quad (2.8.18)$$

$$w(n) = f(2n+1) \qquad 0 \le n \le N-1 \qquad (2.8.19)$$

These subsequences $v(n)$, $w(n)$ can be used to calculate the DFT $F(k)$:

$$F(k) = \sum_{n=0}^{N-1} v(n) W_{2N}^{2nk} + \sum_{n=0}^{N-1} w(n) W_{2N}^{(2n+1)k} \qquad (2.8.20)$$

If the relations (2.8.20) and (2.8.17) are combined, the following formula results for the DCT $C(k)$ after simple mathematical manipulations:

$$C(k) = 2Re\left\{ W_{4N}^k \sum_{n=0}^{N-1} v(n) W_N^{nk} \right\} = 2Re\{W_{4N}^k V(k)\} \qquad (2.8.21)$$

Thus, the DCT coefficients can be computed by using the DFT $V(k)$ of length N. It can also be easily proven that the DCT coefficient $C(N-k)$ is given by a similar relation:

$$C(N-k) = -2Im\{W_{4N}^k V(k)\} \qquad (2.8.22)$$

The DFT coefficients $V(k)$ can also be expressed in terms of the DCT coefficients:

$$V(k) = \frac{1}{2} W_{4N}^{-k}[C(k) - iC(N-k)] \quad 0 \le k \le N-1 \qquad (2.8.23)$$

The sequence $v(n)$ can be expressed directly in terms of the original sequence $x(n)$:

$$v(n) = \begin{cases} x(2n) & 0 \le n \le [\frac{N-1}{2}] \\ x(2N-2n-1) & [\frac{N+1}{2}] \le n \le N-1 \end{cases} \qquad (2.8.24)$$

Thus, the following algorithm can be used for the fast calculation of the DCT:

1. Form the sequence $v(n)$ by using (2.8.24).

2. Compute the DFT $V(k)$, $0 \le k \le N-1$, by using an FFT algorithm of length N.

3. Compute $C(k)$, $0 \le k \le [N/2]$, from (2.8.21) and $C(N-k)$ from (2.8.22).

The fast calculation of the inverse DCT can be done by employing (2.8.23, 2.8.24). The following algorithm results:

1. Compute $V(k)$ by using (2.8.23).

2. Compute $v(n)$ from $V(k)$ by using an inverse FFT algorithm of length N.

3. Retrieve $x(n)$ from $v(n)$ by using (2.8.24).

2.9 TWO-DIMENSIONAL DISCRETE COSINE TRANSFORM

The two-dimensional $N \times N$ DCT is a separable transform having forward transform kernels of the form:

$$g(n_1, n_2, k_1, k_2) = 4\cos\frac{(2n_1+1)k_1\pi}{2N}\cos\frac{(2n_2+1)k_2\pi}{2N} \quad (2.9.1)$$

The following definition of the $N_1 \times N_2$ DCT will be used [LIM90]:

$$C(k_1, k_2) = \sum_{n_1=0}^{N_1-1}\sum_{n_2=0}^{N_2-1} 4x(n_1, n_2)\cos\frac{(2n_1+1)k_1\pi}{2N_1}\cos\frac{(2n_2+1)k_2\pi}{2N_2} \quad (2.9.2)$$

for $0 \leq k_1 \leq N_1 - 1$, $0 \leq k_2 \leq N_2 - 1$. The inverse DCT is given by:

$$x(n_1, n_2) = \frac{1}{N_1 N_2}\sum_{k_1=0}^{N_1-1}\sum_{k_2=0}^{N_2-1} w_1(k_1)w_2(k_2)C(k_1, k_2)$$
$$\times \cos\frac{(2n_1+1)k_1\pi}{2N_1}\cos\frac{(2n_2+1)k_2\pi}{2N_2} \quad (2.9.3)$$

Slightly modified 2-d DCT definitions can be found in the literature [GON87]. The weight functions $w_1(k_1), w_2(k_2)$ are given by:

$$w_1(k_1) = \begin{cases} 1/2 & k_1 = 0 \\ 1 & 1 \leq k_1 \leq N_1 - 1 \end{cases} \quad (2.9.4)$$

$$w_2(k_2) = \begin{cases} 1/2 & k_2 = 0 \\ 1 & 1 \leq k_2 \leq N_2 - 1 \end{cases} \quad (2.9.5)$$

Since both the forward and inverse DCT are separable, they can easily be computed by a row-column algorithm of the form:

$$C'(n_1, k_2) = 2\sum_{n_2=0}^{N_2-1} x(n_1, n_2)\cos\frac{(2n_2+1)k_2\pi}{2N_2} \quad (2.9.6)$$

$$C(k_1, k_2) = 2\sum_{n_1=0}^{N_1-1} C'(n_1, k_2)\cos\frac{(2n_1+1)k_1\pi}{2N_1} \quad (2.9.7)$$

The 2-d DCT (2.9.2–2.9.3) is directly related to the 2-d DFT of the signal $f(n_1, n_2)$ given by [MAK80]:

$$f(n_1, n_2) = \begin{cases} x(n_1, n_2) & 0 \leq n_1 \leq N_1 - 1, \\ & 0 \leq n_2 \leq N_2 - 1 \\ x(2N_1 - n_1 - 1, n_2) & N_1 \leq n_1 \leq 2N_1 - 1, \\ & 0 \leq n_2 \leq N_2 - 1 \\ x(n_1, 2N_2 - n_2 - 1) & 0 \leq n_1 \leq N_1 - 1, \\ & N_2 \leq n_2 \leq 2N_2 - 1 \\ x(2N_1 - n_1 - 1, 2N_2 - n_2 - 1) & N_1 \leq n_1 \leq 2N_1 - 1, \\ & N_2 \leq n_2 \leq 2N_2 - 1 \end{cases}$$
$$(2.9.8)$$

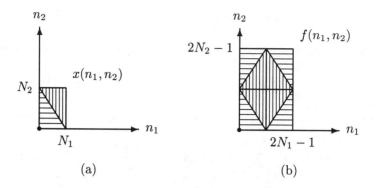

Fig. 2.9.1 (a) Original sequence $x(n_1, n_2)$; (b) expanded sequence $f(n_1, n_2)$.

as will be shown subsequently. This signal is an expanded version of the image $x(n_1, n_2)$. Figure 2.9.1 shows an example of the expansion of the image $x(n_1, n_2)$, according to (2.9.8). It must be noted that the periodic extension of the signal $f(n_1, n_2)$ does not contain sharp discontinuities. This fact is related to the good compression properties of the 2-d DCT, as has already been described in the previous section for the 1-d case. It can be easily proven that the 2-d DFT $F(k_1, k_2)$ of length $2N_1 \times 2N_2$:

$$F(k_1, k_2) = \sum_{n_1=0}^{2N_1-1} \sum_{n_2=0}^{2N_2-1} f(n_1, n_2) W_{2N_1}^{n_1 k_1} W_{2N_2}^{n_2 k_2} \quad (2.9.9)$$

is related to the DCT coefficients $C(k_1, k_2)$ as follows:

$$C(k_1, k_2) = W_{2N_1}^{k_1/2} W_{2N_2}^{k_2/2} F(k_1, k_2) \quad (2.9.10)$$

Thus, a $2N_1 \times 2N_2$ fast Fourier transform can be used for the fast calculation of the $N_1 \times N_2$ DCT.

Another fast algorithm for the 2-d DCT is based on the construction of the following subsequences [MAK80]:

$$v(n_1, n_2) = f(2n_1, 2n_2) \quad (2.9.11)$$
$$w_1(n_1, n_2) = f(2n_1 + 1, 2n_2) \quad (2.9.12)$$
$$w_2(n_1, n_2) = f(2n_1, 2n_2 + 1) \quad (2.9.13)$$
$$w_3(n_1, n_2) = f(2n_1 + 1, 2n_2 + 1) \quad (2.9.14)$$

The subsequences $w_i(n_1, n_2)$, $i = 1, 2, 3$, can be expressed in terms of the subsequence $v(n_1, n_2)$:

$$w_1(n_1, n_2) = v(N_1 - n_1 - 1, n_2) \quad (2.9.15)$$

$$w_2(n_1, n_2) = v(n_1, N_2 - n_2 - 1) \qquad (2.9.16)$$

$$w_3(n_1, n_2) = v(N_1 - n_1 - 1, N_2 - n_2 - 1) \qquad (2.9.17)$$

The subsequence $v(n_1, n_2)$ can be expressed in terms of the original sequence $x(n_1, n_2)$, as follows:

$$v(n_1, n_2) = \begin{cases} x(2n_1, 2n_2) & 0 \le n_1 \le [\frac{N_1-1}{2}], \\ & 0 \le n_2 \le [\frac{N_2-1}{2}] \\ x(2N_1 - 2n_1 - 1, 2n_2) & [\frac{N_1+1}{2}] \le n_1 \le N_1 - 1, \\ & 0 \le n_2 \le [\frac{N_2-1}{2}] \\ x(2n_1, 2N_2 - 2n_2 - 1) & 0 \le n_1 \le [\frac{N_1-1}{2}], \\ & [\frac{N_2+1}{2}] \le n_2 \le N_2 - 1 \\ x(2N_1 - 2n_1 - 1, 2N_2 - 2n_2 - 1) & [\frac{N_1+1}{2}] \le n_1 \le N_1 - 1, \\ & [\frac{N_2+1}{2}] \le n_2 \le N_2 - 1 \end{cases} \qquad (2.9.18)$$

The discrete Fourier transform $V(k_1, k_2)$ of $v(n_1, n_2)$:

$$V(k_1, k_2) = \sum_{n_1=0}^{N_1-1} \sum_{n_2=0}^{N_2-1} v(n_1, n_2) W_{N_1}^{n_1 k_1} W_{N_2}^{n_2 k_2} \qquad (2.9.19)$$

has length $N_1 \times N_2$. By combining (2.9.9–2.9.17), it can be proven that the DCT coefficients $C(k_1, k_2)$ can be expressed in terms of $V(k_1, k_2)$:

$$C(k_1, k_2) = 2Re\{W_{4N_1}^{k_1}[W_{4N_2}^{k_2} V(k_1, k_2) + W_{4N_2}^{-k_2} V(k_1, N_2 - k_2)]\}$$

$$= 2Re\{W_{4N_2}^{k_2}[W_{4N_1}^{k_1} V(k_1, k_2) + W_{4N_1}^{-k_1} V(N_1 - k_1, k_2)]\} \qquad (2.9.20)$$

Thus, the following fast algorithm can be used for the calculation of the 2-d DCT.

1. The sequence $v(n_1, n_2)$ is formed by employing (2.9.18).

2. The DFT $V(k_1, k_2)$ is calculated by using the $N_1 \times N_2$ FFT algorithm (2.9.19).

3. The DCT coefficients are calculated as in (2.9.20).

The resulting DCT algorithm is shown in Figure 2.9.2.

```
int DCT(mat,COS,SIN,N,M)
matrix mat;
vector COS, SIN;
int N,M;
/* Subroutine for the computation of the DCT
```

```
   mat      : input, output matrix
   COS,SIN  :matrices containing cos(2*PI*i/4*N)
             and sin(2*PI*i/4*N)
   N,M: image size */
{
  int i,j,k,l,c,N2,M2;
  float temp,A,B;
  matrix mat1,mat2;

  mat1=matf2(N+1,M+1);
  mat2=matf2(N+1,M+1);
  /*Matrix mat2 contains the imaginary part of input matrix.*/
  c=clearmatrix(mat2,0,0,0,N,M);
  if(c!=0) return(c);
  /* Data expansion from input to matrix mat1. */
  for(k=0;k<N/2;k++)
   for(l=0;l<M/2;l++)
    mat1[k][l]=mat[2*k][2*l];
  for(k=N/2;k<N;k++)
   for(l=0;l<M/2;l++)
    mat1[k][l]=mat[2*N-2*k-1][2*l];
  for(k=0;k<N/2;k++)
   for(l=M/2;l<M;l++)
    mat1[k][l]=mat[2*k][2*M-2*l-1];
  for(k=N/2;k<N;k++)
   for(l=M/2;l<M;l++)
    mat1[k][l]=mat[2*N-2*k-1][2*M-2*l-1];

  /* Compute 2-d DFT */
  c=fft2d(mat1,mat2,mat1,mat2,0,0,0,0,0,N,M);
  if(c!=0) return(c);

  for(i=0;i<N;i++)
   { mat1[i][M]=mat1[i][0]; mat2[i][M]=mat2[i][0]; }

  /* DCT coefficients are derived from DFT coefficients. */
  for(i=0;i<N;i++)
   for(j=0;j<M;j++)
    { A=COS[j]*mat1[i][j]+SIN[j]*mat2[i][j]+
        COS[j]*mat1[i][M-j]-SIN[j]*mat2[i][M-j];
      B=COS[j]*mat2[i][j]-SIN[j]*mat1[i][j]+
        SIN[j]*mat1[i][M-j]+COS[j]*mat2[i][M-j];
      mat[i][j]=2*(COS[i]*A+SIN[i]*B);
    }
  mfreef2(mat1,N+1); mfreef2(mat2,N+1);
```

```
    return(0);
}

int init(COS,SIN,N)
vector COS,SIN;
int N;
{/* Subroutine to compute cos(2*PI*i/4*N) and sin(2*PI*i/4*N)
    which are used for the evaluation of (W4N)^+-i
    COS,SIN: vectors to store cosine and sine functions
    N: dimension of vectors COS,SIN. */

    int i;
    float PI=3.1415926535;

    for(i=0;i<N;i++) { COS[i]=cos((PI*i)/(2*N));
                SIN[i]=sin((PI*i)/(2*N));}
    return(0);
}
```

Fig. 2.9.2 Fast forward 2-d DCT algorithm.

The DFT $V(k_1, k_2)$ can be expressed in terms of the DCT coefficients $C(k_1, k_2)$ as follows:

$$V(k_1, k_2) = \frac{1}{4} W_{4N_1}^{-k_1} W_{4N_2}^{-k_2} \{ [C(k_1, k_2) - C(N_1 - k_1, N_2 - k_2)] \\ - i[C(N_1 - k_1, k_2) + C(k_1, N_2 - k_2)] \} \quad (2.9.21)$$

This relation can be used for the fast computation of the inverse DCT. If the DCT coefficients $C(k_1, k_2)$ are known, the DFT $V(k_1, k_2)$ can be computed from (2.9.21). The sequence $v(n_1, n_2)$ can be calculated by using an inverse $N_1 \times N_2$ 2-d FFT algorithm. Finally, the signal $x(n_1, n_2)$ can be recovered by employing (2.9.18). The fast inverse DCT algorithm is shown in Figure 2.9.3.

```
int IDCT(mat,COS,SIN,N,M)
matrix mat;
vector COS,SIN;
int N,M;

/* Subroutine to compute the inverse Discrete Cosine Transform
    mat: input, output matrix
    COS,SIN: matrices containing
    cos(2*PI*i/4*N) and sin(2*PI*i/4*N) respectively */
```

```
{
 int i,j,k,l,c;
 float A,B,X,Y;
 matrix mati,mat1,mat2;

 mati=matf2(N+1,M+1);
 mat1=matf2(N,M);
 mat2=matf2(N,M);

 /* Initializations */
 for(i=0;i<=N;i++)
   mati[i][M]=0;
 for(i=0; i<=M; i++)
   mati[N][i]=0;
 for(i=0;i<N;i++)
  for(j=0;j<M;j++)
   mati[i][j]=mat[i][j];

 /* Compute DFT coefficients from DCT coefficients */
 for(i=0;i<N;i++)
  for(j=0;j<M;j++)
   {
    A=mati[i][j]-mati[N-i][M-j]; B=mati[N-i][j]+mati[i][M-j];
    X=COS[i]*COS[j]-SIN[i]*SIN[j];
    Y=SIN[i]*COS[j]+COS[i]*SIN[j];
    mat1[i][j]=(A*X+B*Y)/4.0; mat2[i][j]=(A*Y-B*X)/4.0;
   }

 /* Compute inverse DFT. */
 c=fft2d(mat1,mat2,mat1,mat2,1,0,0,0,0,N,M);
 if(c!=0) return(c);

 /* Output matrix is produced */
 for(i=0;i<N/2;i++)
  for(j=0;j<M/2;j++)
   mat[2*i][2*j]=mat1[i][j];

 for(i=N/2;i<N;i++)
  for(j=0;j<M/2;j++)
   mat[2*N-2*i-1][2*j]=mat1[i][j];

 for(i=N/2;i<N;i++)
  for(j=M/2;j<M;j++)
   mat[2*N-2*i-1][2*M-2*j-1]=mat1[i][j];
```

```
    for(i=0;i<N/2;i++)
     for(j=M/2;j<M;j++)
      mat[2*i][2*M-2*j-1]=mat1[i][j];

  mfreef2(mati,N+1); mfreef2(mat1,N); mfreef2(mat2,N);
return(0);
}
```

Fig. 2.9.3 Fast inverse 2-d DCT algorithm.

2.10 DISCRETE WAVELET TRANSFORM

The discrete wavelet transform (DWT) is a fast, linear, invertible and orthogonal operation, just like the Discrete Fourier transform (DFT). The basic idea lying under the discrete wavelet transform is to define a time-scale representation of a signal (unlike short time fourier transform (STFT) which defines a time-frequency signal representation) by decomposing it onto a set of basis functions, called wavelets. Wavelets are obtained from a single prototype wavelet, called mother wavelet, by dilations and contractions, that is, scalings, and shifts [RIO91]. DWT is suitable for the analysis of non-stationary signals since it allows simultaneous localization in time and in scale, unlike STFT which uses fixed time-frequency resolution and thus allows localization only in time or in frequency.

Application of the DWT results in a multilevel decomposition of the input signal into high and low frequency components in different resolutions according to the number of levels employed, as shown in Figure 2.10.1a. Let $H(\omega)$ and $G(\omega)$:

$$H(\omega) = \sum_k h_k e^{-jk\omega} \qquad (2.10.1)$$

$$G(\omega) = \sum_k g_k e^{-jk\omega} \qquad (2.10.2)$$

be a lowpass and highpass filter, respectively, satisfying the orthogonality condition:

$$|H(\omega)|^2 + |G(\omega)|^2 = 1 \qquad (2.10.3)$$

necessary for reconstruction capabilities of the transform. The filters $H(\omega)$ and $G(\omega)$ are also known as quadrature mirror filters. A signal $x[n]$ can be decomposed recursively according to [QIAN96, VET95]:

$$c_{j-1,k} = \sum_n h_{n-2k} c_{j,n} \qquad (2.10.4)$$

$$d_{j-1,k} = \sum_n g_{n-2k} c_{j,n} \qquad (2.10.5)$$

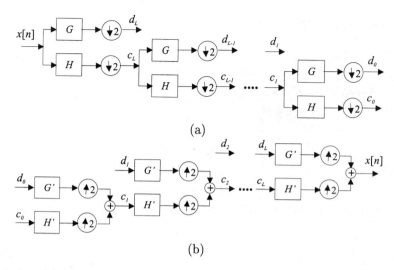

Fig. 2.10.1 (a) Multi-level 1d wavelet decomposition; (b) Multi-level 1d wavelet reconstruction.

which correspond to convolutions followed by downsampling by 2, as can be seen in Figure 2.10.1a. h_k and g_k are the impulse responses of the lowpass and highpass filters, respectively. Index j spans the number of decomposition levels and lies in the range $[0, L+1]$, where $L+1$ represents the index of the high resolution level of the transform and 0 represents the index of the low resolution level. $c_{L+1,k}$ is equal to the input signal $x[k]$. The coefficients $c_{0,k}, d_{0,k}, d_{1,k},..., d_{L-1,k}, d_{L,k}$ are called the DWT wavelet coefficients of $x[n]$. $c_{0,k}$ is the lowest resolution component of $x[n]$ containing lowpass, 'smooth' information and $d_{j,k}$ are the detail coefficients of $x[n]$ at various bands of frequencies. The signal $x[n]$ can now be reconstructed from its DWT coefficients by considering the recursive formula:

$$c_{j,n} = \sum_k h_{n-2k} c_{j-1,k} + \sum_k g_{n-2k} d_{j-1,k} \tag{2.10.6}$$

The reconstruction process is illustrated in Figure 2.10.1b and defines the inverse discrete wavelet transform (IDWT). This time, upsampling precedes filtering at each level of the transform. It is obvious that, in the discrete case, DWT and IDWT can be implemented by two-channel tree-structured filter banks (Figure 2.10.1).

The discrete wavelet transform does not have a single set of basis functions. There are many families of wavelets, the most known of them being the Haar and the Daubechies wavelets [MAL98]. For example, the frequency responses $H(\omega)$ and $G(\omega)$ of the Haar wavelet filters are given by:

$$H(\omega) = \frac{1}{2} + \frac{1}{2}e^{-j\omega} \tag{2.10.7}$$

$$G(\omega) = \frac{1}{2} - \frac{1}{2}e^{-j\omega} \qquad (2.10.8)$$

However, the most frequently used wavelets are the Daubechies ones.

The two-dimensional discrete wavelet transform (2D DWT) and its inverse (2D IDWT) are extensions of the one-dimensional transform. They are simply implemented by using one-dimensional DWTs and IDWTs along each dimension n and m separately:

$$DWT_{n \times m}[x[n,m]] = DWT_n[DWT_m[x[n,m]]] \qquad (2.10.9)$$

In this way, separable two-dimensional filters are only considered. An example of two-dimensional wavelet decomposition is illustrated in Figure 2.10.2. A four level decomposition is shown. Each level is characterized by three detail coefficient components representing the horizontal, vertical and diagonal edges of the input image. The lowest level consists of the low-resolution lowpass version of the initial image.

4-level 2d Wavelet Decomposition

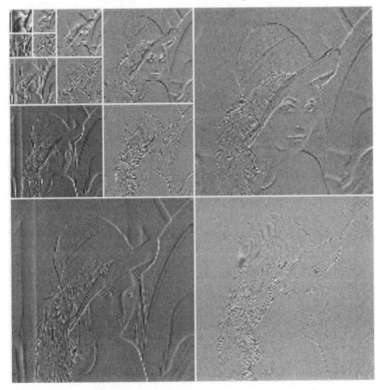

Fig. 2.10.2 Example of 4-level 2D wavelet decomposition.

The discrete wavelet transform is very useful and has found many application areas such as multiresolution signal analysis, subband coding for speech and image compression, study of human vision, radar, earthquake prediction. It is also very likely that the discrete wavelet transform will be incorporated in the JPEG2000 image compression standard.

References

AHM74. N. Ahmed, T. Natarajan, K.R. Rao, "Discrete cosine transform," *IEEE Transactions on Computers*, vol. C-23, pp. 90-93, Jan. 1974.

AHM75. N. Ahmed, K.R. Rao, *Orthogonal transforms for digital signal processing*, Springer Verlag, 1975.

BRI74. E.O. Brigham, *The fast Fourier transform*, Prentice-Hall, 1974.

BUR85. C.S. Burrus, T.W. Parks, *DFT/FFT and convolution algorithms*, Wiley, 1985.

CLA85. R.J. Clark, *Transform coding of images*, Academic Press, 1985.

DUD84. D.E. Dudgeon, R.M. Mersereau, *Multidimensional digital signal processing*, Prentice-Hall, 1984.

EKL72. J.O. Eklundh, "A fast computer method for matrix transposing," *IEEE Transactions on Computers*, vol. C-21, pp. 801-803, July 1972.

GON87. R.C. Gonzalez, P. Wintz, *Digital image processing*, Addison-Wesley, 1987.

HAR77. D.B. Harris, J.H. McClellan, D.S.K. Chan, H.W. Schuessler, "Vector radix fast Fourier transform," *Proc. IEEE Int. Conf. Acoustics, Speech, and Signal Processing*, pp. 548-551, 1977.

JAI89. A.K. Jain, *Fundamentals of digital image processing*, Prentice-Hall, 1989.

JAY84. N.S. Jayant, P. Noll, *Digital coding of waveforms*, Prentice-Hall, 1984.

JAW94. B. Jawerth and W. Sweldens, "An Overview of Wavelet Based Multiresolution Analyses," *SIAM Rev.*, vol. 36, no. 3, pp. 377-412, 1994.

KON95. K. Konstantinides, V. Bhaskaran, *Image and video compression standards*, Kluwer Academic, 1995.

LEV85. M.D. Levine, *Vision in man and machine*, McGraw-Hill, 1985.

LIM90. J.S. Lim, *Two-dimensional signal and image processing*, Prentice-Hall, 1990.

MAK80. J. Makhoul, "A fast cosine transform in one and two dimensions," *IEEE Transactions on Acoustics, Speech, and Signal Processing*, vol. ASSP-28, no. 1, pp. 27-34, Feb. 1980.

MAL98. S. Mallat, "A Wavelet Tour of Signal Processing," Academic Press, 1998.

MAR87. S.L. Marple, *Digital spectral analysis*, Prentice-Hall, 1987.

NUS81. H.J. Nussbaumer, *Fast Fourier transform and convolution algorithms*, Springer Verlag, 1981.

OPP89. A. Oppenheim, R.W. Schafer, *Discrete-time signal processing*, Prentice-Hall, 1989.

PIT86. I. Pitas, M.G. Strintzis, "General in-place calculation of discrete Fourier transforms of multidimensional sequences," *IEEE Transactions on Acoustics, Speech, and Signal Processing*, vol. ASSP-34, no. 3, pp. 565-572, June 1986.

PIT87. I. Pitas, M.G. Strintzis, "Algorithms for the reduction of the I-O operations in the calculation of the 2-D DFT," *Signal Processing*, vol. 12, pp. 277-289, 1987.

PRO89. J.G. Proakis, D.G. Manolakis, *Introduction to digital signal processing*, Macmillan, 1988.

QIAN96. S. Qian and D. Chen, "Joint Time-frequency Analysis," Prentice-Hall, Upper Saddle River, NJ, 1996.

RAO90. K.R. Rao, P. Yip, *Discrete cosine transform: Algorithms, advantages, applications*, Academic Press, 1990.

RIO91. O. Rioul amd M. Vetterli, "Wavelets and Signal Processing," *IEEE Signal Processing Magazine*, pp. 14-38, October 1991.

RIV77. G.K. Rivard, "Direct fast Fourier transform of bivariate functions," *IEEE Transactions on Acoustics, Speech, and Signal Processing*, vol. ASSP-25, no. 3, pp. 250-252, June 1977.

VET95. M. Vetterli and J. Kovacevic, "Wavelets and Subband Coding," Prentice-Hall, Upper Saddle River, NJ, 1995.

3

Digital Image Filtering and Enhancement

3.1 INTRODUCTION

Images are acquired by photoelectronic or photochemical methods. The sensing devices tend to degrade the quality of the digital images by introducing noise, geometric deformation and/or blur due to motion or camera misfocus. One of the primary concerns of digital image processing is to increase image quality and to moderate the degradations introduced by the sensing and acquisition devices. *Image restoration* techniques are concerned primarily with the reconstruction or recovery of an image that has been degraded. A priori knowledge about the degradation phenomenon may be used for this purpose. *Digital image enhancement* techniques increase subjective image quality by sharpening certain image features (e.g. edges and boundaries, contrast) and by reducing noise. Digital enhancement and restoration operations can be thought of as two-dimensional digital filters. They can be distinguished in two large classes: *linear digital filters* and *nonlinear digital filters*. Linear digital image filters can be designed and/or implemented either in the spatial domain (*spatial operations*) or in the frequency domain (*transform operations*). The majority of digital restoration filters are linear ones and are usually implemented in the frequency domain. There are, of course, several nonlinear digital image restoration techniques as well. Spatial operations (linear or nonlinear) are primarily used for digital image enhancement. These operations are two-dimensional digital filters having *regions of support* that consist of a single pixel (*point operations*) or a pixel neighborhood (*local operations*).

In the following sections we shall describe algorithms for the implementation of digital filters that are used either in digital image restoration or in digital image enhancement. This chapter also contains algorithms that are used to create visual effects for digital image display. Such algorithms are digital image pseudocoloring, halftoning and zooming. They are traditionally considered to be subtopics of digital image enhancement, because they influence the subjective visual quality of an image.

3.2 DIRECT IMPLEMENTATION OF TWO-DIMENSIONAL FIR DIGITAL FILTERS

Two-dimensional linear finite impulse response (FIR) digital filters are linear shift-invariant two-dimensional systems, whose region of support is finite. In many cases, the FIR impulse response $h(n_1, n_2)$ is non-zero only in the rectangle $R_{M_1 M_2} = [0, M_1] \times [0, M_2] = \{(n_1, n_2) : 0 \leq n_1 < M_1, 0 \leq n_2 < M_2\}$. The output $y(n_1, n_2)$ of the FIR digital filter is given by the linear convolution:

$$y(n_1, n_2) = \sum_{k_1=0}^{M_1-1} \sum_{k_2=0}^{M_2-1} h(k_1, k_2) x(n_1 - k_1, n_2 - k_2) \qquad (3.2.1)$$

Frequently, the region of support of the FIR digital filter is $[-\nu_1, \nu_1] \times [-\nu_2, \nu_2]$, where $M_i = 2\nu_i + 1, i = 1, 2$, are odd-valued filter lengths. In this case, the output of an FIR filter is given by:

$$y(n_1, n_2) = \sum_{k_1=-\nu_1}^{\nu_1} \sum_{k_2=-\nu_2}^{\nu_2} h(k_1, k_2) x(n_1 - k_1, n_2 - k_2) \qquad (3.2.2)$$

Several digital image filters employ the second definition. The *moving average* filter is the simplest two-dimensional low-pass FIR digital filter. If its lengths are odd numbers $M_i = 2\nu_i + 1$, $i = 1, 2$, it is defined as follows:

$$y(n_1, n_2) = \frac{1}{M_1 M_2} \sum_{k_1=-\nu_1}^{\nu_1} \sum_{k_2=-\nu_2}^{\nu_2} x(n_1 - k_1, n_2 - k_2) \qquad (3.2.3)$$

Its impulse response is given by $h(n_1, n_2) = 1/M_1 M_2$, for $(n_1, n_2) \in R_{M_1 M_2}$. It is very effective in removing white additive Gaussian noise [PIT90]. However, it tends to blur edges and image details (e.g. lines) and degrades image quality. The implementation of the moving average filter can be performed by using the subroutine of Figure 3.2.1.

```
int movav(a,b, NW,MW, N1,M1,N2,M2)
image a,b;
```

```
int NW,MW;
int N1,M1,N2,M2;
/* Subroutine for 2-d moving average filtering
   a,b: input and output buffers
   NW,MW: window size
   N1,M1: start coordinates
   N2,M2: end    coordinates            */

{ int k,l, i,j,NM2,NW2,MW2, s;
  float NM;
  NM=(float)NW*MW; NM2=NM/2;
  NW2=NW/2; MW2=MW/2;
  for(k=N1+NW2; k<N2-NW2; k++)
  for(l=M1+MW2; l<M2-MW2; l++)
     { s=0;
       for(i=0; i<NW; i++)
        for(j=0; j<MW; j++)
          s += a[k+NW2-i][l+MW2-j];
        b[k][l]=(unsigned char)(((float)s)/NM);
     }
  return(0);
}
```

Fig. 3.2.1 Subroutine for the calculation of the moving average filter output.

The direct implementation of an FIR filter (3.2.1) is very easy. The entire image of size $N_1 \times N_2$ is scanned line by line or column by column to produce the filter output. The relevant program code in C-like language is shown in Figure 3.2.2. It can be easily seen that the filter operates only in the interior $[\nu_1, N_1 - \nu_1) \times [\nu_2, N_2 - \nu_2)$ of the input image of size $N_1 \times N_2$. If filtering of the entire $N_1 \times N_2$ image is desired the image must be extended to the size $(N_1 + M_1 - 1) \times (N_2 + M_2 - 1)$ by introducing suitable border pixel values (usually set equal to zero). Alternatively, linear filters of smaller extent $m_1 \times m_2, 0 < m_1 < M_1, 0 < m_2 < M_2$, can be used to produce the filter output close to the image border.

```
int conv(a,b, hcoe, NW,MW, N1,M1,N2,M2)
image a,b;
win_f hcoe;
int NW,MW;
int N1,M1,N2,M2;

/* Subroutine for 2-d convolution
   a,b: input and output buffers
```

```
    hcoe: 2-d filter coefficient buffer
    NW,MW: window size
    N1,M1: start coordinates
    N2,M2: end   coordinates           */

{ int k,l, i,j,NM2,NW2,MW2;
  float s;

  NW2=NW/2;  MW2=MW/2;

  for(k=N1+NW2; k<N2-NW2; k++)
  for(l=M1+MW2; l<M2-MW2; l++)
     { s=0;
       for(i=0; i<NW; i++)
        for(j=0; j<MW; j++)
         s+=(float)a[k+i-NW2][l+j-MW2]*hcoe[i][j];
       b[k][l]=(unsigned char)s;
     }
  return(0);
}
```

Fig. 3.2.2 Subroutine for the calculation of the linear FIR filter output.

The number of real multiplications needed for the calculation of a pixel of the filter output is $M_1 M_2$. Thus, the total number of multiplications for the entire filter output is $(N_1 - M_1 + 1)(N_2 - M_2 + 1)M_1 M_2$. When the filter size $M_1 \times M_2$ is much smaller than the image size $N_1 \times N_2$, the computational complexity of the direct filter computation is of the order $O(kN^2)$ for $N_1 \simeq N_2 \simeq N$. If the filter size is comparable to the image size, i.e. $N_1 \simeq N_2 \simeq M_1 \simeq M_2 \simeq N$ and the entire filter output is calculated by extending the input image with suitable border, the computational complexity rises to the order $O(N^4)$. In this case, frequency domain approaches are used to lower the computational load required.

In several cases, zero-phase FIR filters are used, satisfying $H(\omega_1,\omega_2) = H^*(\omega_1,\omega_2)$. The corresponding impulse response has the following spatial symmetry:

$$h(n_1, n_2) = h(-n_1, -n_2) \tag{3.2.4}$$

when the impulse response is real (as happens in most practical digital image processing applications). The symmetry (3.2.4) can be used to reduce the number of multiplications almost to one-half [DUD84]:

$$y(n_1, n_2) = \sum_{k_1=-\nu_1}^{\nu_1} \sum_{k_2=1}^{\nu_2} h(k_1, k_2)[x(n_1 - k_1, n_2 - k_2) + x(n_1 + k_1, n_2 + k_2)]$$

$$+ \sum_{k_1=1}^{\nu_1} h(k_1,0)[x(n_1-k_1,n_2)+x(n_1+k_1,n_2)] + h(0,0)x(n_1,n_2) \quad (3.2.5)$$

The existence of other impulse response symmetries [PIT86CAS] can also be exploited in a similar fashion to reduce the computational complexity of the implementation of the FIR digital filter.

3.3 FAST FOURIER TRANSFORM IMPLEMENTATION OF FIR DIGITAL FILTERS

A well-known property of the discrete Fourier transform $X(k_1,k_2)$ of a two-dimensional signal $x(n_1,n_2)$:

$$X(k_1,k_2) = \sum_{n_1=0}^{N_1-1} \sum_{n_2=0}^{N_2-1} x(n_1,n_2) W_{N_1}^{n_1 k_1} W_{N_2}^{n_2 k_2} \quad (3.3.1)$$

is that it supports the *circular convolution* of the signal $x(n_1,n_2)$ with the impulse response $h(n_1,n_2)$:

$$y(n_1,n_2) \stackrel{\triangle}{=} h(n_1,n_2) \circledast\circledast\ x(n_1,n_2) \quad (3.3.2)$$

$$= \sum_{k_1=0}^{N_1-1} \sum_{k_2=0}^{N_2-1} x(k_1,k_2) h(((n_1-k_1))_{N_1}, ((n_2-k_2))_{N_2})$$

where $((n))_N$ denotes the modulo operation $((n))_N = n \mod N$, as has already been described in Chapter 2. This property of the DFT states that circular convolution in the spatial domain is equivalent to multiplication in the DFT domain:

$$y(n_1,n_2) = h(n_1,n_2) \circledast\circledast\ x(n_1,n_2) \longleftrightarrow Y(k_1,k_2) = H(k_1,k_2) X(k_1,k_2) \quad (3.3.3)$$

Therefore, the circular convolution of two sequences x,h can be calculated by the following scheme:

$$y(n_1,n_2) = IDFT[DFT[x(n_1,n_2)]DFT[h(n_1,n_2)]] \quad (3.3.4)$$

where DFT and $IDFT$ denote the direct and inverse discrete Fourier transforms respectively. This computational scheme has advantages because both the direct and the inverse two-dimensional DFTs can be calculated using the two-dimensional fast Fourier transform algorithms. However, the computation of the cyclic convolution is of limited use. The computation of the linear convolution (3.2.1) is much more important in digital image processing applications. It is well known [LIM90], [JAI89] that the discrete time Fourier

126 DIGITAL IMAGE FILTERING AND ENHANCEMENT

transform $X(\omega_1,\omega_2)$ supports linear convolution:

$$y(n_1,n_2) = h(n_1,n_2) ** x(n_1,n_2) \longleftrightarrow Y(\omega_1,\omega_2) = H(\omega_1,\omega_2)X(\omega_1,\omega_2)$$
(3.3.5)

Thus, a scheme similar to that described by (3.3.4) can be used for the calculation of the linear convolution. The direct application of this scheme is not practical, because ω_1,ω_2 are continuous variables and because the inverse discrete time Fourier transform requires the calculation of double complex integrals. However, it is possible to employ the fast calculation scheme (3.3.4) for the computation of a two-dimensional linear convolution as has already been described in Chapter 2. If the signal $x(n_1,n_2)$ and the filter $h(n_1,n_2)$ have regions of support $[0, N_1) \times [0, N_2)$ and $[0, M_1) \times [0, M_2)$ respectively, the linear convolution output $y(n_1,n_2)$ has region of support $[0, N_1 + M_1 - 1) \times [0, N_2 + M_2 - 1)$. Let us suppose that we augment both sequences x, h to dimensions $L_i \geq N_i + M_i - 1, i = 1,2$, respectively by padding them with zeros. In this case, it can be easily proven that the computational scheme (3.3.4) gives the correct linear convolution output. The resulting algorithm for the calculation of the linear convolution is shown in Figure 3.3.1.

```
int convolution(mats1,mats2, matd, ND1,MD1, N1,M1,N2,M2)
matrix mats1,mats2,matd;
int ND1,MD1;
int N1,M1,N2,M2;

/* Subroutine to compute the convolution of two signals
    mats1,mats2 : source matrices
    matd : destination matrix (convolution)
    ND1,MD1 : destination coordinates
    N1,M1 : upper left  corner of source matrices
    N2,M2 : lower right corner of source matrices
    NOTE! size of destination matrix: 2*(N2-N1) x 2*(M2-M1) */

{ int nr,nc,rows2,cols2;
  int ret;
  float a,b,c,d;
  matrix matmere1,matmeim1,matmere2,matmeim2;

  ret=0;
  rows2=2*(N2-N1); cols2=2*(M2-M1);

  /* Create temporary matrices having double the size of ROI */
  matmere1=matf2(rows2,cols2);
  matmeim1=matf2(rows2,cols2);
```

```
matmere2=matf2(rows2,cols2);
matmeim2=matf2(rows2,cols2);

/* Clear temporary matrices */
for(nr=0; nr<rows2; nr++)
for(nc=0; nc<cols2; nc++)
   { matmere1[nr][nc]-0.0, matmeim1[nr][nc]=0.0;
     matmere2[nr][nc]=0.0; matmeim2[nr][nc]=0.0;
   }

/* Copy source matrices to temporary matrices */
for(nr=N1; nr<N2; nr++)
for(nc=M1; nc<M2; nc++)
   { matmere1[nr-N1][nc-M1]=mats1[nr][nc];
     matmere2[nr-N1][nc-M1]=mats2[nr][nc];
   }

/* Perform the FFT transforms */
ret=fft2d(matmere1,matmeim1,
          matmere1,matmeim1, 0, 0,0, 0,0,rows2,cols2);
if(ret) goto R22;
ret=fft2d(matmere2,matmeim2,
          matmere2,matmeim2, 0, 0,0, 0,0,rows2,cols2);
if(ret) goto R22;

/* Multiply the transformed images */
for(nr=0; nr<rows2; nr++)
for(nc=0; nc<cols2; nc++)
   { a=matmere1[nr][nc]; c=matmere2[nr][nc];
     b=matmeim1[nr][nc]; d=matmeim2[nr][nc];
     matmere1[nr][nc]=a*c-b*d; matmeim1[nr][nc]=a*d+b*c;
   }

/* Perform inverse FFT transform */
ret=fft2d(matmere1,matmeim1,
          matmere1,matmeim1, 1, 0,0, 0,0,rows2,cols2);

for(nr=0; nr<rows2; nr++)
for(nc=0; nc<cols2; nc++)
    matd[ND1+nr][MD1+nc]=matmere1[nr][nc];

R22: mfreef2(matmeim2,rows2);
R21: mfreef2(matmere2,rows2);
R12: mfreef2(matmeim1,rows2);
R11: mfreef2(matmere1,rows2);
```

```
    RET: return(ret);
}
```

Fig. 3.3.1 Algorithm for the calculation of the two-dimensional linear convolution by using the two-dimensional FFT.

Any of the 2-d FFT algorithms described in Chapter 2 can be used for the fast calculation of the 2-d linear convolution. All of them operate on sequences whose length is a power of 2. Thus, the transform lengths L_1, L_2 must be powers of 2. If a row–column FFT is used, the total number of real multiplications needed for the calculation of (3.3.4) is:

$$N_{mult} = 6L_1L_2 \log_2(L_1L_2) + 4L_1L_2 \qquad (3.3.6)$$

If we take into account that $L_i \geq N_i + M_i - 1$, $i = 1, 2$, it can be easily seen that the multiplicative complexity of the calculation scheme (3.3.4) is of the order $O(kL^2 \log_2 L)$. Therefore, this scheme offers great computational advantages for large image sizes and for filter lengths M_1, M_2 that are large and comparable to the image sizes N_1, N_2. When the filter size is small compared to the two-dimensional image size, the FFT calculation scheme (3.3.4) does not offer computational advantages and the direct linear convolution calculation can be used instead.

One big problem in using FFTs for the linear two-dimensional convolution calculation is the increase in the memory requirements. Let us suppose that we have to process an image having dimensions $N_1 = N_2 = 2^s$ with a filter having dimensions $M_1 \times M_2$, where $M_1, M_2 \ll 2^s$. In this case, both filter and image arrays must be augmented to sizes $L_1 \times L_2$, where $L_1 = L_2 = 2^{s+1}$. Thus, eight times more RAM is needed for the calculation of linear convolution by using the DFT than for its direct implementation. If the conjugate properties of the two-dimensional DFT are not exploited, 16 times more RAM is needed for the convolution calculation via the FFT. The extra memory is needed for the storage of both real and imaginary parts of the transformed sequences. These memory requirements are relatively high, even by current technological standards. Therefore, alternative techniques must be employed that can combine the efficiency of the FFT implementation with reduced memory requirements.

3.4 BLOCK METHODS IN THE LINEAR CONVOLUTION CALCULATION

Block methods have successfully been used in the calculation of the one-dimensional convolution of long sequences [OPE89]. They can be easily extended to the calculation of the linear two-dimensional convolution of large

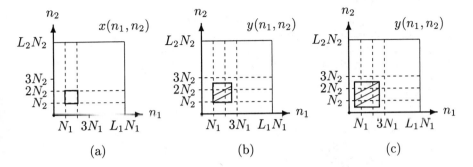

Fig. 3.4.1 Overlap–add method for linear convolution. (a) Non-overlapping blocks; (b) convolution output block, when h is defined over $[0, M_1) \times [0, M_2)$; (c) convolution output block when h is defined over $[-\nu_1, \nu_1] \times [-\nu_2, \nu_2]$.

images when the use of the FFT implementation poses insurmountable memory problems [DUD84], [LIM90].

The *overlap–add* method is based on the distributive property of convolution. A sequence $x(n_1, n_2)$ can be divided into $K_1 \times K_2$ non-overlapping subsequences having dimensions $N_1 \times N_2$ each, as can be seen in Figure 3.4.1:

$$x_{ij}(n_1, n_2) = \begin{cases} x(n_1, n_2) & iN_1 \leq n_1 < (i+1)N_1,\ jN_2 \leq n_2 < (j+1)N_2 \\ 0 & \text{otherwise} \end{cases} \quad (3.4.1)$$

The sequence $x(n_1, n_2)$ can be easily reconstructed from its blocks:

$$x(n_1, n_2) = \sum_{i=1}^{K_1} \sum_{j=1}^{K_2} x_{ij}(n_1, n_2) \quad (3.4.2)$$

The linear convolution output $y(n_1, n_2)$ is the sum of the convolution outputs contributed by the input sequence blocks:

$$\begin{aligned} y(n_1, n_2) &= x(n_1, n_2) ** h(n_1, n_2) \quad (3.4.3) \\ &= \sum_{i=1}^{K_1} \sum_{j=1}^{K_2} (x_{ij}(n_1, n_2) ** h(n_1, n_2)) = \sum_{i=1}^{K_1} \sum_{j=1}^{K_2} y_{ij}(n_1, n_2) \end{aligned}$$

The output convolution blocks $y_{ij}(n_1, n_2)$ are overlapping. If the filter impulse response $h(n_1, n_2)$ is defined over $[0, M_1) \times [0, M_2)$, the output blocks overlap in the way shown in Figure 3.4.1b. If the filter impulse response $h(n_1, n_2)$ is defined over $[-\nu_1, \nu_1] \times [-\nu_2, \nu_2]$, the output convolution blocks overlap along the entire border of the block, as shown in Figure 3.4.1c. The convolution output blocks $y_{ij}(n_1, n_2)$ can be calculated by using two-dimensional FFTs of size at least $(M_1 + N_1 - 1) \times (M_2 + N_2 - 1)$. The algorithm is shown in Figure 3.4.3. The block size $N_1 \times N_2$ controls implementation efficiency. If the block size is small, the overlap–add method requires a small amount of RAM at the expense of increasing computational load. If the block size is large, the overlap–add method is more efficient computationally, at the expense of the memory required. It is relatively difficult to find an optimal block size, since the implementation is machine-dependent. A thorough analysis of the algorithm can be found in [TWO78].

An alternative approach to the linear convolution calculation is the *overlap–save* method. It is based on the following simple property of linear convolution. Let $w(n_1, n_2)$ be a sequence defined over $[0, N_1) \times [0, N_2)$ and convolved with an impulse response $h(n_1, n_2)$ defined over $[0, M_1) \times [0, M_2)$. It is easily proven that the linear convolution output $y'(n_1, n_2)$ is equal to the circular convolution output of extent $N_1 \times N_2$ only within the rectangle $[M_1 - 1, N_1) \times [M_2 - 1, N_2)$ [LIM90]:

$$y'(n_1, n_2) = w(n_1, n_2) **h(n_1, n_2) = w(n_1, n_2) \circledast\circledast\ h(n_1, n_2)$$
$$\text{for } M_1 - 1 \leq n_1 < N_1, M_2 - 1 \leq n_2 < N_2 \quad (3.4.4)$$

as shown in Figure 3.4.2a. The circular convolution is based on $N_1 \times N_2$ periodicity. The rest of the rectangle $[0, N_1) \times [0, N_2)$ contains an aliased version of $y'(n_1, n_2)$, which is not useful. This property can be exploited for the block calculation of the linear convolution in the following way. The linear convolution output is split into non-overlapping blocks of size $N_1 \times N_2$:

$$y(n_1, n_2) = \sum_{i=1}^{K_1} \sum_{j=1}^{K_2} y_{ij}(n_1, n_2) \quad (3.4.5)$$

as shown in Figure 3.4.2c. Each block $y_{ij}(n_1, n_2)$ can be calculated as follows. The corresponding input block $x_{ij}(n_1, n_2)$ of size $N_1 \times N_2$ is augmented to the block $x'_{ij}(n_1, n_2)$ having size $(N_1 + M_1 - 1) \times (N_2 + M_2 - 1)$ by extending it over its left-hand and lower neighbors, respectively, as shown in Figure 3.4.2b. The convolution $x'_{ij}(n_1, n_2) \circledast\circledast\ h(n_1, n_2)$ is calculated by using DFTs of size $(N_1 + M_1 - 1) \times (N_2 + M_2 - 1)$. Since the entire block $x_{ij}(n_1, n_2)$ is non-zero, only the upper right-hand rectangle of the convolution output is correct. It has size $N_1 \times N_2$ and is equal to the linear convolution output block $y_{ij}(n_1, n_2)$. A similar overlap–save method can be constructed for filter impulse response $h(n_1, n_2)$ defined on the rectangle $[-\nu_1, \nu_1] \times [-\nu_2, \nu_2]$. The only difference is that the block $x'_{ij}(n_1, n_2)$ of size $(N_1 + M_1 - 1) \times (N_2 + M_2 - 1)$ overlaps with all its eight neighboring blocks.

Both overlap–save and overlap–add methods have approximately the same computational load, if the block sizes used are approximately equal in both methods. If the block size is carefully chosen, they give very good computational savings in comparison to the direct method. Although both methods are similar, sometimes the overlap–add method is preferred due to its conceptual simplicity.

BLOCK METHODS IN THE LINEAR CONVOLUTION CALCULATION   131

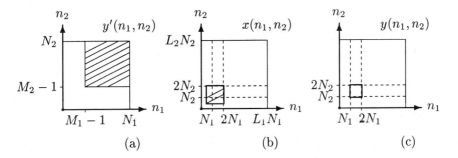

Fig. 3.4.2 (a) Result of the circular convolution of two sequences without zero padding; (b) partition of the input sequence in overlapping blocks; (c) output blocks of the overlap–save method.

```
overlap_add(x,h,y,NL,ML,N1,M1,N2,M2,Nh,Mh)
matrix x,h,y;
int NL,ML,N1,M1,N2,M2,Nh,Mh;
/* Subroutine to perform overlap--add convolution.
   x: input image matrix
   h: filter impulse response matrix
   y: output image matrix
   NL ,ML: block size
   N1, M1: start coordinates
   N2 ,M2: end coordinates
   Nh ,Mh: size of impulse response matrix h */
{
      int i,j,k,l;
      float tr,ti;
      int NS,MS;          /* Number of blocks              */
      int NY,MY;          /* Size of y                     */
      int NR,MR;          /* Size of the FFT blocks        */

      /* Buffers to store intermediate results */
      matrix xbr,xbi,hbr,hbi;

      /* Calculate block buffer sizes NR,MR.
         They must be a power of 2. */
      i = NL+Nh-2;
      j=0;
      while(i>1) { i>>=1; j++; }
      while(j) { i<<=1; j--; }
      NR=(i<<1);

      i = ML+Mh-2;
```

```
j=0;
while(i>1) { i>>=1; j++; }
while(j) { i<<=1; j--; }
MR=(i<<1);

/* Allocate memory for block buffers */
xbr = matf2(NR,MR); xbi = matf2(NR,MR);
hbr = matf2(NR,MR); hbi = matf2(NR,MR);

/* Copy input h to a block buffer. */
for(i=0;i<MR;i++)
for(j=0;j<NR;j++)
 {
  if ( (i<Mh) && (j<Nh) ) hbr[j][i]=h[j][i];
                            else hbr[j][i]=0.0;
  hbi[j][i] = 0.0;
 }

/* Calculate DFT of h. The DFT size is (Nh+NL-1,Mh+ML-1) */
rcfft(hbr,hbi,hbr,hbi,0,0,0,0,0,NR,MR);

/* Clear entire output buffer */
NY=N2+Nh-1;
MY=M2+Mh-1;
for(i=N1;i<NY;i++)
for(j=M1;j<MY;j++)
  y[i][j]=0.0;

/* Calculate the number of blocks NS*MS */
NS = (N2-N1)/NL; MS = (M2-M1)/ML;

/* Start convolution calculation in a column-wise manner.*/
for (i=0;i<NS;i++)
for (j=0;j<MS;j++)
 {
  /* Transfer one block from x to buffer xb. Initialize
     imaginary buffer xi. */
  for (k=0;k<MR;k++)
  for (l=0;l<NR;l++)
   {
    if ((k<ML) && (l<NL))
     xbr[l][k] = x[N1+i*NL+l][M1+j*ML+k] ;
    else xbr[l][k]=0.0;
    xbi[l][k] = 0.0;
   }
```

```
    /* Calculate FFT of (xbr,xbi) */
    rcfft(xbr,xbi,xbr,xbi,0,0,0,0,0,NR,MR);
    /* Multiply (xbr,xbi) by (hbr,hbi) */
    for (k=0;k<NR;k++)
    for (l=0;l<MR;l++)
     {
      tr = xbr[k][l]*hbr[k][l] - xbi[k][l]*hbi[k][l];
      ti = xbr[k][l]*hbi[k][l] + xbi[k][l]*hbr[k][l];
      xbr[k][l]=tr; xbi[k][l]=ti;
     }
    /* Inverse DFT of the product */
    rcfft(xbr,xbi,xbr,xbi,1,0,0,0,0,NR,MR);

    /* The resulting block must now be overlap-added to
       the proper position at output buffer y */
    for (k=0;k<(NL+Nh-1);k++)
    for (l=0;l<(ML+Mh-1);l++)
      y[N1+i*NL+k][M1+j*ML+l] += xbr[k][l] ;
   }

   mfreef2(hbi,NR); mfreef2(hbr,NR);
   mfreef2(xbi,NR); mfreef2(xbr,NR);
   return 0;
}
```

Fig. 3.4.3 Algorithm for overlap–add method for linear convolution calculation.

3.5 INVERSE FILTER IMPLEMENTATIONS

The aim of digital image restoration is the recovery of an image that has been degraded by the sensing instruments. Degradation sources usually produce motion blur, defocus blur and/or noise introduced during sensing. Image sensing is a complicated process and its modelling is cumbersome, as has already been described in Chapter 1. However, approximate models are sufficient in many applications. Such a model is described by the following equation:

$$g(n_1, n_2) = f(n_1, n_2) * * h(n_1, n_2) + n(n_1, n_2) \tag{3.5.1}$$

where f is the original image, g is the observed image, h is the blur point spread function and n is the sensing noise. Noise is usually assumed to be white additive Gaussian. Such an image degraded by horizontal motion blur of the form $h(n_1, n_2) = 1/N_1$, $-\nu_1 \leq n_1 \leq \nu_1$, $n_2 = 0$, $N_1 = 2\nu_1 + 1$,

Fig. 3.5.1 (a) Original image; (b) image corrupted by horizontal blur of length $N_1 = 5$ and by white additive Gaussian noise having variance 20.

and additive white noise is shown in Figure 3.5.1. The problem of image restoration is to estimate f from the observed image g, possibly based on some a priori knowledge about the blur point spread function and on the statistics of the noise n and of the images g, f. Several techniques have been proposed for digital image restoration [AND77], [PRA91], [JAI89], [LIM90]. In this section we shall describe only the implementation issues of the *inverse filter* approach.

Let us suppose that the formation noise n is negligible. The image formation model (3.5.1) takes the following format in the frequency domain:

$$G(\omega_1, \omega_2) = F(\omega_1, \omega_2) H(\omega_1, \omega_2) \qquad (3.5.2)$$

If the degradation function $H(\omega_1, \omega_2)$ is known a priori, the simplest approach to digital image restoration is inverse filtering:

$$\hat{F}(\omega_1, \omega_2) = \frac{G(\omega_1, \omega_2)}{H(\omega_1, \omega_2)} \qquad (3.5.3)$$

One of the biggest disadvantages of the inverse filter is that it cannot be defined in regions (ω_1, ω_2) of the transform domain, where $H(\omega_1, \omega_2)$ is zero. Furthermore, it is very sensitive to the presence of the formation noise n. If $|H(\omega_1, \omega_2)|$ is very small, the magnitude of the inverse filter $1/|H(\omega_1, \omega_2)|$ is large and amplifies the noise existing in the corresponding frequency regions. This problem can be solved by using a pseudoinverse filter, which is a stabilized variation of the inverse filter:

$$H^-(\omega_1, \omega_2) = \begin{cases} \frac{1}{H(\omega_1, \omega_2)} & \text{if } |H(\omega_1, \omega_2)| > \varepsilon \\ 0 & \text{otherwise} \end{cases} \qquad (3.5.4)$$

The most straightforward implementation of the inverse or the pseudoinverse filter is by using the discrete Fourier transform:

$$\hat{f}(n_1, n_2) = IDFT[G(k_1, k_2)/H(k_1, k_2)] \qquad (3.5.5)$$

where $G(k_1, k_2)$ is the DFT of the observed image $g(n_1, n_2)$ and $H(k_1, k_2)$ is the DFT of the blur point spread function $h(n_1, n_2)$. An iterative approach to the implementation of the inverse filter or its variation (3.5.4) is presented in [SCH81], [LIM90]. Let $\hat{F}_k(\omega_1, \omega_2)$ be the estimate of the original image at the iteration k. This estimate produces an approximation $\hat{F}_k(\omega_1, \omega_2) H(\omega_1, \omega_2)$ of the observed image $G(\omega_1, \omega_2)$. A correction term is added to $\hat{F}_k(\omega_1, \omega_2)$ to produce the estimate $\hat{F}_{k+1}(\omega_1, \omega_2)$:

$$\hat{F}_{k+1}(\omega_1, \omega_2) = \hat{F}_k(\omega_1, \omega_2) + \mu[G(\omega_1, \omega_2) - \hat{F}_k(\omega_1, \omega_2) H(\omega_1, \omega_2)] \quad (3.5.6)$$

where μ is a positive convergence parameter. The initial estimate of $F(\omega_1, \omega_2)$ is given by:

$$F_0(\omega_1, \omega_2) = \mu G(\omega_1, \omega_2) \quad (3.5.7)$$

It can be proven that $\hat{F}_k(\omega_1, \omega_2)$ tends to the inverse filter output $G(\omega_1, \omega_2)/H(\omega_1, \omega_2)$, provided that the convergence parameter μ satisfies the inequality:

$$|1 - \mu H(\omega_1, \omega_2)| < 1 \quad (3.5.8)$$

for any point in the transform domain. The advantage of the iterative method (3.5.6) is that it can be stopped after a certain number of iterations if the filtering output is acceptable. Furthermore, the convergence parameter μ can be changed in order to alter the convergence speed. The iterative algorithm can be implemented either in the frequency domain, as shown in (3.5.6), or in the spatial domain:

$$\hat{f}_{k+1}(n_1, n_2) = \hat{f}_k(n_1, n_2) + \mu(g(n_1, n_2) - \hat{f}_k(n_1, n_2) ** h(n_1, n_2)) \quad (3.5.9)$$

The choice of the implementation domain depends on the size of the region of support of $h(n_1, n_2)$. If it is relatively small, the implementation in the spatial domain can be used. If it is large, the transform domain implementation is preferable.

3.6 WIENER FILTERS

Let us suppose that an image $s(n_1, n_2)$ is corrupted by additive zero-mean white noise $n(n_1, n_2)$ producing an observed image $g(n_1, n_2)$:

$$g(n_1, n_2) = s(n_1, n_2) + n(n_1, n_2) \quad (3.6.1)$$

$s(n_1, n_2)$ is assumed to be a two-dimensional zero-mean stationary random signal. We want to estimate image $\hat{s}(n_1, n_2)$ from the noisy observed image $g(n_1, n_2)$, by using a linear filter having impulse response $h_w(n_1, n_2)$:

$$\hat{s}(n_1, n_2) = g(n_1, n_2) ** h_w(n_1, n_2) \quad (3.6.2)$$

The filter impulse response is chosen in such a way that the mean square error:

$$E = E[|\,e(n_1, n_2)\,|^2] = E[|\,s(n_1, n_2) - \hat{s}(n_1, n_2)\,|^2] \qquad (3.6.3)$$

is minimized. $E[\cdot]$ denotes the expectation operator. It can be easily proven that the filter, which minimizes (3.6.3), satisfies the following set of equations [JAI89], [LIM90]:

$$\sum_{k=-\infty}^{\infty} \sum_{l=-\infty}^{\infty} h_w(m-k, n-l) R_{gg}(k, l) = R_{sg}(m, n) \qquad (3.6.4)$$

$R_{gg}(k,l)$ and $R_{sg}(m,n)$ denote autocorrelation and cross-correlation functions, respectively:

$$R_{gg}(k, l) = E[g(m, n) g(m-k, n-l)] \qquad (3.6.5)$$
$$R_{sg}(k, l) = E[s(m, n) g(m-k, n-l)] \qquad (3.6.6)$$

By taking the Fourier transform of both sides of (3.6.4), the following relation gives the filter transfer function:

$$H_w(\omega_1, \omega_2) = \frac{P_{sg}(\omega_1, \omega_2)}{P_{gg}(\omega_1, \omega_2)} \qquad (3.6.7)$$

where $P_{sg}(\omega_1, \omega_2)$, $P_{gg}(\omega_1, \omega_2)$ are the cross-power spectrum of s, g and the power spectrum of g respectively, that is, the Fourier transforms of (3.6.5, 3.6.6). This filter is called the *non-causal Wiener filter*. If the signal $s(n_1, n_2)$ is uncorrelated with the noise $n(n_1, n_2)$, it can be easily proven that the Wiener filter transfer function is given by:

$$H_w(\omega_1, \omega_2) = \frac{P_{ss}(\omega_1, \omega_2)}{P_{ss}(\omega_1, \omega_2) + P_{nn}(\omega_1, \omega_2)} \qquad (3.6.8)$$

The Wiener filter (3.6.8) can be efficiently used for additive noise removal. For frequencies where the signal power spectrum P_{ss} is much stronger than the noise power spectrum P_{nn}, the transfer function approaches unity. These frequencies belong to the filter passband. For frequencies where the noise power spectrum dominates, the filter transfer function approaches $P_{ss}(\omega_1, \omega_2)/P_{nn}(\omega_1, \omega_2)$. Thus, these frequency components are attenuated in proportion to their noise to signal ratio. In most images, the signal to noise ratio is usually high at low spatial frequencies. Therefore, the Wiener filter performs as a low-pass filter.

The Wiener filter can be easily modified to be used in digital image restoration. Let us suppose that the image formation model is described by (3.5.1). The power spectra $P_{fg}(\omega_1, \omega_2)$ and $P_{gg}(\omega_1, \omega_2)$ are given by:

$$P_{fg}(\omega_1, \omega_2) = H^*(\omega_1, \omega_2) P_{ff}(\omega_1, \omega_2) \qquad (3.6.9)$$
$$P_{gg}(\omega_1, \omega_2) = |H(\omega_1, \omega_2)|^2 P_{ff}(\omega_1, \omega_2) + P_{nn}(\omega_1, \omega_2) \qquad (3.6.10)$$

Thus, the Wiener filter takes the following form:

$$H_w(\omega_1,\omega_2) = \frac{H^*(\omega_1,\omega_2)P_{ff}(\omega_1,\omega_2)}{\mid H(\omega_1,\omega_2)\mid^2 P_{ff}(\omega_1,\omega_2) + P_{nn}(\omega_1,\omega_2)} \qquad (3.6.11)$$

A subroutine for the calculation of the frequency response of the Wiener filter by using (3.6.11) is shown in Figure 3.6.1.

```
int wienerfrequency(Fre,Fim,Pss,Hre,Him,Pw,N,M)
matrix Fre,Fim;
matrix Pss,Hre,Him;
float Pw;
int N,M;
{
/* Subroutine to calculate the Wiener filter
   in the frequency domain
   Image formation model : X = H S + W
   Filter output : Y = F X
   where F is the Wiener filter:F = (H* Pss)/(|H|^2 Pss + Pw)
   Fre: real part of the filter (output)
   Fim: imaginary part of the filter (output)
   Pss: power spectral density of the original image
   Hre: real part of the DFT of the blur function
   Him: imaginary part of the DFT of the blur function
   Pw:  variance of the additive Gaussian noise
        on the blurred image
   N,M: size of the matrices */
   int x,y; float fme;
   for(y=0; y<N; y++)
   for(x=0; x<M; x++)
      { fme=Pss[y][x]/((Hre[y][x]*Hre[y][x]+
                Him[y][x]*Him[y][x])*Pss[y][x]+Pw);
        Fre[y][x] =  Hre[y][x] * fme;
        Fim[y][x] = -Him[y][x] * fme;
      }
   return(0);
}
```

Fig. 3.6.1 Calculation of the frequency response of a Wiener filter.

If the noise is zero, that is, $P_{nn}(\omega_1,\omega_2) = 0$, the Wiener filter reduces to the inverse filter, as can be seen from (3.6.11). If both noise and blur are present, the Wiener filter is a compromise between the low-pass filter that reduces the additive noise and the inverse filter which has high-pass characteristics that reduce blur and enhance the noise.

An important problem in the design and implementation of Wiener filters is the estimation of the blur transfer function $H(\omega_1, \omega_2)$ and of the power spectra $P_{ff}(\omega_1, \omega_2)$, $P_{nn}(\omega_1, \omega_2)$. The blur transfer function is usually known a priori or it is estimated from the observed image [AND77], [LIM90]. The average noise level is estimated from the high-frequency parts of the observed image, where the image power is assumed to be negligible. The original image power spectrum $P_{ff}(\omega_1, \omega_2)$ is estimated from the power spectrum of the observed image:

$$\hat{P}_{ff}(\omega_1, \omega_2) = \frac{\hat{P}_{gg}(\omega_1, \omega_2) - \hat{P}_{nn}(\omega_1, \omega_2)}{\mid H(\omega_1, \omega_2) \mid^2} \quad (3.6.12)$$

When the magnitude of the point spread function $\mid H(\omega_1, \omega_2) \mid$ becomes less than a certain threshold, it is modified locally to avoid division by zero in (3.6.12). The easiest modification is to clip it to the threshold level. The threshold choice is rather heuristic and must be low enough to avoid aliasing artifacts in the Wiener filter output. The power spectrum of the observed image $P_{gg}(\omega_1, \omega_2)$ can be estimated by using two-dimensional spectral estimation techniques [LIM90], [MAR87]. Such techniques based either on Fourier transform or on AR modeling are described in Chapter 2. The estimator $P_{gg}(\omega_1, \omega_2)$ can be used directly in the denominator of the Wiener filter (3.6.11). The numerator of the Wiener filter can be calculated by employing (3.6.9) and (3.6.12).

The Wiener filter can be implemented either in the frequency or in the spatial domain. If $H_w(\omega_1, \omega_2)$ is available, it can be sampled to produce the DFT coefficients $H_w(k_1, k_2)$ of the Wiener filter:

$$H_w(k_1, k_2) = H_w(\omega_1, \omega_2) \quad \omega_1 = \frac{2\pi k_1}{L_1}, \quad \omega_2 = \frac{2\pi k_2}{L_2} \quad (3.6.13)$$

The input image $g(n_1, n_2)$ of size $N_1 \times N_2$ ($L_i > N_i$, $i = 1, 2$) can be windowed by using a window function $w(n_1, n_2)$ in order to reduce artifacts. The input image (windowed or not) is augmented to size $L_1 \times L_2$ by padding with zeros and its DFT of size $L_1 \times L_2$ is calculated. After multiplication in the frequency domain, the Wiener filter output is calculated by using an inverse DFT of size $L_1 \times L_2$. This algorithm has been applied to the image shown in Figure 3.6.2a. The Wiener filter output is shown in Figure 3.6.2b. It is clearly seen that motion blur has been removed. However, certain artifacts along the vertical direction appear in the output image.

The impulse response $h_w(n_1, n_2)$ of the Wiener filter can be calculated by sampling $H_w(\omega_1, \omega_2)$ on an $L_1 \times L_2$ grid and by employing the inverse DFT of size $L_1 \times L_2$. In many cases, the effective length of the impulse response is relatively small and $h_w(n_1, n_2)$ can be truncated to size $M_1 \times M_2$, where $M_i \ll L_i, i = 1, 2$. Direct implementation of the linear convolution

Fig. 3.6.2 (a) Image corrupted by horizontal blur of length $N_1 = 5$ and by white additive Gaussian noise having variance 20; (b) Wiener filter output.

$h_w(n_1, n_2) * *g(n_1, n_2)$ can be employed for the calculation of the Wiener filter output.

The extension of Wiener filtering to color images is not straightforward, because color images are multichannel signals whereas monochrome images are single-channel signals. Interchannel correlations exist in color images that carry additional information which can be taken into account in the restoration procedure. Recently, a variety of multichannel Wiener filters was designed both in the spatial and frequency domain [ANG94]. FIR and IIR Wiener filters were developed. Color image restoration based on multi-channel autoregressive (AR) image modeling was examined as well. It was found that the improvement in the quality of color images restored by the designed multi-channel Wiener filters is greater than that of using corresponding independent single-channel Wiener filters.

3.7 MEDIAN FILTER ALGORITHMS

Digital image noise usually appears in the high frequencies of the image spectrum. Therefore, a low-pass digital filter may be used for noise removal. Such a filter is the moving average filter (3.2.3). However, linear low-pass filters tend to smear image details (e.g. lines, corners), whose power is in the high frequencies as well. Furthermore, they tend to blur the image edge for similar reasons, thus degrading the digital image quality. Nonlinear low-pass filters have been proposed in the literature that remove noise effectively and preserve image edges and details [PIT90]. Such a filter class is based on *data ordering*. Let x_i, $i = 1, \ldots, n$, be n observations, whose number $n = 2\nu + 1$ is odd. They can be ordered according to their magnitude as follows:

$$x_{(1)} < x_{(2)} < \cdots < x_{(n)} \tag{3.7.1}$$

$x_{(i)}$ denotes the i-th *order statistic*. $x_{(1)}, x_{(n)}$ are the maximum and the minimum observations respectively [DAV81]. The observation $x_{(\nu+1)}$ lies in the middle and is called the *median* of the observations [TUK77]. It is also denoted by $med(x_i)$. By definition, the median lies in the 'middle' of the observation data. It minimizes the L_1 norm:

$$\sum_{i=1}^{n} |x_i - med| \to min \qquad (3.7.2)$$

According to (3.7.2), the median is the maximum likelihood estimate (MLE) of the location for the Laplacian distribution:

$$f(x) = \frac{1}{2}\exp(-|x|) \qquad (3.7.3)$$

A two-dimensional median filter has the following definition:

$$y(i,j) = med\{\ x(i+r, j+s),\ \ (r,s) \in A \quad (i,j) \in Z^2\} \qquad (3.7.4)$$

where $Z^2 = Z \times Z$ denotes the digital image plane. The set $A \subset Z^2$ defines the filter window. If the input image is of finite extent $N \times M$, $0 \leq i \leq N-1, 0 \leq j \leq M-1$, definition (3.7.4) is valid only in the interior of the output image, that is, for those i, j for which

$$0 \leq i+r \leq N-1,\ 0 \leq j+s \leq M-1,\ (r,s) \in A \qquad (3.7.5)$$

(3.7.5) is not valid at the border of the image. There are two approaches to solve this problem. In the first one, the filter window A is truncated in such a way so that (3.7.5) is valid and definition (3.7.4) can be used again. In the second approach, the input sequence is appended with sufficient samples and (3.7.4) is applied for $0 \leq i \leq N-1, 0 \leq j \leq M-1$. Median filters have some very interesting properties. They have low-pass characteristics and they remove additive white noise [PIT90]. They are very efficient in the removal of noise having a long-tailed distribution (e.g. Laplacian distribution). However, they are not as efficient as the moving average filter (3.2.3) in additive Gaussian noise removal. The median is a robust estimator of location [HAM86]. Therefore, a single outlier (e.g. impulse) can have no effect on its performance, even if its magnitude is very large or very small. The median becomes unreliable only if more than 50% of the data are outliers. On the contrary, the moving average filter is very susceptible to impulses [JUS81], [PIT90]; even a single outlier can destroy its performance. The robustness properties of the median make it very suitable for impulse noise filtering, as can be seen in Figure 3.7.1.

Edge information is very important for human perception. Its preservation and, possibly, its enhancement is a very important subjective feature of the performance of a digital image filter. Edges, by definition, contain high frequencies. The moving average filter, which has low-pass characteristics, smooths them and produces images which are unpleasant to the eye. In contrast, the median filter tends to preserve the edge sharpness (Figure 3.7.1c).

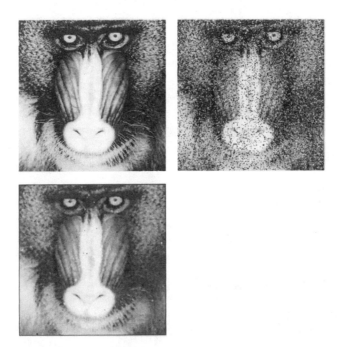

Fig. 3.7.1 (a) Original image; (b) image corrupted by impulsive noise; (c) output of a 3 × 3 median filter.

The robustness properties of the median filter make it very suitable for edge filtering.

The median filter not only smooths noise in homogeneous image regions but tends to produce regions of constant or nearly constant intensity. The shapes of these regions depend on the geometry of the filter window. Usually they are either linear patches (streaks) or blotches. These effects are undesirable, because they are perceived as lines or contours, which do not exist in the original image.

The calculation of the median filter output can be done easily, provided that a fast *sorting* or a fast *selection* algorithm is available. Such algorithms (e.g. MERGESORT, QUICKSORT, BUBBLESORT) are very well known in the literature [KNU73], [HOR84], [PIT90] and will not be repeated here. A subroutine for the median filtering calculation based on a BUBBLESORT algorithm is shown in Figure 3.7.2.

```
int median(a,b, NW,MW, N1,M1,N2,M2)
image a,b;
int NW,MW;
int N1,M1,N2,M2;
```

142   DIGITAL IMAGE FILTERING AND ENHANCEMENT

```
/* Subroutine for 2-d median filtering
   a,b: input and output buffers
   NW,MW: window size
   N1,M1: start coordinates
   N2,M2: end   coordinates        */

{ int k,l, i,j, NM,NM2,NW2,MW2;
  int * t;

  /* Temporary buffer to be used in data sorting */
  t=(int *)malloc((unsigned int)(NW*MW)*sizeof(int));

  NM=NW*MW; NM2=NM/2; NW2=NW/2; MW2=MW/2;

  for(k=N1+NW2; k<N2-NW2; k++)
   for(l=M1+MW2; l<M2-MW2; l++)
     { for(i=0; i<NW; i++)
       for(j=0; j<MW; j++)
        t[MW*i+j]=a[k+i-NW2][l+j-MW2];
       sort(t,NM);
       b[k][l]=t[NM2];
     }

  free(t);
  return(0);
}

sort(t, NM)
int * t;
int NM;
/* Bubblesort */
{ int i, k,j;
  int r;

  i=0;
  while(i==0)
   { i=1;
     for(k=0; k<NM-1; k++)
      if(t[k]>t[k+1]) {i=0; r=t[k+1]; t[k+1]=t[k]; t[k]=r;}
   }
}
```

Fig. 3.7.2 Two-dimensional median filtering algorithm.

Median filtering is a *running* operation. The filter windows centered at pixels (i,j) and $(i,j+1)$ overlap. Window overlapping cannot be used to advantage if classical sorting and selection algorithms, like the one of Figure 3.7.2, are used. The running median algorithm described subsequently is based exactly on window overlapping. This algorithm has been proposed by Huang et al. [HUA79]. Let us suppose that the filter window has $m \times n$ pixels and that it is moving row-wise. When the window is moved from position (i,j) to position $(i,j+1)$, $mn - 2m$ pixels remain unchanged, m must be discarded and m new pixels must be inserted. Huang's algorithm is based on the local gray-level histogram of the nm pixels and on its adaptation, as the window moves. The algorithm is shown in Figure 3.7.3.

```
int run_median(a,b,NW,MW,N1,M1,N2,M2)
image a,b;
int NW,MW;
int N1,M1,N2,M2;

/* Subroutine for running 2-d median filtering
   a,b: input and output buffers
   NW,MW: window size
   N1,M1: start coordinates
   N2,M2: end     coordinates         */

{ int k,l, i,j, NM,NM2,NW2,MW2, mdn,ltmdn,g;
  int * t;   int * leftcol;   int * rightcol;
  int hist[256];

  t=(int *)malloc((unsigned int)(NW*MW)*sizeof(int));
  leftcol=(int *)malloc((unsigned int)(NW)*sizeof(int));
  rightcol=(int *)malloc((unsigned int)(NW)*sizeof(int));

  NM=NW*MW; NM2=NM/2; NW2=NW/2; MW2=MW/2;

  /* Perform running median filtering */
  for(k=N1+NW2; k<N2-NW2; k++)
   {
    /* Initialize histogram */
    for(i=0; i<256; i++)
     hist[i]=0;
    /* Calculate the histogram and the median
       of the first window */
    for(i=0; i<NW; i++)
     for(j=0; j<MW; j++)
      { t[MW*i+j]=a[k+i-NW2][M1+MW2+j-MW2];
```

```
              g=a[k+i-NW2][M1+MW2+j-MW2];
              hist[g]++;
            }
    sort(t,NM);
    mdn=t[NM2];
    b[k][M1+MW2]=mdn;

    /* Calculate ltmdn */
    ltmdn=0;
    for(i=0;i<mdn;i++)
    ltmdn+=hist[i];

    /* Calculate the median for the rest of this line */
    for(l=M1+MW2+1; l<M2-MW2; l++)
     {
       /* Update histogram and calculate ltmdn */
       for(i=0; i<NW; i++) leftcol[i]=a[k+i-NW2][l-MW2-1];
       for(i=0; i<NW; i++) rightcol[i]=a[k+i-NW2][l+MW2];

       for(i=0; i<NW; i++)
        {
          g=leftcol[i]; hist[g]--;
          if(g<mdn) ltmdn--;
          g=rightcol[i]; hist[g]++;
          if(g<mdn) ltmdn++;
        }

       /* Calculate the median */
       if(ltmdn>NM2)
         do {
              mdn--;
              ltmdn-=hist[mdn];
            } while (ltmdn>NM2);
       else while(ltmdn+hist[mdn]<=NM2)
              { ltmdn+=hist[mdn]; mdn++; }
       b[k][l]=mdn;
      }
    }
   free(t); free(leftcol); free(rightcol);
   return(0);
}
```

Fig. 3.7.3 Running median filtering algorithm.

It has the following steps:

1. Calculate the local histogram of the nm pixels in the first window and find the median. Count the number ltmdn of pixels with gray levels less than that of the median.

2. Move to the next window by deleting m pixels and inserting the m new pixels. Update the histogram. Count the number ltmdn of pixels whose gray level is less than the median of step 1.

3. Starting from the median of the previous window, move up/down the histogram bins one at a time if the count ltmdn is not greater/greater than $(mn+1)/2$ and update ltmdn until the median is reached.

4. Stop if the line end has been reached. Otherwise go to step 2.

Huang's algorithm requires only $2m$ comparisons per output point, whereas the QUICKSORT algorithm requires $O(2m^2 \log m)$ comparisons if $m \simeq n$. Thus the running median algorithm is much faster than the QUICKSORT used in median filtering.

The median filters can be extended to the case of angular data. Such an application comes from the area of color image processing: in HSI, HSV, HLS, $L^*C_{ab}^*h_{ab}^*$, $L^*C_{uv}^*h_{uv}^*$ color representation systems, that are used in computer graphics and color image analysis, hue is essentially an angular quantity defined on a circle, as has already been described in Chapter 1. A new class of filters that are suitable for angular images has been proposed in [NIK98]. The basis for the derivation of these operators has been the theory of angular statistics that is comprehensively described in [MAR72]. The *arc distance median (ADM)* [LIU92] can be used in filtering applications. The ADM of a set of points $\theta_1 \ldots \theta_N$ on the unit circle is the direction a that minimizes the following measure of dispersion :

$$d_0 = \frac{1}{N} \sum_{i=1}^{N} arc(\theta_i, a) \quad (3.7.6)$$
$$arc(\theta_i, a) = \pi - |\pi - |\theta_i - a|| \quad (3.7.7)$$

where $arc(\theta_i, a)$ is the smallest of the two angles that are defined by the points a and θ_i on the unit circle. d_0 is called *circular mean deviation*. ADM is equivalent to the sample media direction when this is unique. In case of multiple sample median directions, one of them will be the ADM of the data set.

Separable median filter. A separable two-dimensional median filter of size n results from two successive applications of a one-dimensional median filter of length n along rows and then along columns of an image (or vice versa) [NAR81]:

$$y_{ij} = med(z_{i,j-\nu}, \ldots, z_{ij}, \ldots, z_{i,j+\nu}) \quad (3.7.8)$$
$$z_{ij} = med(x_{i-\nu,j}, \ldots, x_{ij}, \ldots, x_{i+\nu,j}) \quad (3.7.9)$$

The algorithm for separable median filtering is very similar to that of the standard median calculation. The separable median filter of length n has greater output variance than a non-separable one having extent $n \times n$. However, its main advantage is its low computational complexity in comparison to that of the non-separable median filter, because it sorts n numbers two times, whereas the non-separable $n \times n$ median sorts n^2 numbers.

Recursive median filter. One intuitive modification of the median filter is to use the already computed output samples $y_{i-\nu}, \ldots, y_{i-1}$ in the calculation of y_i:

$$y_i = med(y_{i-\nu}, \ldots, y_{i-1}, x_i, \ldots, x_{i+\nu}) \qquad (3.7.10)$$

This is a recursive median filter. Its output tends to be much more correlated than that of the standard median filter. Recursive median filters have higher immunity to impulsive noise than the non-recursive median filters. A variation of the recursive median filter is the separable recursive median filter:

$$y_{ij} = med(y_{i,j-\nu}, \ldots, y_{i,j-1}, z_{ij}, \ldots, z_{i,j+\nu}) \qquad (3.7.11)$$

$$z_{ij} = med(z_{i-\nu,j}, \ldots, z_{i-1,j}, x_{ij}, \ldots, x_{i+\nu,j}) \qquad (3.7.12)$$

The algorithm for recursive median filtering can be constructed by a simple modification of the standard median filter algorithm.

Weighted median filters. The *weighted median* is the estimator T that minimizes the weighted L_1 norm of the form [PIT90]:

$$\sum_{i=1}^{n} w_i |x_i - T| \rightarrow min \qquad (3.7.13)$$

The weighted median filter is described by:

$$y_i = med\{w_{-\nu} \Box x_{i-\nu}, \ldots, w_{\nu} \Box x_{i+\nu}\} \qquad (3.7.14)$$

where $w \Box x$ denotes duplication of x ($x, \ldots, x$, w times). By choosing the filter weights properly, weighted median filters can incorporate time information in their structure (i.e. central pixels in the filter window can be weighted more heavily). An algorithm for weighted median filtering is shown in Figure 3.7.4.

```
int wei_median(a,b,g, NW,MW,N1,M1,N2,M2)
image a,b;
image g;
int NW,MW;
int N1,M1,N2,M2;

/* Subroutine for weighted 2-d median filtering
    a,b: input and output buffers
    g: filter coefficients
    NW,MW: window size
```

      N1,M1: start coordinates
      N2,M2: end   coordinates   */
{ int k,l, i,j, NM,NM2,NW2,MW2,s=0,m=0,n,s2;
  int * t;

  for(i=0; i<NW; i++)
   for(j=0; j<MW; j++)
    s +=g[i][j];

  /* Temporary 1-d buffer of size s
     to be used in data ordering */
  t=(int *)malloc((unsigned int)(s)*sizeof(int));

  NM=NW*MW; NM2=NM/2; NW2=NW/2; MW2=MW/2;
  s2=s/2;
  /* Perform weighted median filtering */
  if(s%2==1)
   {
    /* The sum of weights is an odd number */
    for(k=N1+NW2; k<N2-NW2; k++)
    for(l=M1+MW2; l<M2-MW2; l++)
     {
      m=0;
      for(i=0; i<NW; i++)
       for(j=0; j<MW; j++)
        for(n=1;n<g[i][j]+1;n++){t[m]=a[k+i-NW2][l+j-MW2];m++;}

      sort(t,s);
      b[k][l]=t[s2];
     }
   }
  else
   {
    /* The sum of weights is an even number */
    for(k=N1+NW2; k<N2-NW2; k++)
    for(l=M1+MW2; l<M2-MW2; l++)
     {
      m=0;
      for(i=0; i<NW; i++)
       for(j=0; j<MW; j++)
        for(n=1;n<=g[i][j];n++){t[m]=a[k+i-NW2][l+j-MW2];m++;}

      sort(t,s);
      b[k][l]=(char) ((float)(t[s2-1]+t[s2])/2.0);
```
```

```
        }
    }

    free(t);
    return(0);
}
```

Fig. 3.7.4 Two-dimensional weighted median filtering algorithm.

Multistage median filters. In the theoretical analysis of median filters, especially the deterministic one, several, rather unrealistic, assumptions are made, e.g. that the image consists of constant neighborhoods and edges. In reality the images have fine details, for example, lines and sharp corners, which are very valuable for human vision. These details are usually destroyed by medians having relatively large windows (larger than 5×5). It is the ordering process which destroys the structural and spatial neighborhood information. Thus, several efforts have been made to take into account structural information. Such a modification of the median leads to the multistage median filters [ARC89]:

$$
\begin{align}
y_{ij} &= med(med(z_1, z_2, x_{ij}), med(z_3, z_4, x_{ij}), x_{ij}) \tag{3.7.15} \\
z_1 &= med(x_{i,j-\nu}, \ldots, x_{ij}, \ldots, x_{i,j+\nu}) \tag{3.7.16} \\
z_2 &= med(x_{i-\nu,j}, \ldots, x_{ij}, \ldots, x_{i+\nu,j}) \tag{3.7.17} \\
z_3 &= med(x_{i+\nu,j-\nu}, \ldots, x_{ij}, \ldots, x_{i-\nu,j+\nu}) \tag{3.7.18} \\
z_4 &= med(x_{i-\nu,j-\nu}, \ldots, x_{ij}, \ldots, x_{i+\nu,j+\nu}) \tag{3.7.19}
\end{align}
$$

Multistage median filters can preserve details in horizontal, vertical and diagonal directions, because they use subfilters that are sensitive to these directions. An algorithm for multistage median filtering is shown in Figure 3.7.5.

```
int multistage_median(a,b, NW,MW, N1,M1,N2,M2)
image a,b;
int NW,MW;
int N1,M1,N2,M2;

/* Subroutine for 2-d multistage median filtering
   a,b: input and output buffers
   NW,MW: window size
   N1,M1: start coordinates
   N2,M2: end   coordinates        */
```

```
{ int k,l, i,j, NM,NM2,NW2,MW2;
  float NMT;
  int * t;   int * s;   int * r;   int * q;
  int p[5];
  t=(int *)malloc((unsigned int)(NW)*sizeof(int));
  r=(int *)malloc((unsigned int)(NW)*sizeof(int));
  s=(int *)malloc((unsigned int)(NW)*sizeof(int));
  q=(int *)malloc((unsigned int)(NW)*sizeof(int));
  NM=NW*MW; NM2=NM/2; NW2=NW/2; MW2=MW/2;

  for(k=N1+NW2; k<N2-NW2; k++)
   for(l=M1+MW2; l<M2-MW2; l++)
    {
      /* Compute the median subfilters            */
      for(j=0;j<MW;j++)
       {
        t[j]=a[k][l+j-MW2]; s[j]=a[k+j-NW2][l];
        r[j]=a[k-j+NW2][l-j+MW2]; q[j]=a[k-j+NW2][l+j-MW2];
       }
      sort(t,NW); sort(s,NW); sort(r,NW); sort(q,NW);
      p[0]=t[NW2]; p[1]=s[NW2]; p[2]=r[NW2];
      p[3]=q[NW2]; p[4]=a[k][l];
      /* Compute filter output                    */
      sort(p,5);
      b[k][l]=p[2];
    }
  free(t); free(r); free(s); free(q);
  return(0);
}
```

Fig. 3.7.5 Two-dimensional multistage median filtering algorithm.

3.8 DIGITAL FILTERS BASED ON ORDER STATISTICS

The median filter and its modifications that have been described in the previous section are only special cases of a large filter class that is based on order statistics. Such filters are based on L-estimators [HAM86] or are extensions of median filters. They are designed to meet various criteria, for example, edge preservation, robustness, adaptation to noise statistics or preservation of image details. In the following, some filters of this class are described in brief and algorithms for their calculation are given.

Rank-order filters. An r-th rank-order filter of the signal x_i is the r-th order statistic [HEY82]:

$$y_i = r\text{-th order statistic of } \{x_{i-\nu}, \ldots, x_i, \ldots, x_{i+\nu}\} \qquad (3.8.1)$$

The r-th rank-order filter introduces a strong bias in the estimation of the mean when the rank is small or large. In this case, the filter tends to perform as a maximum or minimum filter respectively. The bias is even stronger when the input data have a long-tailed distribution. The algorithm for rank-order filtering is very similar to that of median filtering and is not repeated here.

Max/min filters. The maximum $x_{(n)}$ and the minimum $x_{(1)}$ are the two extremes of the rank-order filters. They are closely related to two morphological operations called *dilation* and *erosion* of a function by a set respectively, as will be described in Chapter 7 [SER82], [SER88]. Max/min filtering is a running operation. An approach to one-dimensional running max/min calculation is the following [PIT89].

Let us denote by y_i the output of the max filter at the position i. We suppose that the filter window contains the samples $x_{i-n+1}, \ldots, x_i$. If y_{i-1} is known and y_i must be computed, the window is moved only by one element (x_{i-n} is deleted and x_i is inserted in the window). If x_i is greater than or equal to y_{i-1}, the new element is the maximum, that is, $y_i = x_i$. If x_i is less than or equal to y_{i-1}, the maximum is not affected by the introduction of the new element. In this case, if x_{i-n} is less than the maximum y_{i-1}, the new maximum is the same as the old one: $y_i = y_{i-1}$. If x_{i-n} is equal to the maximum y_{i-1}, a new max selection $y_i = \max(x_i, \ldots, x_{i-n+1})$ must be performed. This running max selection algorithm is described by the following formula:

$$y_i = \begin{cases} x_i & \text{if } x_i \geq y_{i-1} \\ y_{i-1} & \text{if } x_i < y_{i-1} \text{ and } x_{i-n} < y_{i-1} \\ \max(x_i, .., x_{i-n+1}) & \text{if } x_i < y_{i-1} \text{ and } x_{i-n} = y_{i-1} \end{cases} \qquad (3.8.2)$$

A theoretical analysis of the average number of comparisons required by this algorithm is given in [PIT88]. The average number of comparisons is approximately only 3 and does not increase with the filter window size n. Therefore, it is much better than the straightforward max selection algorithm, which requires $n-1$ comparisons. The two-dimensional version of this algorithm is shown in Figure 3.8.1. In this algorithm, the running max calculation is applied along rows and columns independently.

```
int run_max(a,b,c,NW,MW,N1,M1,N2,M2)
image a,b,c;
int NW,MW;
int N1,M1,N2,M2;
```

```c
/* Subroutine for fast 2-d max filtering
   a,b: input and output buffers
   c: buffer for intermediate results
   NW,MW: window size
   N1,M1: start coordinates
   N2,M2: end   coordinates          */

{ int k,l, i,j, NM,NM2,NW2,MW2,bm;

  NM=NW*MW; NM2=NM/2; NW2=NW/2; MW2=MW/2;

  /* Calculate the maximum row-wise */
  for(k=N1; k<N2; k++)
    {
      /* Calculate first output point */
      bm=a[k][M1];
      for(i=1; i<MW; i++)
       if(bm<a[k][M1+i]) bm=a[k][M1+i];
      c[k][M1+MW2]=bm;
      /* Calculate the max for the rest of this line */
      for(l=M1+MW2+1; l<M2-MW2; l++)
        {
          if(a[k][l+MW2]>=bm) bm=a[k][l+MW2];
          else
            {
              if(a[k][l-MW2-1]==bm)
                {
                  bm=a[k][l-MW2];
                  for(i=1; i<MW; i++)
                    if(bm<a[k][l+i-MW2]) bm=a[k][l+i-MW2];
                }
            }
          c[k][l]=bm;
        }
    }
  /* Calculate the max column-wise */
  for(l=M1+MW2; l<M2-MW2; l++)
    {
      /* Calculate first output point */
      bm=c[N1][l];
      for(i=1; i<NW; i++)
       if(bm<c[N1+i][l]) bm=c[N1+i][l];
      b[N1+NW2][l]=bm;
      /* Calculate the max for the rest of this column */
      for(k=N1+NW2+1; k<N2-NW2; k++)
```

```
    {
      if(c[k+NW2][1]>=bm) bm=c[k+NW2][1];
      else
        {
          if(c[k-NW2-1][1]==bm)
            {
              bm=c[k-NW2][1];
              for(i=1; i<NW; i++)
                if(bm<c[k+i-NW2][1]) bm=c[k+i-NW2][1];
            }
        }
      b[k][1]=bm;
    }
  }
  return(0);
}
```

Fig. 3.8.1 Two-dimensional running max filtering algorithm.

The maximum filter can effectively remove negative impulses (black spots) in an image, whereas the minimum filter can remove positive impulses (white spots). Both fail in the removal of mixed impulse noise, because minimum and maximum filters tend to enhance the negative and positive spikes respectively. However, cascades of max and min filters can remove mixed impulse noise effectively. However, their performance is generally inferior to that of the median filter. Both the maximum and the minimum filters have good edge preservation properties. Their disadvantage is that they tend to enhance the bright and the dark regions of the image respectively. Maximum and minimum filters are very popular, especially in the context of mathematical morphology, despite their disadvantages. The main reason is their computational simplicity and the existence of fast specialized processors for their calculation. The maximum/minimum filters are related to the L_p-mean filters given by [PIT86ASSP]:

$$y_i = \left(\sum_{j=1}^{n} a_j x_{i-\nu+j-1}^p \right)^{1/p} \qquad p \in Z \qquad (3.8.3)$$

The L_p-mean filter tends to the maximum filter when p tends to infinity. The L_{-p}-mean filter tends to the minimum filter when p tends to infinity. Therefore, if the effects due to the highly nonlinear nature of max and min operators are not desirable (e.g. the enhancement of dark or bright regions), L_p-mean or L_{-p}-mean filters of moderate p can be used instead. L_p-mean filters have relatively good noise smoothing and edge preservation properties.

An overview of applications of L_p means in image filtering and edge detection can be found in [KOT99]. L_2 mean filters are proven to be optimal for both multiplicative Rayleigh speckle and signal-dependent Gaussian speckle [KOT92UI, KOT94TIP]. Therefore, they are very useful for ultrasonic image filtering.

α-trimmed mean filters. It has already been stated that the moving average filter suppresses additive white Gaussian noise better than the median filter, whereas the second preserves better edges and rejects impulses. Therefore, a good compromise between the two is highly desirable. Such a filter is the α-trimmed mean filter [BED84]:

$$y_i = \frac{1}{n(1-2\alpha)} \sum_{j=\alpha n+1}^{n-\alpha n} x_{(j)} \qquad (3.8.4)$$

where $x_{(j)}$, $j = 1, \ldots, n$, are the order statistics of $x_{i-\nu}, \ldots, x_i, \ldots, x_{i+\nu}$. Thus, the α-trimmed mean filter rejects the smaller and the larger observation data. This data rejection depends on the coefficient α, $0 \leq \alpha < 0.5$. If $\alpha = 0$, no data are rejected and the filter performs as a moving average filter. If α is close to 0.5, all data but the median are rejected. Therefore, the α-trimmed mean filter can be used as a compromise between the median filter and the moving average filter. The rejected data are usually outliers. This fact explains the robustness of this filter. The performance of the α-trimmed mean filter is poor for short-tailed distributions. It is known that the *midpoint MP*:

$$MP = \frac{1}{2}(x_{(1)} + x_{(n)}) \qquad (3.8.5)$$

is a very good estimator of location for short-tailed distributions (e.g. uniform distribution).

A different approach to trimmed filters is to exclude the samples $x_{i+r,j+s}$ in the filter window, which differ considerably from the local median $med\{x_{ij}\}$ or from the central pixel x_{ij}. This is the *modified trimmed mean* (MTM) filter:

$$y_{ij} = \frac{\sum\sum_A a_{rs} x_{i+r,j+s}}{\sum\sum_A a_{rs}} \qquad (3.8.6)$$

The summations cover the entire filter window A. The filter coefficients are chosen as follows:

$$a_{rs} = \begin{cases} 1 & |x_{i+r,j+s} - med\{x_{ij}\}| \leq q \\ 0 & \text{otherwise} \end{cases} \qquad (3.8.7)$$

The amount of trimming depends on the parameter q. Data deviating strongly from the local median are trimmed out. Since such data are usually outliers, the modified trimmed mean filter has good robustness properties. A variation of (3.8.6–3.8.7) employs two filter windows of different sizes and is called the *double window modified trimmed mean* (DW MTM) filter. It is proven to have good robustness and edge preservation properties. Another modification is the *modified nearest neighbor* (MNN) filter, whose coefficients are given by:

$$a_{rs} = \begin{cases} 1 & |x_{i+r,j+s} - x_{ij}| \leq q \\ 0 & \text{otherwise} \end{cases} \qquad (3.8.8)$$

This filter trims out pixels deviating strongly from the central pixel. Therefore, the MNN filter has good edge preservation properties.

L-filters. An important generalization of the median is the *L-filter* (also called the *order statistic filter*) [BOV83]:

$$y_i = \sum_{j=1}^{n} a_j x_{(j)} \qquad (3.8.9)$$

The moving average, median, r-th rank-order, α-trimmed mean and midpoint filters are special cases of (3.8.9), if the coefficients a_j, $j = 1, \ldots, n$, are chosen appropriately. The filter coefficients a_j, $j = 1, \ldots, n$, can be chosen to satisfy an optimality criterion that is related to the probability distribution of the input noise. $\mathbf{a}$ is the coefficient vector $[a_1, \ldots, a_n]^T$ to be calculated. Let us consider the additive noise model $x_{ij} = s_{ij} + n_{ij}$, under the assumption that the signal s_{ij} is constant. A *location invariance* constraint can be imposed on the estimator (3.8.9):

$$\sum_{j=1}^{n} a_j = \mathbf{a}^T \mathbf{e} = 1 \qquad (3.8.10)$$

where $\mathbf{e} = [1, \ldots, 1]^T$. The filter coefficients can be chosen in such a way that the error norm is minimized:

$$\text{MSE} = E\left[(s_i - y_i)^2\right] = E\left[\left(\sum_{j=1}^{n} a_j x_{(j)} - s_i\right)^2\right] = \mathbf{a}^T \mathbf{R} \mathbf{a} \qquad (3.8.11)$$

where $\mathbf{R}$ is the $n \times n$ correlation matrix of the vector of the ordered noise variables $\mathbf{n} = [n_{(1)}, \ldots, n_{(n)}]^T$. The minimization of the MSE under the constraint (3.8.10) gives the following coefficient vector:

$$\mathbf{a} = \frac{\mathbf{R}^{-1} \mathbf{e}}{\mathbf{e}^T \mathbf{R}^{-1} \mathbf{e}} \qquad (3.8.12)$$

The filter coefficients depend entirely on the correlation matrix $\mathbf{R}$, that is, on the input noise probability function. If the input noise distribution is Gaussian, the filter coefficients are given by $a_j = 1/n$, $j = 1, \ldots, n$. As expected, the optimal L-filter for the Gaussian noise is the moving average, the optimal L-filter for the Laplacian distribution is close to the median and the optimal L-filter for the uniform distribution is the midpoint. The ability of the L-filter to have optimal coefficients for a variety of input distributions makes it suitable for a large number of applications. An algorithm for two-dimensional L-filtering is shown in Figure 3.8.2.

```
int l_filter(a,b, h, NW,MW, N1,M1,N2,M2)
image a,b;
vector h;
int NW,MW;
int N1,M1,N2,M2;

/* Subroutine for 2-d L-filtering
   a,b: input and output buffers
   h: buffer of filter coefficients
   NW,MW: window size
   N1,M1: start coordinates
   N2,M2: end   coordinates            */

{ int k,l, i,j, NM,NM2,NW2,MW2;
  int * t;
  float s;
  t=(int *)malloc((unsigned int)(NW*MW)*sizeof(int));
  NM=NW*MW; NM2=NM/2; NW2=NW/2; MW2=MW/2;
  for(k=N1+NW2; k<N2-NW2; k++)
  for(l=M1+MW2; l<M2-MW2; l++)
      { for(i=0; i<NW; i++)
         for(j=0; j<MW; j++)
           t[MW*i+j]=a[k+i-NW2][l+j-MW2];
        sort(t,NM);
        s=0.0;
        for (i=0;i<NM;i++)
          s += h[i]*t[i];
        if(s<0.0) s=0.0;
          else if(s>255.0) s=255.0;
        b[k][l]=(unsigned char)s;
      }
  free(t);
  return(0);
}
```

Fig. 3.8.2 Two-dimensional L-filter algorithm.

A further advantage of the L-filter over the median filter is that it has no streaking effects, provided that its coefficients are not similar to those of the median filter. Its disadvantage over both the median and the moving average filter is that it has greater computational complexity, because its calculation requires additions, multiplications and comparisons.

Optimal L-filters for multiplicative Rayleigh noise that models the speckle appearing in ultrasound B-mode images have been designed in [KOT92UI]. Segmentation-based L-filtering, that is, filtering processes combining segmentation and optimum L-filtering in ultrasonic images have been proposed in [KOF96]. In this approach, a self-organizing neural network segments the image in approximately homogeneous regions and an optimum L-filter is designed for each image region. L-filter design has been extended to the multichannel case in [KOT94TSP] as well.

Adaptive versions of the L-filters can be constructed as well. The adaptation of L-filter coefficients by using the least mean squares (LMS) has been studied in [KOT92SP, KOT96TIP, KOT99OE]. A review of four adaptation methodologies, namely, the LMS algorithm, the recursive least squares (RLS) algorithm, the backpropagation algorithm and simulated annealing techniques in the design of L-filters, median hybrid filters, order statistic LMS filters, weighted median filters and morphological filters can be found in [KOT94CDS].

3.9 SIGNAL ADAPTIVE ORDER STATISTIC FILTERS

The nonlinear filters described in the previous section are usually optimized for a specific type of noise and sometimes for a specific type of signal. However, this is not usually the case in many applications of nonlinear filtering, especially in image processing. Images can be modelled as two-dimensional stochastic processes, whose statistics vary in the various image regions. Images are non-stationary processes. Furthermore, noise statistics, for example, noise standard deviation and even noise probability density function, vary from application to application. Sometimes, noise characteristics vary in the same application from one image to the next. Such cases are channel noise in image transmission and atmospheric noise (e.g. cloud noise) in satellite images. In these environments, non-adaptive filters cannot perform well, because their characteristics depend on the noise and signal characteristics, which are unknown. Therefore, adaptive filters are the natural choice in such cases. Adaptive filter performance depends on the accuracy of the estimation of certain signal and noise statistics, namely the signal mean and standard deviation and the noise standard deviation. The estimation is usually local, that is, relatively small windows are used to obtain the signal and noise characteristics. Such an adaptive filter can be employed for additive white noise:

$$x_{ij} = s_{ij} + n_{ij} \qquad (3.9.1)$$

The linear *minimal mean square error* (MMSE) estimate of s_{ij} is given by the following formula [LEE80]:

$$\hat{s}_{ij} = \left(1 - \frac{\sigma_n^2}{\sigma_x^2}\right) x_{ij} + \frac{\sigma_n^2}{\sigma_x^2} \hat{m}_x \quad (3.9.2)$$

where σ_n, σ_x, $\hat{m}_x$ are local estimates of noise standard deviation, signal standard deviation, and signal mean, respectively. In homogeneous image regions, the noise standard deviation is approximately equal to the signal standard deviation. Therefore, for these regions the adaptive MMSE filter (3.9.2) is reduced to the local estimate of the signal mean $\hat{s}_{ij} \simeq \hat{m}_x$. At edge regions, the signal standard deviation is much greater than the noise signal deviation ($\sigma_n \ll \sigma_x$). In these regions, no filtering is performed at all ($\hat{s}_{ij} = x_{ij}$). Thus, the adaptive MMSE filter preserves edges, although it does not filter the noise in edge regions. The performance of the adaptive MMSE filter depends on the choice of the local measures of signal mean and standard deviation and of the noise standard deviation. The local arithmetic mean and sample standard deviation have been used for the estimation of the signal mean and standard deviation, respectively. The local median or midpoint has been proposed for the estimation of the signal mean as well. An algorithm for adaptive MMSE filtering is described in Figure 3.9.1.

```
int local_adapt_filt(a,b, sn, NW,MW, N1,M1,N2,M2)
image a,b;
float sn;
int NW,MW;
int N1,M1,N2,M2;

/* Subroutine for 2-d adaptive filtering
   based on local statistics
   a,b: input and output buffers
   sn: noise variance
   NW,MW: window size
   N1,M1: start coordinates
   N2,M2: end   coordinates         */

{ int k,l, i,j,NM2,NW2,MW2, bm,ime;
  long s;
  float sx,snx,NM;

  NM=(float)NW*MW; NM2=NM/2; NW2=NW/2; MW2=MW/2;

  for(k=N1+NW2; k<N2-NW2; k++)
  for(l=M1+MW2; l<M2-MW2; l++)
```

```
{
    /* Find the local mean */
    s=0;
    for(i=0; i<NW; i++)
    for(j=0; j<MW; j++)
      s+=a[k+i-NW2][l+j-MW2];
    bm=(int)((float)s/NM);

    /* Find the local standard deviation */
    s=0;
    for(i=0; i<NW; i++)
    for(j=0; j<MW; j++)
      s+=(a[k+i-NW2][l+j-MW2]-bm)*(a[k+i-NW2][l+j-MW2]-bm);
    sx=(float)s/NM;
    /* Perform adaptation */
    if(sx==0.0) ime=bm;
    else { snx=sn/sx;
           if (snx>1.0) snx=1.0;
           ime=(int)((1.0-snx)*a[k][l]+snx*bm);
         }
    if(ime<0) b[k][l]=0; else
    if(ime>255) b[k][l]=255; else
    b[k][l]=ime;
  }
  return(0);
}
```

Fig. 3.9.1 Two-dimensional adaptive MMSE algorithm.

Another approach that uses adaptive data trimming has also been proposed. The resulting filter is called the *adaptive double window modified trimmed mean* (DW MTM) filter and it is used in signal-dependent noise filtering [DIN87].

Another reason for using adaptive filters is the fact that edge information is very important for the human eye and must be preserved. Certain filters, for instance, the moving average filter, perform well in homogeneous image regions but fail close to edges. The opposite is true for other filters, for example, for the median filter. A combined filter which performs differently in the image edges than in the image plateaus can be used in such a case. These filters are also called *decision-directed filters*, because they employ an edge detector to decide if an edge is present or not. Order statistics are efficient tools in edge detection. The range W and the quasi range $W_{(i)}$ can be used as edge detectors [PIT86PAMI]:

$$W = x_{(n)} - x_{(1)} \qquad (3.9.3)$$

The range edge detector can be calculated easily. However, it has poor robustness to impulses. The *dispersion edge detector*:

$$J_i = W + W_{(1)} + W_{(2)} + \ldots + W_{(i)} \tag{3.9.4}$$

has better robustness properties than the range edge detector. A comparison of the median filter output at neighboring windows can also be used as the edge detector. If the difference of the two outputs is larger than a threshold an edge is declared. Another approach to edge detection is to use *rank tests* [HAJ67]. The *Wilcoxon test* and the *median test* have been proposed [BOV86].

Decision-directed filters can take into account both edge information and impulsive noise information [PIT86SP]. Impulses, when detected, can be removed from the estimation of the local mean, median, and standard deviation. Furthermore, when an edge is detected, the window of the filter can become smaller so that edge blurring is minimized. Such an impulse-sensitive filter is the *adaptive window edge detection* (AWED) filter [PIT90]. The filter initially starts with a 7×7 or 5×5 window. The local image histogram in the filter window is examined. If impulses are detected, they are rejected and the local image standard deviation calculation is based on the rest of the pixels in the window. If the local standard deviation is low enough, a homogeneous image region is assumed and the moving average filter is used. If the local standard deviation is large (above a certain threshold) an edge is declared. If the window size is 3×3, the median filter is used for image filtering, but if the window size is greater than 3×3, it is reduced and the whole procedure is repeated. The window size is increased at each pixel if no edge has been detected.

An adaptive version of the α-trimmed mean filter has also been proposed [RES88]. Order statistics are used as estimators of the data distribution tail. Based on this estimator a decision is made on whether to use the midpoint filter, or the complementary α-trimmed mean filter, or the moving average filter, or the α-trimmed mean filter, or the median filter.

Another approach related to decision-directed filtering is the *two-component model*. An image **x** can be considered to consist of two parts: a low-frequency part $\mathbf{x}_L$ and a high-frequency part $\mathbf{x}_H$:

$$\mathbf{x} = \mathbf{x}_L + \mathbf{x}_H \tag{3.9.5}$$

The low-frequency part is dominant in the homogeneous image regions, whereas the high-frequency part is dominant in the edge regions. The two-component image model allows different treatment of its components. Therefore, it can be used for adaptive image filtering and enhancement, provided that the two components can be separated. A low-pass and a high-pass filter can be used for the separation of the two components. In most cases, the moving average filter or the median filter is used as estimators $\hat{m}_x$ of the low-frequency component, whereas the high-frequency component is given by:

$$x_{Hij} = x_{ij} - \hat{m}_x \tag{3.9.6}$$

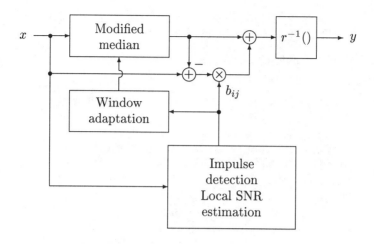

Fig. 3.9.2 Block diagram of the SAM filter.

An adaptive filter based on the two-component image model is shown in Figure 3.9.2. It is called the *signal-adaptive median* (SAM) filter [BER87]. Its output signal is given by:

$$y_{1ij} = \hat{m}_x + b_{ij}(x_{ij} - \hat{m}_x) \qquad (3.9.7)$$
$$y_{ij} = r^{-1}(y_{1ij}) \qquad (3.9.8)$$

where $r(\cdot)$ is the nonlinearity existing in image formation. The performance of the adaptive filter (3.9.7–3.9.8) depends on the choice of the coefficient b_{ij}, which can be done in an optimal way for various noise types (e.g. additive and multiplicative noise). This coefficient also detects edge information. Thus, b_{ij} can be used for the adaptation of the filter window. We can start filtering by using the initial window size 5×5 or 7×7. If the coefficient b_{ij} becomes greater than an appropriate threshold b_t close to image edges, the window size is decreased until the coefficient becomes less than the threshold or until the window reaches the size 3×3. Otherwise the window size is increased to its maximum size. If impulsive noise is present, the impulses can be detected and removed from the filter window. The local median of the remaining pixels can be used as an estimate of the signal mean. The SAM filter has an excellent performance in noise filtering, edge and image detail preservation.

An extension of the SAM filter is the so-called *morphological signal-adaptive median* ($MSAM$) filter [TSEK98]. Two major modifications in the standard SAM filter structure are introduced by the $MSAM$ filter aiming at alleviating its disadvantages/limitations:

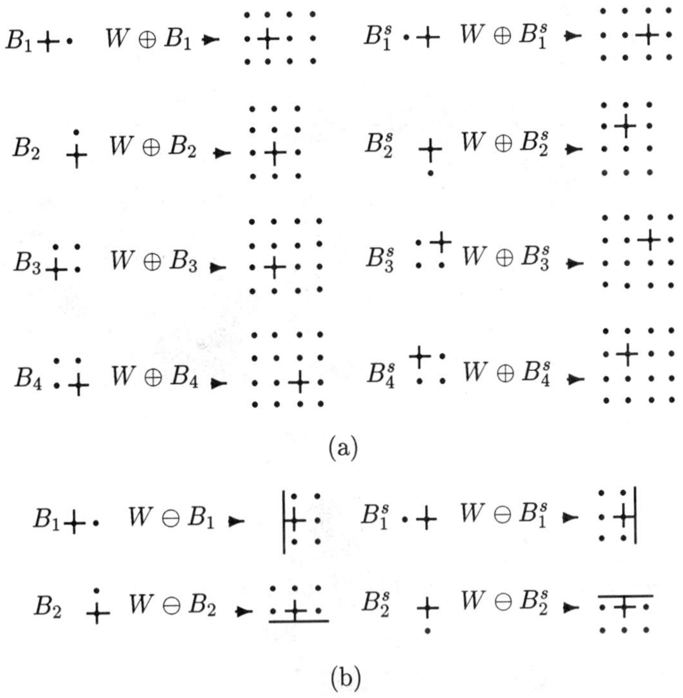

Fig. 3.9.3 Anistotropic windows used in $MSAM$ filter.

- The impulse detection mechanism of the SAM filter is enhanced so that it detects not only impulses of a constant amplitude but randomly valued impulses as well.
- In contrast to the SAM filter, which employs isotropic filter window adaptation, the $MSAM$ filter implements anisotropic window adaptation based on binary morphological erosions/dilations with predefined structuring elements shown in Figure 3.9.3.

The $MSAM$ filter performs well against contaminated Gaussian or impulsive noise corruption cases. It does not require the a priori knowledge of a noise-free image, but only of certain noise characteristics which can easily be estimated. It adapts its behavior based on a local SNR measure thus achieving edge preservation and noise smoothing in homogeneous regions. Moreover, the coefficient b_{ij} can be related to the segmentation information obtained prior to the filtering process [KOT94TIP].

SAM filter can be further extended to the class of signal-adaptive maximum likelihood filters that employ the maximum likelihood estimator of location for the distribution of the noise which corrupts the image. Signal-adaptive maximum likelihood filters for speckle suppression in ultrasonic images have been designed in [KOT94TIP]. A representative real ultrasonic image of liver

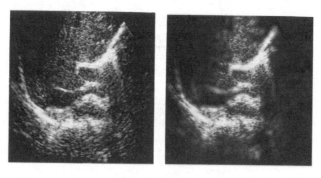

Fig. 3.9.4 (a) Real ultrasonic image of liver recorded using 3 MHz probe; (b) Output of signal-adaptive maximum likelihood filter.

recorded using 3 MHz probe is shown in Figure 3.9.4a. The output of the signal-adaptive maximum likelihood filter that exploits a local signal-to-noise ratio measure to adapt the filter window is shown in Figure 3.9.4b.

3.10 HISTOGRAM AND HISTOGRAM EQUALIZATION TECHNIQUES

A useful approach to digital image processing is to consider image intensities $f(i,j)$ as being random variables having probability density function (pdf) $p_f(f)$. Image pdf carries valuable global information about the image content. However, the pdf is generally not available and must be estimated from the image itself by using the empirical pdf usually called the *histogram*. Let us assume that the digital image has L discrete gray levels (usually from 0 to 255) and that n_k, $k = 0, \ldots, L-1$, is the number of pixels having intensity k. The histogram $\hat{p}_f(f)$ is given by the relation:

$$\hat{p}_f(f_k) = \frac{n_k}{n} \quad k = 0, 1, \ldots, L-1 \qquad (3.10.1)$$

where n is the total number of image pixels. The image histogram can be calculated easily, as can be seen in Figure 3.10.1.

```
int hist(a, h, N1,M1,N2,M2)
image a;
vector h;
int N1,M1,N2,M2;

/* Subroutine to calculate the histogram of an image
```

```
   a: image buffers
   h: histogram buffer
   N1,M1: start coordinates
   N2,M2: end   coordinates    */

{ int i,j; float num;
  unsigned long * lh;
  lh=(unsigned long *)
      malloc((unsigned int)NSIG*sizeof(unsigned long));
  for(i=0; i<NSIG; i++) lh[i]=0L;

  for(i=N1; i<N2; i++)
  for(j=M1; j<M2; j++)
      lh[(unsigned int)a[i][j]]++;
  num=(float)(N2-N1)*(float)(M2-M1);
  for(i=0; i<NSIG; i++) h[i]=lh[i]/num;
  free(lh);
  return(0);
}
```

Fig. 3.10.1 Algorithm for histogram calculation.

The image histogram carries important information about the image content. If its pixel values are concentrated in the low image intensities, as can be seen in Figure 3.10.2a, the image is 'dark'. A 'bright' image has a histogram that is concentrated in the high image intensities, as seen in Figure 3.10.2b. The histogram of Figure 3.10.2c reveals that the image contains two objects with different intensities (or, possibly, one object clearly distinguished from its background). If the image histogram is concentrated on a small intensity region, the image contrast is poor and the subjective image quality is low. Image quality can be enhanced by modifying its histogram. This can be

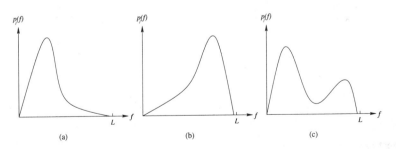

Fig. 3.10.2 (a) Histogram of a dark image; (b) histogram of a bright image; (c) histogram of an image containing two regions with different distributions.

performed by a technique called *histogram equalization*. Let us suppose that an image f is transformed by a nonlinear point-wise function $g = T(f)$. The pdf $p_g(g)$ of the transformation output is given by [PAP65]:

$$p_g(g) = \frac{p_f(f_1)}{\left|\frac{dT(f_1)}{df}\right|} \qquad (3.10.2)$$

provided that the equation $g = T(f)$ possesses only one solution f_1. Based on this relation, it can be easily proven [PAP65], [JAI89] that the transformation function:

$$T(f) = \int_0^f p_f(w)\,dw \qquad (3.10.3)$$

produces an output image g whose pdf is uniform, $p_g(g) = 1$. Generally, the output image has higher contrast and better subjective quality than the original image f. However, it has to be noted that histogram equalization tends also to amplify noise. The transformation function $T(f)$ can be approximated by the following relation:

$$g_k = T(f_k) = \sum_{j=0}^{k} \hat{p}_f(f_j) = \sum_{j=0}^{k} \frac{n_j}{n} \qquad (3.10.4)$$

A histogram equalization algorithm that employs (3.10.4) is shown in Figure 3.10.3.

```
int cdfhist(h,t)
vector h;
vector t;

/* Subroutine to calculate the cdf of a histogram
   It is used in histogram equalization
   h: histogram buffer
   t: transformation buffer */

{ int i;

  t[0]=h[0];
  for(i=1; i<NSIG; i++)
      t[i]=t[i-1]+h[i];
  return(0);
}


int histeq(a,b, h, N1,M1,N2,M2)
image a,b;
vector h;
```

```
int N1,M1,N2,M2;

/* Subroutine for histogram equalization
   a: input buffer
   b: output buffer
   h: histogram cdf function
   N1,M1: start coordinates
   N2,M2: end coordinates */

{ int i,j;
  float * f;

  f=(float *)malloc((unsigned int)NSIG*sizeof(float));

  for(i=0; i<NSIG; i++) f[i]=h[i]*255.0;
  for(i=N1; i<N2; i++)
  for(j=M1; j<M2; j++)
      b[i][j]=(unsigned char)f[(unsigned int)a[i][j]];

  free(f);
  return(0);
}
```

Fig. 3.10.3 Algorithm for the calculation of the transformation function and for histogram equalization.

This algorithm has been applied to the confocal microscopy image of a neuron, shown in Figure 3.10.4a. This image has a very poor contrast. The equalized image has a much better contrast, as can be seen in Figure 3.10.4b.

In certain cases, histogram modification rather than histogram equalization is desired. The image f having histogram $p_f(f)$ must be transformed to an image g having a predefined histogram $p_g(g)$. This can be done in a two-step procedure. Let us suppose that we produce an image $s = T(f)$ by using histogram equalization:

$$s = T(f) = \int_0^f p_f(w)dw \tag{3.10.5}$$

Let us also suppose that the desired image g is available, is equalized and produces an image $z = G(g)$:

$$z = G(g) = \int_0^g p_g(w)dw \tag{3.10.6}$$

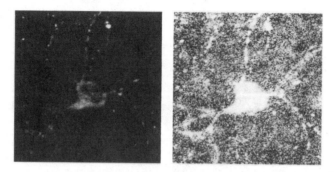

Fig. 3.10.4 (a) Confocal microscopy image of a neuron; (b) equalized image. (*Courtesy of Prof. A. Colchester, Neuromedia Group, University of Canterbury, UK.*)

The intermediate images s, z can be set equal to each other. Thus, the desired image g can be produced in two steps. In the first step, the image s is produced by using (3.10.5). The image g is produced in the second step by using:

$$g = G^{-1}(z) = G^{-1}(s) \qquad (3.10.7)$$

The inverse function $G^{-1}(z)$ can be easily obtained from (3.10.6). These two steps can be combined in one transformation function:

$$g = G^{-1}[T(f)] \qquad (3.10.8)$$

Both transformation functions $T(f)$, $G(g)$ can be easily calculated by using a modification of the algorithm described in Figure 3.10.3.

3.11 PSEUDOCOLORING ALGORITHMS

In many applications, the human eye is more sensitive to color changes rather than to intensity changes for inspection applications (e.g. security inspection, quality control, remote sensing). Therefore, it is natural to encode the intensity of black and white (BW) images by using color information. This process is called *pseudocoloring*. The result of pseudocoloring is usually pleasant to the human eye. Thus, it can be used for artistic applications (e.g. creation of visual effects).

Pseudocoloring is a digital image transformation of the form:

$$\mathbf{c}(x,y) = \mathbf{T}(f(x,y)) \qquad (3.11.1)$$

where $f(x,y)$ is a black and white image and $\mathbf{c}(x,y)$ is a color image. It is usually expressed in terms of its red, green and blue components $\mathbf{c} = [c_R, c_G, c_B]^T$. The transformation function $\mathbf{T}$ produces a three-channel output. Since the result of pseudocoloring is usually judged only subjectively, the choice of the transformation function **T** is rather heuristic. In the following, we shall describe two different ways to find pseudocoloring transformations.

The first method is based on intensity quantization (slicing). Let us suppose that we want to use N different colors in pseudocoloring. If the BW image possesses L different gray levels, the transformation function can take the following form:

$$\mathbf{T}(f(k,l)) = \begin{cases} \mathbf{c}_i & \text{if } i[\frac{L}{N}] \leq f(k,l) < (i+1)[\frac{L}{N}] \\ & i = 0, 1, \ldots, N-2 \\ \mathbf{c}_{N-1} & \text{if } (N-1)[\frac{L}{N}] \leq f(k,l) < L \end{cases} \quad (3.11.2)$$

where $[\frac{L}{N}]$ denotes the integer part of $\frac{L}{N}$. The colors $\mathbf{c}_i$, $i = 0, \ldots, N-1$, can be arbitrarily chosen. Intensity slicing segments the BW image in N different regions. Each region is assigned a different color. More sophisticated image segmentation algorithms can be used that will be presented in Chapter 6. Pseudocoloring (3.11.2) produces acceptable results if the BW image histogram is close to the uniform distribution. If the histogram is concentrated in a small intensity region, that is, if the image has poor contrast, intensity slicing (3.11.2) produces poor results. The solution to this problem is non-uniform intensity quantization. Powerful techniques are known for this non-uniform quantization [JAY84]. We shall describe here an alternative approach that is simple and easy to implement. Histogram equalization $g = G(f)$ can be performed and intensity quantization can be applied on the equalized image g. The equalization function $G(f)$ is given by (3.10.4). This method is equivalent to the following non-uniform transformation function:

$$\mathbf{T}(f(k,l)) = \begin{cases} \mathbf{c}_i & \text{if } i[\frac{L}{N}] \leq G(f(k,l)) < (i+1)[\frac{L}{N}] \\ & i = 0, 1, \ldots, N-2 \\ \mathbf{c}_{N-1} & \text{if } (N-1)[\frac{L}{N}] \leq G(f(k,l)) < L \end{cases} \quad (3.11.3)$$

This method is sensitive to noise, because histogram equalization tends to enhance it. Another modification of image intensity pseudocoloring is based on arbitrary image thresholding. In this case, a set of $N-1$ thresholds f_i, $i = 1, \ldots, N-1$, is defined by the user. The transformation function takes the form:

$$\mathbf{T}(f(k,l)) = \begin{cases} \mathbf{c}_0 & 0 \leq f(k,l) < f_1 \\ \mathbf{c}_i & f_i \leq f(k,l) < f_{i+1}, \ 1 \leq i \leq N-2 \\ \mathbf{c}_{N-1} & f_{N-1} \leq f(k,l) < L \end{cases} \quad (3.11.4)$$

It must be noted that all transformations (3.11.1–3.11.4) are pointwise non-linear operators. Thus, their implementation is very easy. If computation speed is of vital importance (e.g. in video or animation applications), they can be implemented by using look-up tables, as shown in Chapter 1 (subroutine `nltran()`).

Another pseudocoloring method is based on a filtering approach [GON87], [LIM90]. Let us suppose that a low-pass, a bandpass and a high-pass linear FIR filter are available having impulse responses $h_L(k,l), h_B(k,l)$ and $h_H(k,l)$ respectively. Each of these filters can operate on a monochrome image $f(k,l)$ producing a corresponding output. Each output is assigned to one color image component:

$$c_R(k,l) = f(k,l) * *h_H(k,l) \qquad (3.11.5)$$

$$c_G(k,l) = f(k,l) * *h_B(k,l) \qquad (3.11.6)$$

$$c_B(k,l) = f(k,l) * *h_L(k,l) \qquad (3.11.7)$$

These color image components form the pseudocolored image. The color image tends to have more blue in the homogeneous BW image regions and more red close to lines and edges. The ideal frequency responses of the low-pass, bandpass and high-pass filters used in (3.11.5–3.11.7) have the following form:

$$H_L(\omega_1,\omega_2) = \begin{cases} 1 & \text{if } D(\omega_1,\omega_2) \leq D_L \\ 0 & \text{if } D(\omega_1,\omega_2) > D_L \end{cases} \qquad (3.11.8)$$

$$H_B(\omega_1,\omega_2) = \begin{cases} 1 & \text{if } D_L \leq D(\omega_1,\omega_2) < D_H \\ 0 & \text{otherwise} \end{cases} \qquad (3.11.9)$$

$$H_H(\omega_1,\omega_2) = \begin{cases} 0 & \text{if } D(\omega_1,\omega_2) \leq D_H \\ 1 & \text{if } D(\omega_1,\omega_2) > D_H \end{cases} \qquad (3.11.10)$$

where $D(\omega_1,\omega_2) = \sqrt{\omega_1^2 + \omega_2^2}$ is the distance of a point in the frequency domain from the origin (0,0). Several methods exist for the design of linear FIR and IIR filters that approximate the ideal responses (3.11.8–3.11.10) [LIM90], [DUD84] but their description is outside the scope of this book. The linear convolutions (3.11.5–3.11.7) can be implemented either in the spatial domain or in the frequency domain by the techniques described earlier in this chapter.

3.12 DIGITAL IMAGE HALFTONING

Most black-and-white images possess a relatively large number of gray scales (typically 256). Such images can be easily displayed on computer graphics monitors. However, they cannot be printed, because almost all printing technologies can print and display only binary images. Therefore, any grayscale image that is to be printed must be transformed first to a binary image. This can be easily done owing to an interesting property of human vision. The human eye tends to perform a local integration of black dots and white dots. If the black dots and white dots are small and are viewed from a normal reading distance, the dots in various percentages tend to create subjectively perceived grayscales. A *halftone* image is a binary image, whose dots create the impression of grayscales. The process of transforming a grayscale image to a halftone one is called *halftoning*. As expected, the resolution of the halftone image is much less than that of the grayscale image, if both images have the same size. Resolution is sacrificed in order to create grayscales through halftoning. If image resolution is very important, the size of the halftone image must be much higher than that of the grayscale image.

Several techniques have been devised for digital image halftoning [ULI87]. Many of them are based on binary thresholding. Let $f(k,l)$, $g(k,l)$ be the grayscale and the binary image, respectively. If a threshold T is given, the binary output image can be calculated as follows:

$$g(k,l) = \begin{cases} 1 & \text{if } f(k,l) \geq T \\ 0 & \text{otherwise} \end{cases} \quad (3.12.1)$$

The threshold T can be chosen by inspecting the histogram of image f. It can be chosen in such a way that $5\% - 50\%$ of the image pixels are above the threshold. Binary thresholding (3.12.1) is global, since the threshold T is the same over the entire image. Global thresholding may give 'pleasant' binary output if the threshold is chosen appropriately. However, most images have regions whose statistics differ considerably. Thus, a locally adaptive threshold is a better choice. Locally adaptive thresholding is again described by (3.12.1). The only difference is that the threshold $T(k,l)$ changes with the pixel location (k,l). There are several ways to perform threshold adaptation. The first one is to calculate the local threshold by inspecting the local image histogram in a local $M \times M$ neighborhood of the pixel $f(k,l)$. A second possibility is to choose the threshold to be equal to the local midpoint:

$$T(k,l) = \frac{1}{2}(\max\{f(k,l)\} + \min\{f(k,l)\}) \quad (3.12.2)$$

where $\max\{f(k,l)\}$ and $\min\{f(k,l)\}$ denote the local maximum and minimum within the $M \times M$ neighborhood.

Although binary thresholding is the first step to halftoning, it does not produce halftone images, because, generally speaking, it does not produce image regions with varying dot density. A better halftone technique is based on the generation of *grayscale binary fonts*. Lets us suppose that the grayscale image has L gray levels and that it must be transformed to a halftone image having N perceived gray levels. Usually N is much less than the number of gray levels L. For reasons of simplicity we choose N to be of the form $N = n \times n + 1$. We can create N matrices $\mathbf{F}_i = 0, 1, \ldots, N$ of size $n \times n$ each, containing 1s. Such a series of matrices is shown in Figure 3.12.1 for $N = 2 \times 2 + 1 = 5$. Let us suppose that the grayscale image $f(k,l)$ has size $N_1 \times N_2$. The binary array for the halftone image $g(k,l)$ has size $nN_1 \times nN_2$. For every pixel of the input image f, the appropriate gray-level matrix $\mathbf{F}$ is

$$\begin{bmatrix} 0 & 0 \\ 0 & 0 \end{bmatrix} \quad \begin{bmatrix} 0 & 1 \\ 0 & 0 \end{bmatrix} \quad \begin{bmatrix} 1 & 0 \\ 0 & 1 \end{bmatrix} \quad \begin{bmatrix} 1 & 1 \\ 0 & 1 \end{bmatrix} \quad \begin{bmatrix} 1 & 1 \\ 1 & 1 \end{bmatrix}$$

Fig. 3.12.1 Gray-level binary fonts.

chosen:

$$F = \begin{cases} \mathbf{F}_i & \text{if } i[\frac{L}{N}] \leq f(k,l) < (i+1)[\frac{L}{N}], \quad i = 0, 1, \ldots, N-2 \\ \mathbf{F}_{N-1} & \text{if } (N-1)[\frac{L}{N}] \leq f(k,l) < L \end{cases}$$

(3.12.3)

and is copied at the appropriate position of the output image $g(k,l)$. If the output is restricted to have size $N_1 \times N_2$, the grayscale image f must be subsampled by n in both dimensions, before the application of the halftoning procedure. Such an algorithm is described in Figure 3.12.2.

```
int halftone(a,b, cn, N1,M1,N2,M2)
image a,b;
int cn;
int N1,M1,N2,M2;

/* Subroutine to halftone an image
    a: source buffer
    b: destination buffer
    cn: number of halftones
    N1,M1: start coordinates source buffer
    N2,M2: end   coordinates source buffer */

{ int i,j,k,l,count,c;
  /* Clear destination buffer */
  clear(b,0,N1,M1,N2,M2);
  c=(int)(0.001+sqrt((double)cn));
  cn=256/cn;
  for(i=N1; i<N2-c+1; i+=c)
  for(j=M1; j<M2-c+1; j+=c)
     { count=a[i][j]/cn;
       for(k=i; k<i+c && count!=0; k++)
       for(l=j; l<j+c && count!=0; l++)
          { b[k][l]=1; count--; }
     }
  return(0);
}
```

Fig. 3.12.2 Algorithm for digital halftoning by using gray-level binary fonts.

$$D^2 = \begin{bmatrix} 0 & 2 \\ 3 & 1 \end{bmatrix} \qquad D^4 = \begin{bmatrix} 0 & 8 & 2 & 10 \\ 12 & 4 & 14 & 6 \\ 3 & 11 & 1 & 9 \\ 15 & 7 & 13 & 5 \end{bmatrix}$$

Fig. 3.12.3 Dither matrices D^2, D^4.

Gray-level font halftoning is conceptually simple and easy to implement. Its major drawback is that it creates false lines and contours in homogeneous image regions, due to the patterns created by the gray-level binary fonts. Therefore, it cannot be used without modifications.

Pseudorandom thresholding can solve the problem of false contours. If random noise is added to the image and thresholding is applied afterwards, binary noise is introduced in the halftone image. This noise destroys the false patterns created by the halftone procedure. The process of adding random noise and thresholding is equivalent to thresholding the image by a random threshold. Matrices containing pseudorandom thresholds are used in halftoning and are called *dither* matrices [ULI87], [SCH89], [FOL90]. A dither matrix of size $n \times n$ is denoted by D^n. Such dither matrices are shown in Figure 3.12.3 for $n = 2, 4$. Recursive algorithms have been developed for dither matrix calculation, when the size $n \times n$ of the matrix D^n is a power of 2. Such a recursive relation is the following [FOL90]:

$$D^n = \begin{bmatrix} 4D^{\frac{n}{2}} + D_{00}^2 U^{\frac{n}{2}} & 4D^{\frac{n}{2}} + D_{01}^2 U^{\frac{n}{2}} \\ 4D^{\frac{n}{2}} + D_{10}^2 U^{\frac{n}{2}} & 4D^{\frac{n}{2}} + D_{11}^2 U^{\frac{n}{2}} \end{bmatrix} \qquad (3.12.4)$$

where U^n is an $n \times n$ matrix containing 1s and D_{ij}^2 is the (i,j) element of matrix D^2 shown in Figure 3.12.3. Dither matrices have integer elements as threshold values. Their dynamic range is generally different from that of the dynamic range of the grayscale image. Therefore, they must be scaled before being applied to the input image. Let us suppose that we want to use an $n \times n$ dither matrix in halftoning. Both input and output images are assumed to have size $N_1 \times N_2$. The halftoning is described by the following equations [SCH89]:

$$g(k,l) = \begin{cases} 1 & \text{if } f(k,l) > T(k,l) \\ 0 & \text{otherwise} \end{cases} \qquad (3.12.5)$$

$$T(k,l) = D^n(k \bmod n, l \bmod n) \qquad (3.12.6)$$

The dither algorithm is described in Figure 3.12.5. The modulo operation in the threshold calculation (3.12.6) ensures that all thresholds included in matrix D^n are used in an ordered manner. The above-mentioned algorithms have been applied on the grayscale image shown in Figure 3.12.4a. The output of subroutine `halftone()` for $N = 3 \times 3 + 1 = 10$ binary fonts is shown in

172 DIGITAL IMAGE FILTERING AND ENHANCEMENT

Fig. 3.12.4 (a) Original image; (b) output of halftoning by using 10 binary fonts; (c) output of dithering for $n = 8$.

Figure 3.12.4b. The false contours created are evident in the output image. The result of dithering for a dither matrix D^8 is shown in Figure 3.12.4c. Ordered dithering produces subjectively pleasant images. As expected, the dithering procedure (3.12.5–3.12.6) reduces the resolution of the output image. Resolution can be increased by increasing the size of the input and output images up to $nN_1 \times nN_2$ and interpolation techniques can be used to fill the input image. The same dithering algorithm (3.12.5–3.12.6) can be applied to the interpolated grayscale image.

```
int dither(a,b,size,N1,M1,N2,M2)
image a,b;
int size;
int N1,M1,N2,M2;

/* Subroutine to perform dithering
   by using a dither matrix of size size.
   a: input grayscale image buffer
   b: output binary image buffer
   size: size of dither matrix
   N1,M1: start coordinates
```

           N2,M2: end   coordinates   */
{
  int i,j,k,cs, dmax, dmin;
  image dither_matrix;
  float frac;

  /* Construct dither matrix */
  dither_matrix=matuc2(size,size);

  for(i=0; i<size; i++)
   for(j=0; j<size; j++)
      dither_matrix[i][j]=0;
  cs=2;
  for(i=0;i<(int)(log((double)size)/log(2.0));i++)
   { for(j=0;j<cs/2;j++)
        for(k=0;k<cs/2;k++)
          {
            dither_matrix[j][k]*=4;
            dither_matrix[j+cs/2][k]=dither_matrix[j][k]+3;
            dither_matrix[j][k+cs/2]=dither_matrix[j][k]+2;
            dither_matrix[j+cs/2][k+cs/2]=dither_matrix[j][k]+1;
          }
      cs=2*cs;
    }

  /* Scale dither matrix elements */
  dmax=dmin=0;
  for(i=0; i<size; i++)
   for(j=0; j<size; j++)
     { if(dither_matrix[i][j]>dmax)
          dmax=(int)dither_matrix[i][j];
       if(dither_matrix[i][j]<dmin)
          dmin=(int)dither_matrix[i][j];
     }

  frac=255.0/(float)(dmax-dmin);

  for(i=0; i<size; i++)
   for(j=0; j<size; j++)
     dither_matrix[i][j]=dmin+
       (unsigned char)(frac*(float)(dither_matrix[i][j]-dmin));

  /* Apply dithering */
  clear(b,0,0,0,NMAX,MMAX);
```
```

```
    for(i=N1; i<N2; i++)
     for(j=M1; j<M2; j++)
      { if(a[i][j]>dither_matrix[i%(int)size][j%(int)size])
         b[i][j]=255; else b[i][j]=0;
      }

  mfreeuc2(dither_matrix,size);
  return(0);
}
```

Fig. 3.12.5 Dithering algorithm.

3.13 IMAGE INTERPOLATION ALGORITHMS

A digital image $f(n_1, n_2)$ may be thought of as a sampled region of an analog image $f(x, y)$ having continuous coordinates x, y:

$$f(n_1, n_2) = f(x, y)|_{x=n_1 T_1, y=n_2 T_2} \qquad (3.13.1)$$

as has already been described in Chapter 1. T_1, T_2 are the sampling intervals along the x, y axes. If the analog image $f(x, y)$ is band-limited:

$$F(\Omega_1, \Omega_2) = 0 \text{ for } |\Omega_1| \geq \tfrac{\pi}{T_1}, |\Omega_2| \geq \tfrac{\pi}{T_2} \qquad (3.13.2)$$

and the sampling frequencies are above the Nyquist frequencies, it can be recovered from the sampled image $f(n_1, n_2)$ by using the sinc interpolation formula [DUD84]:

$$f(x,y) = \sum_{n_1=-\infty}^{\infty} \sum_{n_2=-\infty}^{\infty} f(n_1, n_2) \frac{\sin \frac{\pi}{T_1}(x - n_1 T_1)}{\frac{\pi}{T_1}(x - n_1 T_1)} \frac{\sin \frac{\pi}{T_2}(y - n_2 T_2)}{\frac{\pi}{T_2}(y - n_2 T_2)} \qquad (3.13.3)$$

A truncated version of (3.13.3) having summations over $-M_1 \leq n_1 \leq M_1$, $-M_2 \leq n_2 \leq M_2$ can be used for digital image interpolation. However, *sinc* interpolation is seldom used in practical applications, due to increased computational complexity in sinc function calculation. Interpolation (3.13.3) is essentially a convolution operation with a low-pass filter $h(x, y)$ (sinc function). Other low-pass filters can be used instead. The rectangular window function:

$$h(x,y) = \begin{cases} 1 & -\frac{T_1}{2} \leq x \leq \frac{T_1}{2}, \ -\frac{T_2}{2} \leq y \leq \frac{T_2}{2} \\ 0 & \text{elsewhere in } [-T_1, T_1] \times [-T_2, T_2] \end{cases} \qquad (3.13.4)$$

produces a *zero-order hold* interpolation. The value $f(x, y)$ is chosen to be equal to the geometrically closest pixel on the image array. Let us suppose that the image $f(n_1, n_2)$ has dimensions $N_1 \times N_2$ and that we want to produce an interpolated image $f_i(n_1, n_2)$ having size $2N_1 \times 2N_2$. This can be done by the following interpolation formula:

$$f_i(n_1, n_2) = f([n_1/2], [n_2/2]) \qquad (3.13.5)$$

where $[x]$ denotes the integer part of x. Repetitive application of (3.13.5) leads to zooming by a factor of $2^n \times 2^n$. Digital image zooming is a zero-order hold interpolation. A zooming algorithm is shown in Figure 3.13.1.

```
int zoom(a,b, t, ND1,MD1, N1,M1,N2,M2)
image a,b;
int t;
int ND1,MD1;
int N1,M1,N2,M2;

/* Subroutine to zoom an image
   a: source buffer
   b: destination buffer
   t: zoom factor
   ND1,MD1: start coordinates destination buffer
   N1,M1: start coordinates source buffer
   N2,M2: end   coordinates source buffer     */

{ int i,j,cs,ce,i1,j1,k,l;

  if(t%2==1) {cs=-t/2; ce=-cs;} else {ce=t/2; cs=-ce+1;}
  for(i=N1; i<N2; i++)
     { i1=(i-N1)*t+ND1-cs;
       for(j=M1; j<M2; j++)
          { j1=(j-M1)*t+MD1-cs;
            for(k=cs; k<ce+1; k++)
            for(l=cs; l<ce+1; l++)
               b[k+i1][l+j1]=a[i][j];
          }
     }
  return(0);
}
```

Fig. 3.13.1 Zooming algorithm.

Zero-order interpolation produces patches having equal intensity. Thus, it is used sometimes to create video effects. In several applications, uniform patches are not acceptable, especially if the patch size and the corresponding zooming factor are large. First-order (linear) interpolation can be used to produce smoother interpolated images. The interpolated image is given by the following relation:

$$f(x,y) = (1 - \Delta_1)(1 - \Delta_2)f(n_1, n_2) + (1 - \Delta_1)\Delta_2 f(n_1, n_2 + 1)$$
$$+ \Delta_1(1 - \Delta_2)f(n_1 + 1, n_2) + \Delta_1\Delta_2 f(n_1 + 1, n_2 + 1) \quad (3.13.6)$$

where:

$$\Delta_1 = \frac{x - n_1 T_1}{T_1} \quad (3.13.7)$$

$$\Delta_2 = \frac{y - n_2 T_2}{T_2} \quad (3.13.8)$$

If we want to interpolate an image $f(n_1, n_2)$ by doubling both its dimensions N_1, N_2 to $2N_1, 2N_2$, the following procedure can be used. The rows and columns of $f(n_1, n_2)$ are copied on an array of size $2N_1 \times 2N_2$ interlaced with rows and columns having zero pixel values. Linear interpolation along rows gives:

$$f_i(2n_1, 2n_2 + 1) = \frac{1}{2}[f(n_1, n_2) + f(n_1, n_2 + 1)]$$
$$0 \leq n_1 \leq N_1 - 1, 0 \leq n_2 \leq N_2 - 1 \quad (3.13.9)$$

Linear interpolation along columns:

$$f_i(2n_1 + 1, 2n_2) = \frac{1}{2}[f_i(2n_1, 2n_2) + f_i(2n_1 + 2, 2n_2)]$$
$$0 \leq n_1 \leq N_1 - 1, 0 \leq n_2 \leq N_2 - 1 \quad (3.13.10)$$

produces the desired two-dimensional linearly interpolated image. Direct application of (3.13.6) gives the same results. Two-dimensional linear interpolation is equivalent to the convolution of the image:

$$f'(n_1, n_2) = \begin{cases} f(\frac{n_1}{2}, \frac{n_2}{2}) & \text{if } n_1 = 2k, n_2 = 2l \\ 0 & \text{otherwise} \end{cases} \quad (3.13.11)$$

of size $2N_1 \times 2N_2$ with a low-pass filter having an impulse response described by the matrix:

$$\mathbf{H} = \begin{bmatrix} \frac{1}{4} & \frac{1}{2} & \frac{1}{4} \\ \frac{1}{2} & 1 & \frac{1}{2} \\ \frac{1}{4} & \frac{1}{2} & \frac{1}{4} \end{bmatrix} \quad (3.13.12)$$

Direct implementation of this convolution is preferable. If integer arithmetic is used, no multiplications are needed because filter coefficients are powers of 2. Thus, multiplications can be performed by right shift operations. The convolution kernel $\mathbf{H}$ can be used to obtain image interpolation by higher factors [JAI89]. Let us suppose that we want to create an interpolated image of size $pN_1 \times pN_2$. First we interlace the image $f(n_1, n_2)$ with zeros:

$$f'(n_1, n_2) = \begin{cases} f(\frac{n_1}{p}, \frac{n_2}{p}) & \text{if } n_1 = pk, n_2 = pl \\ 0 & \text{otherwise} \end{cases} \quad (3.13.13)$$

An interpolation of order p can be obtained by convolving the image $f'(n_1, n_2)$ with the convolution matrix $\mathbf{II}$ p times. The cubic spline interpolation is obtained for $p = 3$ [JAI89]. Linear interpolation produces acceptable images in most cases. However, if better performance is desired, polynomial interpolation schemes must be used. Such methods are treated extensively in the graphics literature [FOL90].

The inverse procedure to image interpolation is image decimation:

$$g(n_1, n_2) = f(an_1, bn_2) \quad (3.13.14)$$

where $a, b > 1$. If a, b are integers, the decimation routine is very simple, as shown in Figure 3.13.2.

```
int decim(a,b, t, ND1,MD1, N1,M1,N2,M2)
image a,b;
int t;
int ND1,MD1; int N1,M1,N2,M2;
/* Subroutine to decimate an image
   a: source buffer
   b: destination buffer
   t: decimation factor
   ND1,MD1: start coordinates destination buffer
   N1,M1:   start coordinates source buffer
   N2,M2:   end coordinates source buffer    */
{ int i,j,k=-1,l=-1;
  for(i=N1; i<N2; i+=t)
    { k++; l=-1;
      for(j=M1; j<M2;j+=t) { l++; b[k+ND1][l+MD1]=a[i][j]; }
    }
  return(0);
}
```

Fig. 3.13.2 Decimation algorithm.

3.14 ANISOTROPIC DIFFUSION

Anisotropic diffusion [PER90] is an efficient nonlinear technique for simultaneously performing contrast enhancement and noise reduction, and for deriving consistent deterministic scale-space image descriptions. Isotropic diffusion is based on the heat diffusion formula and smooths the entire image, the image intensity being equivalent to the image plate temperature. The anisotropic diffusion smooths homogeneous image regions but retains image edges, since it does not allow heat dissipation across image edges.

The anisotropic diffusion can be defined as:

$$I_t = div(c(x,y,t)\nabla I) = c(x,y,t)\Delta I + \nabla c \cdot \nabla I \qquad (3.14.1)$$

where we indicate with div the divergence operator and with ∇ and Δ respectively the gradient and Laplacian operator, with respect to the space variables. Equation (3.14.1) can be discretized on a square lattice-structure, with brightness values associated to the vertices and conduction coefficients to the arcs [PER90]. A 4 - nearest-neighbors discretization of the Laplacian operator can be used:

$$I_{i,j}^{t+1} = I_{i,j}^t + \lambda [c_N \cdot \nabla_N I + c_S \cdot \nabla_S I + c_E \cdot \nabla_E I + c_W \cdot \nabla_W I]_{i,j}^t \qquad (3.14.2)$$

where $0 \le \lambda \le 1/4$ ensures numerical stability, N, S, E, W are the mnemonic subscripts for North, South, East, West and the symbol ∇ (not to be confused with the gradient operator ∇) indicates nearest-neighbor differences:

$$\nabla_N I^{i,j} \equiv I_{i-1,j} - I_{i,j}$$
$$\nabla_S I^{i,j} \equiv I_{i+1,j} - I_{i,j}$$
$$\nabla_E I^{i,j} \equiv I_{i,j+1} - I_{i,j}$$
$$\nabla_W I^{i,j} \equiv I_{i,j-1} - I_{i,j} \qquad (3.14.3)$$

Iterating this scheme can be thought as moving towards coarser resolutions in scale-space. The conduction coefficients are updated at every iteration as a function of the brightness gradient:

$$c_{N_{i,j}}^t = g\left(\|\, (\nabla I)_{i+(1/2),j}^t \,\|\right)$$
$$c_{S_{i,j}}^t = g\left(\|\, (\nabla I)_{i-(1/2),j}^t \,\|\right)$$
$$c_{E_{i,j}}^t = g\left(\|\, (\nabla I)_{i,j+(1/2)}^t \,\|\right)$$
$$c_{W_{i,j}}^t = g\left(\|\, (\nabla I)_{i,j-(1/2)}^t \,\|\right) \qquad (3.14.4)$$

The value of the gradient can be computed on different neighborhood structures, thus achieving different compromises between accuracy and locality. The simplest choice consists in approximating the norm of the gradient at each arc location with the absolute value of its projection along the direction of the arc [PER90]:

$$c_{N_{i,j}}^t = g\left(|\, \nabla_N I_{i,j}^t \,|\right)$$
$$c_{S_{i,j}}^t = g\left(|\, \nabla_S I_{i,j}^t \,|\right)$$
$$c_{E_{i,j}}^t = g\left(|\, \nabla_E I_{i,j}^t \,|\right)$$
$$c_{W_{i,j}}^t = g\left(|\, \nabla_W I_{i,j}^t \,|\right) \qquad (3.14.5)$$

This implementation scheme is rather simple computationally. Different functions can be used for $g(\cdot)$ giving rather similar results. The following function can be used:

$$g(\nabla I) = \frac{1}{1 + (\frac{\|\nabla I\|}{K})^2} \tag{3.14.6}$$

The constant K can be manually fixed. This parameter specifies a stationarity threshold for the intensity function above which anisotropic diffusion occurs. Anisotropic diffusion has found important applications as a preprocessing step in medical imaging (MRI, CT) before image segmentation. It has also being used for filling cracks in digital crack restoration of old paintings [GIA98].

3.15 IMAGE MOSAICING

Mosaicing has been developed for the detailed reconstruction of large images by successively acquiring image patches in a given row-column or column-row order. The final digital image is assembled by mosaicing its patches. The classical approach adopted in image mosaicing is based on selecting, in a semiautomatic way, a set of corresponding points on each pair of image patches containing overlapping parts. Automatic matching of images has been proposed as well. The images are assembled according to an algorithm which find similarities in the overlapping regions.

An algorithm based on matching two regions of the same size from the two given images has been recently presented [BOR97, BOR97A]. This approach is similar to the block matching algorithms, extensively used for estimating the optical flow in image sequences. A search area is defined in the overlapping parts of the image patches and the best match for a given area from a neighboring image is found. The best match corresponds to the minimum of a cost function. In the general case, when we assume that images are not distorted, the cost function is the mean square error (MSE) between the image luminance of patch regions of the same size. The computational complexity depends on the total number of component images and on the overlapping area size. Knowing with a certain approximation the size of the search region can reduce the computational burden. In the case of an automatic acquisition system controlled by computer, the overlapping area size can be estimated quite well.

When mosaicing, we have to deal with possible differences in the illumination between neighboring images. In order to assure a smooth transition between regions with different illumination we weight the contributions of pixels coming from different images. The weights are proportional with the distances from the given pixel to the margin of the respective image. The weighting procedure ensures a smooth luminance transition between the mosaiced images.

The moisaicing algorithm has been applied on several images. We present an application on an infrared image of a Byzantine icon. The 2 × 3 image patches are represented in Figure 3.15.1a. The resulting mosaiced image is shown in Figure 3.15.1b. We can observe that the mosaicing result is quite good, despite the fact that features are fainted in infrared images. Mosaicing has several applications in painting restoration, medical imaging and remote sensing.

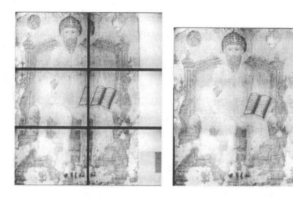

Fig. 3.15.1 (a) Original set of images; (b) resulting mosaiced image.

3.16 IMAGE WATERMARKING

One of the most important digital image features is easy storage, manipulation and transmission, a fact that gave rise to a wide range of new applications (digital television, digital video disc, digital image databases, electronic publishing etc.). However, this feature has an important side effect; it allows easy data piracy, that is, unauthorized copying, reproduction and transmission of information. Due to this fact, protection of intellectual property rights, (copyright protection) of stored/transmitted digital images has become a very important issue.

The solution to the problem of copyright protection seems to lie in a technique called *data hiding* or *steganography*. Steganography deals with methods of embedding data within a medium (in our case in images) in an imperceptible way. Data hiding techniques in images explore the so-called noise or distortion masking property of the Human Visual System (HVS) that refers to the decrease in the perceived intensity of a visual stimulus when this is superimposed over another stimulus.

The invisibly embedded signal, which is usually referred to as *watermark*, completely characterizes the person who applied it on the image and, therefore, can be used as a proof of digital media ownership in case of a copyright dispute. Obviously the secure and unambiguous identification of the legal owner of an image requires that each individual or organization that produces owns or transmits digital images uses a different watermark. Watermarks serving the purpose of copyright protection should be undeletable by an attacker, easily and securely detectable by the owner and resistant to standard image processing operations (e.g. filtering, compression).

Another application envisaged for data hiding is *authentication* or *tamper-proofing*, a functionality equivalent to data integrity verification. Authentication methods should signal an authentication violation only when significant modifications of the visual content occur and also pinpoint the image regions where alterations took place. Contrary to copyright protection applications, data inserted for authentication purposes should be fragile, that is, they should be modified when the image is manipulated.

In a watermarking scheme one can distinguish between three fundamental stages: watermark generation, embedding and detection. Watermark generation aims at producing the watermark pattern, using an owner and/or image dependent key. Watermark embedding can be considered as a superposition of the watermark signal on the original image. Finally, watermark detection aims at verifying whether the image under consideration hosts a certain watermark. For a more detailed discussion of the general watermarking framework the interested reader can consult [VOY98A], [VOY99]. Watermarking techniques can be categorized in a number of classes on the basis of their distinct features. An overview can be found in [NIK99].

A number of watermarking techniques require that the original image is available during the detection phase. Such schemes are sometimes referred to as private schemes or image escrow schemes. They are robust to a wide range of image manipulations that include filtering, compression and geometric distortions such as cropping, scaling, rotation etc. Despite their obvious advantages, methods based on the original image do not fit well in certain applications e.g. automatic Internet search. Image watermarking methods that do not require the original image for watermark detection are called oblivious or blind methods and do not suffer from the above-mentioned disadvantages of private methods. However, their robustness to image modifications and attacks is limited in comparison to private methods.

Another classification scheme for watermarking techniques can be devised by taking into account the domain where the watermark signal is embedded. Certain methods do data embedding in the spatial domain by modulating the intensity of certain pixels, while other techniques modify the magnitude of coefficients in an appropriate transform domain, that is, the DCT, DFT or DWT domain that are described in Chapter 2.

Watermarking techniques can also be classified on the basis of the watermark signal dependency on the original image. Image dependency is necessary if image characteristics are to be exploited in order to obtain invisible watermarks using the masking properties of the HVS.

A typical blind, image independent, spatial domain image watermarking technique generates a 2-D watermark pattern $W(i,j)$ of the same dimensions with the host image $I(i,j)$. Such a watermark image generated by chaotic

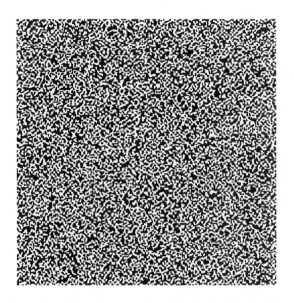

Fig. 3.16.1 Watermark produced by a chaotic function

techniques is shown in Figure 8.6.4 [VOY98]. Then it embeds the watermark in the image in an additive:

$$I'(i,j) = I(i,j) + aW(i,j) \qquad (3.16.1)$$

or multiplicative way:

$$I'(i,j) = I(i,j) + aI(i,j)W(i,j) \qquad (3.16.2)$$

Finally, detection is performed using correlation based techniques by evaluating a sum of the form:

$$C = \sum_{ij} W(i,j)I(i,j) \qquad (3.16.3)$$

The value of this sum will be large when a watermark is embedded in the image and close to zero when no watermark is embedded in the image ($I = I'$). The correlator output is compared against a threshold T in order to decide if the image is watermarked or not. Hypothesis testing can be also used for watermark detection.

One of the first and most cited watermarking schemes that makes use of the original image at the detection phase is proposed in [COX97]. According to this scheme the watermark sequence x_i is placed on the n-th highest magnitude coefficients u_i of the DCT transformed image using the following formula:

$$u'_i = u_i + au_i x_i \qquad (3.16.4)$$

that is, in a multiplicative way. The watermark coefficients x_i are zero mean, unit variance independent identically distributed (i.i.d.) samples that follow a Gaussian distribution. Watermark detection is done by correlating the watermark signal with an estimate of the embedded watermark that is derived by subtracting the original image from the watermarked one. The utilization of the original image by the detection scheme robustifies the proposed technique against a wide range of attacks. A variation of the above technique that incorporates perceptual masking is presented in [POD98].

The Fourier transform followed by the so-called Fourier-Mellin transform (FMT), that is, the Fourier transform applied on a log-polar coordinate system, can be proven to be rotation, scale and translation invariant. In [RUA98] the authors exploit this property in order to construct watermarks that are invariant to such signal manipulations. The watermark signal is embedded in the transform domain. For watermark detection, the original image is subtracted from the watermarked image and the residual is transformed to the FT-FMT domain where correlation with the watermark signal takes place.

A multiresolution watermarking technique is proposed in [XIA97]. The host image is decomposed into subbands using a two-step Discrete Wavelet Transform. The watermark sequence, which has the form of zero-mean unit variance Gaussian noise, is added to the largest coefficients that do not belong to the lowest resolution band. Decoding requires the original image and is done in a hierarchical way.

In the method proposed in [PIT98], the image I is split into two random subsets A, B and the intensity of pixels in A is increased by a constant embedding factor k. Watermark detection is performed by evaluating the difference of the mean values of the pixels in sets A, B. This difference is expected to be equal to k for a watermarked image and equal to zero for an image that is not watermarked. Hypothesis testing can be used to decide on the existence of the watermark. The above algorithm is vulnerable to lowpass operations. Extensions to this algorithm are proposed in [NIK98A]. The watermark signal can be appropriately shaped using an optimization procedure that calculates the appropriate embedding value for each pixel so that the energy of the watermark signal is concentrated at low frequencies, while taking also into account invisibility constraints.

In the method proposed in [SOL99] watermark embedding is performed in the magnitude of the DFT image coefficients:

$$M'(k_1, k_2) = M(k_1, k_2) + aW(k_1, k_2) \qquad (3.16.5)$$

where $M(k_1, k_2)$ is the magnitude of the original image DFT coefficients. The watermark consists of a 2-D zero-mean random bi-valued sequence. The frequency region in which the watermark is embedded is a ring covering the middle frequencies. The ring is separated in S sectors and in homocentric circles of radius $r \in [R_1, R_2]$. Each circular sector is assigned the same value 1 or -1. Watermark detection does not involve the original image and is carried out using a correlator detector. The ring shape of the watermark along with the properties of DFT make the method robust to compression as well as translation, rotation, cropping and scaling.

In [VOY98] bi-valued digital signals generated by properly tuned chaotic dynamical systems are used as watermark patterns. Chaotic signals possess certain properties (cryptographical security, controlled lowpass characteristics) that make them suitable for watermarking applications. A 1-D chaotic watermark signal can be produced using a one-dimensional discrete time chaotic map:

$$z(n+1) = \mathbf{F}(z(n), \lambda) \qquad (3.16.6)$$

where $n = 0, 1, 2, ...$ denotes map iterations. The mapping is chosen so that strongly chaotic signals are produced, a fact that is necessary to obtain watermarks with the desired characteristics. A bi-valued sequence $s(n) \in \{-1, 1\}$ is obtained by "thresholding" $z(n)$. Parameter λ controls the frequency characteristics of $s(n)$. A 2-D watermark W_0 is then formed by appropriately wrapping the 1-D signal on the 2-D image. A function $\mathcal{F}$ that modifies the original watermark signal W_0 according to the host image is used to produce image-dependent watermarks. Watermarks generated by this procedure can be embedded using general superposition and multiplication operators. The detection module does not require the original image and implies the use of a suitable test statistic.

References

AND77. H.C. Andrews, B.R. Hunt, *Digital image restoration*, Prentice-Hall, 1977.

ANG94. G. Angelopoulos, I. Pitas, "Multichannel Wiener filters in color image restoration," *IEEE Transactions on Circuits and Systems for Video Technology*, vol. 4, no. 1, pp. 83-87, February 1994.

ARC89. G.R. Arce, R.E. Foster, "Detail preserving ranked-order based filters for image processing," *IEEE Transactions on Acoustics, Speech, and Signal Processing*, vol. ASSP-37, no. 1, pp. 83-98, Jan. 1989.

BED84. J.B. Bednar, T.L. Watt "Alpha-trimmed means and their relationship to the median filters," *IEEE Transactions on Acoustics, Speech, and Signal Processing*, vol. ASSP-32, no. 1, pp. 145-153, Feb. 1984.

BER87. R. Bernstein, "Adaptive nonlinear filters for simultaneous removal of different kinds of noise in images," *IEEE Transactions on Circuits and Systems*, vol. CAS-34, no. 11, pp. 1275-1291, Nov. 1987.

BOR97. A.G. Bors, W. Puech, I. Pitas, J.M. Chassery, "Perspective Distortion Analysis for Mosaicing Images Painted on Cylindrical Surfaces," *IEEE International Conference on Acoustics, Speech, and Signal Processing (ICASSP'97)*, Munich, Germany, vol. 4, pp. 3049-3052, 20-24 Apr. 1997.

BOR97A. A.G. Bors, W. Puech, I. Pitas, J.M. Chassery, "Mosaicing of Flattened Images From Straight Homogeneous Generalized Cylinders," *Lecture Notes in Computer Science*, vol. 1296, 7th International Conference on Computer Analysis of Images and Patterns, G. Sommer, K. Daniilidis, J. Pauli (Eds.), Kiel, Germany, pp. 122-129, 1997.

BOV83. A.C. Bovik, T.S. Huang, D.C. Munson, "A generalization of median filtering using linear combinations of order statistics," *IEEE Transactions on Acoustics, Speech, and Signal Processing*, vol. ASSP-31, no. 6, pp. 1342-1349, Dec. 1983.

BOV86. A.C. Bovik, T.S. Huang, D.C. Munson, "Nonparametric tests for edge detection in noise," *Pattern Recognition*, vol. 19, no. 3, pp. 209-219, 1986.

COX97. I. Cox, J. Kilian, T. Leighton, and T. Shamoon, "Secure spread spectrum watermarking for multimedia," *IEEE Trans. Image Processing*, vol. 6, no. 12, pp. 1673-1687, 1997.

DAV81. H.A. David, *Order statistics*, Wiley, 1981.

DIN87. R. Ding, A.N. Venetsanopoulos, "Generalized homomorphic and adaptive order statistic filters for the removal of impulsive and signal-dependent noise," *IEEE Transactions on Circuits and Systems*, vol. CAS-34, no. 8, pp. 948-955, Aug. 1987.

DUD84. D.E. Dudgeon, R.M. Mersereau, *Multidimensional digital signal processing*, Prentice-Hall, 1984.

FOL90. J.D. Foley, A. van Dam, J.F.Hughes, *Computer graphics: Principles and practice*, Addison-Wesley, 1990.

GIA98. I. Giakoumis, I. Pitas, "Digital restoration of painting cracks," *Proc. IEEE Intl. Symp. on Circuits and Systems (ISCAS98)*, pp. 357, 1998.

GON87. R.C. Gonzalez, P. Wintz, *Digital image processing*, Addison-Wesley, 1987.

HAJ67. J. Hajek, Z. Sidak, *Theory of rank tests*, Academic Press, 1967.

HAM86. F. Hampel, E. Ronchetti, P. Rousseevw, W. Stahel, *Robust statistics: An approach based on influence functions*, Wiley, 1986.

HEY82. G. Heygster, "Rank filters in digital image processing," *Computer Vision, Graphics and Image Processing*, vol. 19, pp. 148-164, 1982.

HOR84. E. Horowitz, S. Sahni, *Fundamentals of computer algorithms*, Computer Science Press, 1984.

HUA79. T.S. Huang, G.J. Yang, G.Y. Tang, "A fast two-dimensional median filtering algorithm," *IEEE Transactions on Acoustics, Speech, and Signal Processing*, vol. ASSP-27, no. 1, pp. 13-18, Feb. 1979.

JAI89. A.K. Jain, *Fundamentals of digital image processing*, Prentice-Hall, 1989.

JAY84. N.S. Jayant, P. Noll, *Digital coding of waveforms*, Prentice Hall, 1984.

JUS81. B.I. Justusson, "Median filtering: Statistical properties," in *Two-dimensional digital signal processing II*, T.S. Huang (editor), Springer Verlag, 1981.

KNU73. D.E. Knuth, *The art of computer programming*, vol. 3, Addison-Wesley, 1973.

KOF96. E. Kofidis, S. Theodoridis, C. Kotropoulos, and I. Pitas, "Nonlinear adaptive filters for speckle suppression in ultrasonic images," *Signal Processing*, vol. 52, no. 3, pp. 357-372, August 1996.

KOT92SP. C. Kotropoulos, and I. Pitas, "Constrained LMS adaptive L-filters," *Signal Processing*, vol. 26, no. 3, pp. 335-358, March 1992.

KOT92UI. C. Kotropoulos, and I. Pitas, "Optimum nonlinear signal detection and estimation in the presence of ultrasonic speckle," *Ultrasonic Imaging*, vol. 14, no. 3, pp. 249-275, July 1992.

KOT94CDS. C. Kotropoulos and I. Pitas, "Adaptive nonlinear filters for digital signal/image processing," in *Control and Dynamic Systems* (Prof. C.T. Leondes, ed.), vol. 67, pp. 263-318, Academic Press, 1994.

KOT94TIP. C. Kotropoulos, X. Magnisalis, I. Pitas, and M.G. Strintzis, "Nonlinear ultrasonic image processing based on signal-adaptive filters and self-organizing neural networks," *IEEE Trans. on Image Processing*, vol. 3, no. 1, pp. 65-77, January 1994.

KOT94TSP. C. Kotropoulos, and I. Pitas, "Multichannel L-filters based on marginal data ordering," *IEEE Trans. on Signal Processing*, vol. 42, no. 10, pp. 2581-2595, October 1994.

KOT96TIP. C. Kotropoulos, and I. Pitas, "Adaptive LMS L-filters for noise suppression in images," *IEEE Trans. on Image Processing*, vol. 5, no. 12, pp. 1596-1609, December 1996.

KOT99. C. Kotropoulos, M. Pappas, and I. Pitas, "Nonlinear mean filters and their application in image filtering and edge detection," in *Nonlinear Image Processing* (S. Mitra and G. Sicuranza, Eds.), Academic Press, 2000.

KOT99OE. C. Kotropoulos, and I. Pitas, "Adaptive multichannel marginal L-filters," *Optical Engineering*, vol. 38, no. 4, pp. 688-704, April 1999.

LEE80. J.S. Lee, "Digital image enhancement and noise filtering by local statistics," *IEEE Transactions on Pattern Analysis and Machine Intelligence*, vol. PAMI-2, no. 2, pp. 165-168, March 1980.

LIM90. J.S. Lim, *Two-dimensional signal and image processing*, Prentice-Hall, 1990.

LIU92. R. Liu, K. Singh, " Ordering directional data: Concepts of data depth on circles and spheres," *Annals of Statistics*, vol 20, no. 3, pp. 1468-1484, 1992.

MAR72. K.V. Mardia, *Statistics of directional data*, Academic Press, 1972.

MAR87. S.L. Marple, *Digital spectral analysis*, Prentice-Hall, 1987.

NAR81. P.M. Narendra, "A separable median filter for image noise smoothing," *IEEE Transactions on Pattern Analysis and Machine Intelligence*, vol. PAMI-3, no. 1, pp. 20-29, Jan. 1981.

NIK96. N. Nikolaidis and I. Pitas, "Copyright protection of images using robust digital signatures," *IEEE International Conference on Acoustics, Speech and Signal Processing*, vol. 4, pp. 2168-2171, May 1996.

NIK98. N. Nikolaidis and I. Pitas, "Nonlinear processing and analysis of angular signals," *IEEE Trans. on Signal Processing*, vol. 46, no. 12, pp. 3181-3194, December 1998.

NIK98A. N. Nikolaidis, I. Pitas, "Robust image watermarking in the spatial domain," *Signal Processing*, vol. 66, no. 3, pp. 385-403, May 1998.

NIK99. N. Nikolaidis, I. Pitas, "Digital Image Watermarking: an Overview," *Proc. ICMCS*, vol. 1, pp. 1-6, Italy 1999.

OPE89. A. Oppenheim, R.W. Schafer, *Discrete-time signal processing*, Prentice-Hall, 1989.

PAP65. A. Papoulis, *Probability, random variables and stochastic processes*, McGraw-Hill, 1965.

PER90. P. Perona, J. Malik, "Scale-space and edge detection using anisotropic diffusion," *IEEE Trans. on Pattern Analysis and Machine Intelligence*, vol. 12, no. 7, pp. 629-639, July 1990.

PIT86ASSP. I. Pitas, A.N. Venetsanopoulos, "Nonlinear mean filter in image processing," *IEEE Transactions on Acoustics, Speech, and Signal Processing*, vol. ASSP-34, pp. 573-584, June 1986.

PIT86CAS. I. Pitas, A.N. Venetsanopoulos, "The use of symmetries in the design of multidimensional digital filters," *IEEE Transactions on Circuits and Systems*, vol. CAS-33, no. 10, pp. 863-874, Oct. 1986.

PIT86PAMI. I. Pitas, A.N. Venetsanopoulos, "Edge detectors based on nonlinear filters," *IEEE Transactions on Pattern Analysis and Machine Intelligence*, vol. PAMI-8, pp. 538-560, June 1986.

PIT86SP. I. Pitas, A.N. Venetsanopoulos, "Nonlinear order statistic filter for image filtering and edge detection," *Signal Processing*, vol. 10, pp. 395-413, June 1986.

PIT88. I. Pitas, A.N. Venetsanopoulos, "A new filter structure for the implementation of certain classes of image operations," *IEEE Transactions on Circuits and Systems*, vol. CAS-35, no. 6, pp. 636-647, June 1988.

PIT89. I. Pitas, "Fast algorithms for running ordering and max/min calculation," *IEEE Transactions on Circuits and Systems*, vol. CAS-36, no. 6, pp. 795-804, June 1989.

PIT90. I. Pitas, A.N. Venetsanopoulos, *Nonlinear digital filters: Principles and applications*, Kluwer Academic, 1990.

PIT98. I. Pitas, "A Method for Watermark Casting in Digital Images," *IEEE Trans. on Circuits and Systems on Video Technology*, vol. 8, no. 6, pp. 775-780, October 1998.

PIT98A. I. Pitas, "A Method for Watermark Casting in Digital Images," *IEEE Trans. on Circuits and Systems on Video Technology*, vol. 8, no. 6, pp. 775-780, October 1998.

POD98. C. Podilchuk, W. Zeng, "Image-Adaptive Watermarking Using Visual Models," *IEEE Journal on Selected Areas in Communications*, vol. 16, no. 4, pp. 525-539, May 1998.

PRA91. W.K. Pratt, *Digital image processing*, Wiley, 1991.

RES88. A. Restrepo, A.C. Bovik, "Adaptive trimmed mean filters for image restoration," *IEEE Transactions on Acoustics, Speech, and Signal Processing*, vol. ASSP-36, no. 8, pp. 1326-1337, 1988.

RUA98. J.O. Ruanaidh, T. Pun, "Rotation scale and translation invariant spread spectrum digital image watermarking," *Signal Processing*, special issue on Copyright Protection and Access Control, vol. 66, no. 3, pp. 303-318, May 1998.

SCH81. R.W. Schafer, R.M. Mersereau, M.A. Richards, "Constrained iterative restoration algorithms," *Proc. IEEE*, vol. 69, pp. 432-450, Apr. 1981.

SCH89. R.J. Schalkof, *Digital image processing and computer vision*, Wiley, 1989.

SER82. J. Serra, *Image analysis and mathematical morphology*, Academic Press, 1982.

SER88. J. Serra (editor), *Image analysis and mathematical morphology: Theoretical advances*, vol. 2, Academic Press, 1988.

SOL99. V. Solachidis, I. Pitas, "The use of circular symmetric watermarks in the 2-D DFT domain," *Proc. IEEE Int. Conf. on Acoustics, Speech and Signal Processing*, (ICASSP'99), vol. 6, pp. 3469-3472, USA, 1999.

TSEK98. S. Tsekeridou, C. Kotropoulos and I. Pitas, "Adaptive order statistic filters for the removal of noise from corrupted images," *SPIE Optical Engineering Journal*, vol. 37, no. 10, pp. 2798-2816, October 1998.

TUK77. J.W. Tukey, *Exploratory data analysis*, Addison-Wesley, 1977.

TWO78. R.E. Twogood, M.P. Ekstrom, S.K. Mitra, "Optimal sectioning procedure for the implementation of 2-d digital filters," *IEEE Transactions on Circuits and Systems*, vol. CAS-25, no. 5, pp. 260-269, May 1978.

ULI87. R. Ulichney, *Digital halftoning*, MIT Press, 1987.

VOY98. G. Voyatzis and I. Pitas, "Chaotic Watermarks for Embedding in the Spatial Digital Image Domain," *IEEE Int. Conference on Image Processing (ICIP'98)*, vol. II, pp. 432-436, October 1998.

VOY98A. G. Voyatzis, I. Pitas, "Protecting Digital-Image Copyrights: A Framework," *IEEE Computer Graphics & Applications*, vol. 19, no. 1, January/February 1999.

VOY99. G. Voyatzis and I. Pitas, "The use of watermarks in the protection of digital multimedia products," *Proc. of the IEEE*, vol. 87, no. 7, pp. 1197-1207, July 1999.

XIA97. X.G. Xia, C.G. Boncelet, G.R. Arce. "A multiresolution watermark for digital images," *Proceedings of ICIP'97*, vol. I, pp. 548-551, Atlanta, October 1997.

4
Digital Image Compression

4.1 INTRODUCTION

Digital image coding and compression is concerned with the minimization of the memory needed to represent and store a digital image. It is well known that raw digital images occupy a large amount of memory. For example, a 1024×1024 color image requires 3 MB of memory for its storage. The amount of memory required creates problems in massive digital image storage for image archiving in a variety of applications, for example, office automation (documents), medical imaging, video storage and remote sensing (satellite images). Fast compression and especially fast decompression algorithms for image retrieval are of primary importance in image data base applications. Digital image transmission is another area where coding and compression techniques are needed. Compressed images require less transmission time. The use of compression techniques leads to substantial economic savings in fax transmission, videoconferencing, medical image transmission, and so forth. Fast compression/decompression algorithms and architectures are of primary importance for applications where real-time performance is required, for example, in videoconferencing and high-definition TV (HDTV). Since most communication channels are noisy, the robustness of coding/compression techniques in the presence of noise is important as well.

Digital image compression techniques can be divided into two large classes: *lossless* and *lossy* compression. Lossless compression is employed in applications where raw image data are difficult to obtain or contain vital information that may be destroyed by compression, for example, in medical diagnostic imaging. Lossy compression can be used when raw image data can be easily reproduced or when the information loss can be tolerated at the receiver site. A typical case is in videoconferencing and digital TV applications where the final 'receiver' is the human eye. It is well known that the eye can tolerate certain image imperfections. For example, slight losses of chromatic information can be easily tolerated in the case of color images. However, other imperfections, for example, edge blurring and image sequence flickering, are unacceptable. Thus, human visual perception characteristics can be taken into account when designing an image coder in order to reduce visible image degradations.

All digital image compression techniques are based on the exploitation of *information redundancy* that exists in most digital images. The redundancy stems from the statistics of the image data (e.g. strong spatial correlation). Image redundancy can be described in various ways. Each of them leads to a particular class of digital image compression algorithms. *Statistical redundancy* is directly related to the image data probability distribution and can be treated by information theory techniques using image entropy concepts. Its removal results in lossless image compression techniques (Huffman coding, run-length coding) that are used in facsimile transmission. Another way of describing image redundancy is by using *predictability* in local image neighborhoods. Local prediction models take advantage of the strong spatial correlation and try to decorrelate the pixels in a local neighborhood. The transmission or storage of the prediction model and of the decorrelated (error) image results in image compression. Most predictive compression schemes result in lossy compression. Finally, image compression can be achieved by information packing through *image transforms*. This is obtained by exploiting the fact that certain orthogonal transforms (e.g. discrete cosine transform, Karhunen–Loeve transform, discrete sine transform, Walsh–Hadamard transform) can concentrate image energy in a few transform coefficients. Thus, coding of transform coefficients can lead to substantial data compression. Transform coding algorithms are mainly used in lossy compression schemes. Transform compression can achieve great data reduction and it is currently one of the best coding techniques known.

4.2 HUFFMAN CODING

Let us suppose that an image has been digitized to a size of $N \times M$ pixels by using B bits per pixel. The digital image is already coded in the *pulse code modulation* (PCM) system. Raw digital image transmission, i.e. PCM transmission, requires NMB bits for the entire digital image. In this case, each of the 2^B image intensity levels is transmitted by using B bits. The average number of bits per pixel can be reduced by assigning binary codes of different bit length to the various image intensities. Let $p(i)$ be the probability density function (pdf) of the image intensities for $0 \leq i < 2^B$. The pdf $p(i)$

can be estimated by calculating the digital image histogram. Once the pdf is known, we can assign short codewords to intensities having a high probability of occurrence and larger codewords to less frequent image intensity levels. This coding scheme is called *entropy coding*. Let us suppose that a codeword of length $L(i)$ is assigned to the intensity level i, $0 \leq i < 2^B$. The average codeword length is given by:

$$\bar{L} = \sum_{i=0}^{2^B-1} L(i)p(i) \qquad (4.2.1)$$

The lengths $L(i)$ must be chosen in such a way that $\bar{L}$ is minimized. Information theory [GAL68] gives a lower bound on the average codeword length $\bar{L}$:

$$\bar{L} \geq H(B) \qquad (4.2.2)$$

where $H(B)$ is *image entropy*:

$$H(B) = -\sum_{i=0}^{2^B-1} p(i) \log_2 p(i) \qquad (4.2.3)$$

If the image histogram $\hat{p}(i)$ is close to the uniform distribution, image entropy attains its maximum. In this case, image redundancy is small and the average codeword length $\bar{L}$ is relatively large. In fact, if $p(i)$ is a uniform distribution $p(i) = 2^{-B}$, $0 \leq i \leq 2^B - 1$, entropy $H(B)$ is equal to B and the average codeword length is equal to B as well. In this case, no benefit can result from entropy coding. Entropy coding decreases $\bar{L}$ if the intensity levels are highly predictable, i.e. if some intensity levels have large probabilities $p(i)$. This happens in images having bimodal or multimodal histograms.

If the image intensity levels are coded by using variable length codewords, the codewords are concatenated to form a binary data stream. This stream must be decoded at the receiver site. Thus, the combinations of the concatenated codewords must be decipherable. The Huffman code possesses this property [GAL68]. Its average codeword length is very close to image entropy:

$$H(B) \leq \bar{L} \leq H(B) + 1 \qquad (4.2.4)$$

In the Huffman code, no codeword can be the prefix of another codeword. This guarantees that the encoded binary data stream is decipherable. The resulting code has a tree form. The Huffman code can be constructed by using a tree, as illustrated in Figure 4.2.1. Let us suppose that the image has eight intensity levels. The number of bits for PCM coding is $B = 3$. We assume that the probabilities $p(i)$, $i = 0, \ldots, 2^{B-1}$, are known and we create a column of intensity levels with descending probabilities $p(i)$ as shown in Figure 4.2.1a. The intensities of this column constitute the leaves of the Huffman code tree. The tree is constructed in $B - 1$ steps. At each step the two tree

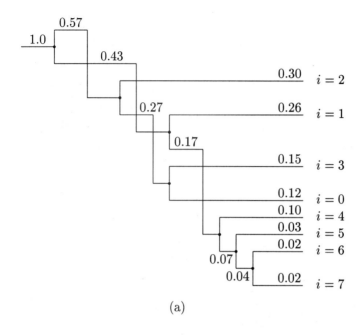

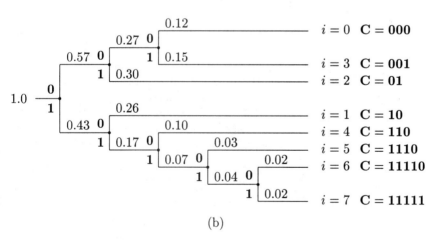

Fig. 4.2.1 (a) Construction of Huffman code tree; (b) tree rearrangement.

HUFFMAN CODING   195

nodes having minimal probabilities are connected to form an intermediate node. The probability assigned to this node is the sum of the probabilities of the two branches. This procedure is repeated until all branches (intensity levels) are used and the probability sum is 1. The code tree is unscrambled to eliminate branch crossovers, as shown in Figure 4.2.1b. The codewords are constructed by traversing the decoding tree from the root to its leaves. At each level we assign 0 to the top branch and 1 to the bottom branch. This procedure is repeated until we reach all tree leaves. Each leaf corresponds to a unique intensity level. The codeword for this intensity consists of the 0s and the 1s that exist in the path from the root to this specific leaf. The resulting decoding tree and codebook are shown in Figure 4.2.1b. As expected, the intensity levels $i = 1, 2$ are assigned small codewords, because they have high probabilities of occurrence. A program which constructs a Huffman codebook is shown in Figure 4.2.2. Subroutine `huftree()` uses the symbol probability vector to create a coding tree whose nodes have the form:

```
typedef struct symbol * SPOINTER;
struct symbol { float prob;
        SPOINTER comp[2]; };
```

Originally, only tree leaves exist in a symbol array called `leaf` and pointed by `*lpnr`. Each element of this array contains symbol probability. At each stage, it uses the subroutine `find2min()` to find two tree nodes having the minimal probability. These nodes are merged to a new node by using subroutine `merge()`. When tree construction finishes, the subroutine returns the tree root. Subsequently, this tree is accessed by using `treeaccess()` in order to produce the codeword of each tree leaf (symbol). The entire code table is returned in the array `code`.

```
int huftree(pdf,rpnr,lpnr,NUMOFSYMBOLS)
vector pdf;
SPOINTER *rpnr;
struct symbol **lpnr;
int NUMOFSYMBOLS;
/* Subroutine for constructing a Huffman decoding tree
   pdf : vector of grayscale probabilities
   rpnr: pointer to root of decoding tree
   lpnr: pointer to array of decoding tree leaves
   NUMOFSYMBOLS : number of image symbols */
{
 int i,j;
 SPOINTER *nodes;
 int min[2];
 /* Prepare the construction of Huffman decoding tree */
 /* 'leaf' is an array of elements of type symbol that correspond
```

196   DIGITAL IMAGE COMPRESSION

```
         to Huffman tree leaves.
         'nodes' is an array of elements of type SPOINTER. One of them
         will point to the Huffman tree root after tree construction.
         The rest of them will be null. */
 if((*lpnr=(struct symbol *)
             malloc(NUMOFSYMBOLS*sizeof(struct symbol)))==NULL)
        return(-10);
 if((nodes=(SPOINTER *)
             malloc(NUMOFSYMBOLS*sizeof(SPOINTER)))==NULL)
        return(-10);
 /* Initialize 'leaf'.
    Make elements of 'nodes' to point to the elements of leaf.*/
 i=initialize(nodes,lpnr,pdf,NUMOFSYMBOLS);
 if(i) return(i);
 /* Construct Huffman decoding tree */
 for(i=NUMOFSYMBOLS-1;i>0;i--)
  {
    if(find2min(nodes,min,NUMOFSYMBOLS)) return(-310);
    if((nodes[(min[0]>min[1] ? min[1] : min[0])]
         = merge(nodes,min)) == NULL) return(-10);
    nodes[(min[0]>min[1] ? min[0] : min[1])] = NULL;
  }
 /* Check how many elements of 'nodes' are NULL. If more
 than one element of 'nodes' are not NULL or if all elements of
 'nodes' are NULL return decoding error. */
 *rpnr=NULL;
 for(i=0;i<NUMOFSYMBOLS;i++)
   if (nodes[i]!=NULL)
    {
       if (*rpnr==NULL) *rpnr=nodes[i];
       else return(-310);
    }
  if (*rpnr==NULL) return(-310);
  return(0);
 }


 int initialize(nodes,lpnr,pdf,NUMOFSYMBOLS)
 SPOINTER *nodes;
 struct symbol **lpnr;
 vector pdf;
 int NUMOFSYMBOLS;
 /* Subroutine for initialization of 'leaf' and 'nodes' elements
       nodes: array of elements of type SPOINTER. One of them
       will point to the Huffman tree root after tree
       construction.
```

lpnr: pointer to array of elements of type symbol that
   correspond to Huffman tree leaves.
   pdf: grayscale pdf
   NUMOFSYMBOLS : number of image symbols */
{
 int i;

 /* If pdf values are 0.0 add a small constant */
 for(i=0;i<NUMOFSYMBOLS;i++)
   {
    if(pdf[i]==0.0) (*lpnr)[i].prob=0.000001;
       else (*lpnr)[i].prob=pdf[i];
    (*lpnr)[i].comp[0]=NULL; (*lpnr)[i].comp[1]=NULL;
   }
 for(i=0;i<NUMOFSYMBOLS;i++) nodes[i] = &(*lpnr)[i];
 return(0);
}

int find2min(nodes,min,NUMOFSYMBOLS)
SPOINTER *nodes;
int min[2];
int NUMOFSYMBOLS;
/* Subroutine that finds the two elements pointed to by
   'nodes' that have the minimal probability.
   nodes: array of elements of type SPOINTER.
   min: output table containing elements pointing to
   the position of two elements of 'nodes' having minimal
   probability.
   NUMOFSYMBOLS : number of image symbols */
{
 int i;
 i=NUMOFSYMBOLS-1;
 while(nodes[i]==NULL && i>=0) i--;
 if (i<0) return(-310);
 min[0]=i;
 for(i=NUMOFSYMBOLS-1;i>=0;i--)
    if (nodes[i]!=NULL)
       if(nodes[i]->prob < nodes[min[0]]->prob) min[0]=i;
 i=NUMOFSYMBOLS-1;
 while((nodes[i]==NULL && i>=0) || i==min[0]) i--;
 if (i<0) return(-310);
 min[1]=i;
 for(i=NUMOFSYMBOLS-1;i>=0;i--)
    if (nodes[i]!=NULL && i!=min[0])

```
    if(nodes[i]->prob < nodes[min[1]]->prob) min[1]=i;
 return(0);
}

SPOINTER merge(nodes,min)
SPOINTER *nodes;
int min[2];
/* Subroutine to merge two elements of nodes.
   nodes: array of elements of type SPOINTER.
   min: output table containing elements pointing to the
   position of two elements of nodes having
   minimal probability. */

{
 SPOINTER p;
 if((p=(SPOINTER) malloc(sizeof(struct symbol)))==NULL)
    return(NULL);
 p->prob = nodes[min[0]]->prob + nodes[min[1]]->prob;
 p->comp[0] = min[0]>min[1] ? nodes[min[1]] : nodes[min[0]];
 p->comp[1] = min[0]>min[1] ? nodes[min[0]] : nodes[min[1]];
 return(p);
}

int treeaccess(root,stackpointer,stack,leaf,code)
SPOINTER root;
int *stackpointer;
char *stack;
struct symbol *leaf;
unsigned char * * code;
/* Subroutine to perform preorder access of Huffman tree
   root: tree root (input)
   stackpointer: number of bits in stack
   stack: stack of bits
   leaf: array of tree leaves
   code: table of strings containing codewords (output) */

{
 int i,j;

 if (root==NULL) return(-1);
 /* If a leaf is reached, store codeword contained
    in stack on table code */
 if(root->comp[0]==NULL)
  {
   i=(int)(root-leaf);
```

```
    if((code[i]=(unsigned char *)
     malloc((*stackpointer+1)*sizeof(char)))==NULL)return(-1);
    for(j=0;j<=*stackpointer;j++) code[i][j]=stack[j];
  }
  /* Otherwise access the two children of the current node */
  else
  {
    /* Add a zero to 'stack' for branch '0' */
    stack[*stackpointer] = 48;
    (*stackpointer)++;
    stack[*stackpointer] = 0;
    /* Visit the child that corresponds to branch '0' */
    i=treeaccess(root->comp[0],stackpointer,stack,leaf,code);
    if(i) return(i);
    /* Add a unit to 'stack' for branch '1' */
    stack[(*stackpointer)-1] = 49;
    /* Visit the child that corresponds to branch '1' */
    i=treeaccess(root->comp[1],stackpointer,stack,leaf,code);
    if(i) return(i);
    /* End with current node */
    (*stackpointer)--;
    stack[*stackpointer] = 0;
  }
  return(0);
}
```

Fig. 4.2.2 Subroutine for the construction of a Huffman code table.

The size of the Huffman codebook is $L = 2^B$ and the longest codeword may have up to L bits. Thus, some codewords may be extremely long, in the case of relatively large L. A practical modification of the Huffman code is to truncate it to a suitable length $L_1 < L$. The first L_1 intensity levels are Huffman coded. The remaining intensity levels are coded by a prefix code followed by a fixed-length code. This coding scheme is called *truncated Huffman code*.

Another modification of the Huffman code comes from the following representation of the intensity level i, $0 \leq i \leq L - 1$:

$$i = qL_1 + r \qquad 0 \leq q \leq \left\lfloor \frac{L-1}{L_1} \right\rfloor \qquad 0 \leq r \leq L_1 - 1 \qquad (4.2.5)$$

The first L_1 intensities $0 \leq i \leq L_1 - 1$ are Huffman coded. The remaining intensities are coded by a prefix (representing the quotient q) followed by the terminator code. The terminator code is the same Huffman code used for the intensities, $0 \leq i \leq L_1 - 1$. It represents the remainder r of (4.2.5).

200 DIGITAL IMAGE COMPRESSION

The histogram of image sequences (e.g. video images) evaluated over a relatively large time period tends to the uniform distribution, despite the fact that histograms of single frames may deviate considerably from uniformity. Thus, Huffman coding is not the best choice for raw image data coding and does not result in considerable data savings. However, it is very useful in combination with other coding schemes, for example, predictive coding, transform coding of grayscale images and run-length coding of binary images.

4.3 RUN-LENGTH CODING

Let us suppose that $(x_1, x_2, \ldots, x_M)$ are image pixels in an image line, as shown in Figure 4.3.1. Each image line consists of k segments having length l_i and gray level g_i, $1 \leq i \leq k$. Thus, it can be represented by the couples (g_i, l_i), $1 \leq i \leq k$:

$$(x_1, \ldots, x_M) \longrightarrow (g_1, l_1), (g_2, l_2), \ldots, (g_k, l_k) \qquad (4.3.1)$$

where:

$$g_1 = x_1 \quad g_k = x_M \qquad (4.3.2)$$

$$\sum_{i=1}^{k} l_i = M \qquad (4.3.3)$$

Each couple (g_i, l_i) is called a *gray-level run*. Representation (4.3.1) results in considerable compression if the gray-level runs are relatively large. The savings are even larger when the image is binary. If we assume that a line starts with a white run, the line can be represented as follows:

$$(x_1, \ldots, x_M) \longrightarrow l_1, l_2, \ldots, l_k \qquad (4.3.4)$$

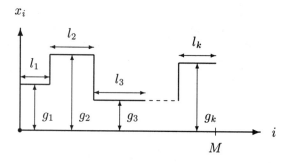

Fig. 4.3.1 Graphical representation of an image line.

Thus, only the run lengths must be encoded. The resulting coding scheme is one-dimensional and can be used for binary image compression. An *end of line* (EOL) codeword indicates the start of an image and the end of a line. The end of an image can be represented by a number (usually 6) of consecutive EOL codewords. The resulting image format is shown in Figure 4.3.2. The run-length coding has been standardized by CCITT and is included in Group 3 coding schemes used primarily for binary image transmission over public telephone networks for facsimile applications. The typical image size is A4 (210 mm×298 mm). The typical vertical sampling rates are 3.85 lines/mm (normal resolution) and 7.7 lines/mm (high resolution). The horizontal sampling rate is 200 dpi (dots per inch). This results in 1728 pixels per line for an A4 page. The horizontal sampling rate can be increased to 400–1000 dpi for certain applications. If the data of a line are very short, special *fill* bits consisting of zeros are concatenated to form a minimum line length. The recommended standard minimum is 96 bits per line for a transmission rate of 4800 bps (bits per second). The fill data concern transmitter–receiver synchronization and are discarded by the receiver. Therefore, they can be omitted if the run-length code is used for digital binary image storage. Runlengths have a specific probability distribution for each image. This pdf can be used to construct a Huffman code for run-length coding. Ideally, a specific Huffman code should be constructed for each binary image, but in practice this is unacceptable due to the delays caused by codebook construction and transmission. CCITT has chosen eight test documents that are typical in facsimile transmission and constructed a modified Huffman code based on the document statistics [JAI89]. The resulting codebook is shown in Table 4.3.1. Different codewords are used for black and white runs. The first 64 runs $[0, \ldots, 63]$ are Huffman coded and use a *terminating* codeword for transmission. The lengths between 64 and 1728 are transmitted by using a *make-up codeword* (MUC) followed by a terminating code. MUC represents an integer of the form $64k$, $k = 1, \ldots, 27$, that is equal to or less than the run-length l to be encoded. The terminating code that follows represents the number $l - 64k$. An extended Huffman code table is also provided to code document sizes up to A3. This size has 2560 pixels per line. Finally, the codebook also contains a special codeword for EOL.

| EOL | Data | EOL | Data | Fill | EOL | ⋯ | EOL | Data | EOL | EOL | EOL | EOL | EOL |

Fig. 4.3.2 Format of an image coded by a run-length code.

Table 4.3.1 Modified Huffman codebook for run-length coding (CCITT).

Run length	Terminating codewords White run	Terminating codewords Black run
0	00110101	0000110111
1	000111	010
2	0111	11
3	1000	10
4	1011	011
5	1100	0011
6	1110	0010
7	1111	00011
8	10011	000101
9	10100	000100
10	00111	0000100
11	01000	0000101
12	001000	0000111
13	000011	00000100
14	110100	00000111
15	110101	000011000
16	101010	0000010111
17	101011	0000011000
18	0100111	0000001000
19	0001100	00001100111
20	0001000	00001101000
21	0010111	00001101100
22	0000011	00000110111
23	0000100	00000101000
24	0101000	00000010111
25	0101011	00000011000
26	0010011	000011001010
27	0100100	000011001011
28	0011000	000011001100
29	00000010	000011001101
30	00000011	000001101000
31	00011010	000001101001

Run length	Terminating codewords White run	Terminating codewords Black run
32	00011011	000001101010
33	00010010	000001101011
34	00010011	000011010010
35	00010100	000011010011
36	00010101	000011010100
37	00010110	000011010101
38	00010111	000011010110
39	00101000	000011010111
40	00101001	000001101100
41	00101010	000001101101
42	00101011	000011011010
43	00101100	000011011011
44	00101101	000001010100
45	00000100	000001010101
46	00000101	000001010110
47	00001010	000001010111
48	00001011	000001100100
49	01010010	000001100101
50	01010011	000001010010
51	01010100	000001010011
52	01010101	000000100100
53	00100100	000000110111
54	00100101	000000111000
55	01011000	000000100111
56	01011001	000000101000
57	01011010	000001011000
58	01011011	000001011001
59	01001010	000000101011
60	01001011	000000101100
61	00110010	000001011010
62	00110011	000001100110
63	00110100	000001100111

Run length	Make-up codewords White run	Make-up codewords Black run
64	11011	0000001111
128	10010	000011001000
192	010111	000011001001
256	0110111	000001011011
320	00110110	000000110011
384	00110111	000000110100
448	01100100	000000110101
512	01100101	0000001101100
576	01101000	0000001101101
640	01100111	0000001001010
704	011001100	0000001001011
768	011001101	0000001001100
832	011010010	0000001001101
896	011010011	0000001110010
960	011010100	0000001110011
1024	011010101	0000001110100
1088	011010110	0000001110101
1152	011010111	0000001110110
1216	011011000	0000001110111
1280	011011001	0000001010010
1344	011011010	0000001010011

Run length	Make-up codewords White run	Make-up codewords Black run
1408	011011011	0000001010100
1472	010011000	0000001010101
1536	010011001	0000001011010
1600	010011010	0000001011011
1664	011000	0000001100100
1728	010011011	0000001100101
1792	00000001000	00000001000
1856	00000001100	00000001100
1920	00000001101	00000001101
1984	000000010010	000000010010
2048	000000010011	000000010011
2112	000000010100	000000010100
2176	000000010101	000000010101
2240	000000010110	000000010110
2304	000000010111	000000010111
2368	000000011100	000000011100
2432	000000011101	000000011101
2496	000000011110	000000011110
2560	000000011111	000000011111
EOL	000000000001	000000000001

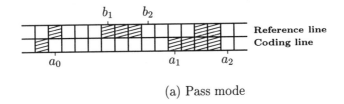

(a) Pass mode

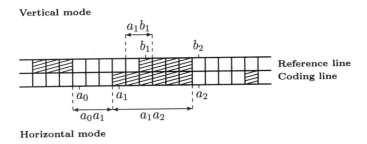

(b) Vertical and horizontal mode

Fig. 4.4.1 Transition elements in modified READ coding.

4.4 MODIFIED READ CODING

Run-length coding is a one-dimensional scheme that cannot take into account vertical correlations among run transitions in consecutive image lines. This correlation is exploited in the modified READ (*Relative Element Address Designate*) code. It is a two-dimensional coding scheme that codes a binary image line with reference to the previous line. The first image line can be coded in the conventional one-dimensional run-length coding scheme. One-dimensional coding is performed every k consecutive lines in order to stop errors propagating from one line to the rest of the image. The recommended values for k are 2 and 4 for documents scanned in normal resolution and high resolution respectively. The modified READ code has been standardized by CCITT for Group 3 facsimile transmission.

The modified READ coding scheme uses three transition elements a_0, a_1, a_2 in the current line and two transition elements b_1, b_2 in the reference line, as shown in Figure 4.4.1. The reference line has already been coded. The initial position of a_0 is the (imaginary) white transition pixel to the left of the first pixel in the current line. a_1, a_2 is the next pair of transitions to the right of a_0 in the present line; b_1, b_2 is the next pair of transitions to the right of a_0 in the reference line. The transition b_1 has opposite color to the transition a_0. The current position to be coded is a_1 and the coding will be done with reference to b_1, b_2 and a_0. Three possibilities (coding modes) can be used, as follows. The code table is given in Table 4.4.1.

Table 4.4.1 Code table of modified READ code.

Mode	Changing elements to be coded	Notation	Codeword
Pass	b_1, b_2	P	0001
Horizontal	$a_0 a_1, a_1 a_2$	H	$001 + M(a_0 a_1) + M(a_1 a_2)$
Vertical	a_1 just under b_1, $a_1 b_1 = 0$	$V(0)$	1
	a_1 to the right of b_1, $a_1 b_1 = 1$	$V_R(1)$	011
	a_1 to the right of b_1, $a_1 b_1 = 2$	$V_R(2)$	000011
	a_1 to the right of b_1, $a_1 b_1 = 3$	$V_R(3)$	0000011
	a_1 to the left of b_1, $a_1 b_1 = 1$	$V_L(1)$	010
	a_1 to the left of b_1, $a_1 b_1 = 2$	$V_L(2)$	000010
	a_1 to the left of b_1, $a_1 b_1 = 3$	$V_L(3)$	0000010
	EOL		000000000001
	1-d coding of next line		EOL+'1'
	2-d coding of next line		EOL+'0'

Pass mode. Let us suppose that b_2 is to the left of pixel a_1 (to be coded), as shown in Figure 4.4.1a. This means that the local black and white runs of the reference and the current line do not match. The pass mode uses only one codeword (0001), as can be seen in Table 4.4.1. Before doing the next coding, the pixel a_0 is set just below b_2 and b_1, b_2 are updated.

Vertical mode. Let us suppose that b_2 is to the right of a_1. Furthermore, we suppose that b_1 is within 3 pixels' distance from a_1 (to the right or to the left). Thus, a_1 can be coded with reference to b_1. The distance $a_1 b_1$ can take seven possible values: 0, right 1 2 3, left 1 2 3. In the vertical mode we assign the corresponding codeword from Table 4.4.1 (cases $V(0), V_R(i), i = 1, 2, 3, V_L(i), i = 1, 2, 3$). We update the reference pixel a_0 by making it equal to the current position a_1 ($a_0 = a_1$), update a_1, b_1, b_2 and proceed to the next coding step.

Horizontal mode. If pass mode or vertical mode is rejected, this means that a_1 is far from b_1, i.e. their distance $|a_1 b_1|$ is larger than 3 pixels. In this case, the position a_2 is detected and lengths $a_0 a_1$ and $a_1 a_2$ are coded by using the codebook of the modified Huffman code used in run-length coding (Table 4.3.1). The codeword is $001 + M(a_0 a_1) + M(a_1 a_2)$, where + denotes concatenation and $M(\cdot)$ denotes the Huffman table. During horizontal mode, both positions a_1 and a_2 are encoded. Therefore, the new reference position is $a_0 = a_2$. The positions a_1, b_1, b_2 are updated before the start of a new coding step.

All three modes stop when an end of line is reached, i.e. when $a_0 = 1729$ (for A4 pages). If a number $k = 2, 4$ of lines is coded in a two-dimensional way, the next line is coded in a one-dimensional way (run-length coding). Both run-length coding and modified READ coding are primarily binary image compression schemes, although some modifications and extensions can be used for gray-level image coding as well. In the following, we shall give a compression scheme that can be used equally well for binary and for grayscale image compression.

4.5 LZW COMPRESSION

The LZW algorithm was proposed by Lempel-Ziv and Welch as a general-purpose compression scheme [WEL84]. It can be used for the compression of any binary data file. It is currently one of the most popular digital image compression schemes and is incorporated in several *de facto* image storage standards (e.g. TIFF, GIF standard image files) [RIM90]. Its popularity is due to the fact that it is lossless, fast, effective and can operate on images of any bit depth.

LZW compression is based on the construction of a code table that maps frequently encountered bit strings to output codewords. The digital image is treated as a one-dimensional bit string. The coded image is treated as a bit string as well. The algorithm produces the output string and updates the code table simultaneously. Thus, the codebook can adapt to the particular characteristics of the image to be compressed. The code table can have up to 4096 codewords and thus each codeword can be up to 12 bits long. The first 256 codewords $[0,\ldots,255]$ contain the numbers $0\ldots 255$. The codeword #256 contains a special *Clear code*. Its use will be described later on in this section. The codeword #257 contains an *End of information* (EOI) code. It is used to denote the end of file of an image or the end of an image strip (if the image is segmented into strips). The codewords from #258 to #4095 contain bit strings that are frequently encountered in the image. The structure of the LZW codebook is shown in Figure 4.5.1. The first 512 table elements are represented by 9 bits each. The codewords from #512 to #1023 are 10 bits long. Similarly, we switch to 11 bit codewords at #1024 and to 12 bit codewords at #2048. The part of the codebook from #258 to #4095 is constructed 'on the fly'.

The encoding algorithm is relatively simple, as can be seen in Figure 4.5.2 [TIF88]. It uses the items: `code table T`, `prefix string P` and `current string S`. The characters read from the input image are stored on a byte C.

```
#define    TRUE    1
#define    FALSE   0

typedef struct llink * LPOINTER;
struct llink {
  int n;
  LPOINTER next;
};
```

0	0
⋮	⋮
255	255
Clear code	256
EOI code	257
string	258
string	259
⋮	⋮
string	4095

Fig. 4.5.1 LZW codebook.

```
int lzw(im,fw,Ns,Nc,N,M)
image im;
FILE *fw;
int Ns,Nc,N,M;
/* Subroutine for LZW encoding
    im     : image array to be encoded
    fw     : output file
    Ns     : number of image symbols (usually 256)
    Nc     : number of code strings  (usually 4096)
    N      : number of rows
    M      : number of columns */
{
    int i,j,h,tablepointer,c,oldc,maxcode,codesize,ENDOUT,
        Plength, * T, * Tlength;
    unsigned int RemainingBits,BitsRead,item;
    LPOINTER *llinks;
    unsigned char C;
    /* T:           code table
       Tlength:     array of code string lengths
       llinks:      array of pointers to all code strings
                    having the same prefix
       tablepointer: pointer to the last location of T
```

```
       Plength :     length of the current prefix P
       C:            current pixel
       c:            position of (P+C) in T
       oldc:         position of P in T
       codesize:     current number of bits sent to output
                     per pixel (it varies from 9 to 12)
       maxcode:      2^(codesize) */

/* Allocate memory for code table */
if (( T=(int *)malloc(Nc*sizeof(int)) )==NULL)
   return(-10);
if (( Tlength=(int *)malloc(Nc*sizeof(int)) )==NULL)
   return(-10);
if (( llinks=(LPOINTER *)malloc(Nc*sizeof(LPOINTER)) )==NULL)
   return(-10);

/* Initializations */
tablepointer=INT_MAX;
oldc=INT_MAX;
RemainingBits=16; BitsRead=0;
item=0;
ENDOUT=FALSE;
init_lzw(Ns,Nc,&tablepointer,T,Tlength,&Plength,
         &maxcode,&codesize,llinks);
/* Send clear code to output. */
send2output(fw,(unsigned int)Ns,codesize,&RemainingBits,
            &BitsRead,ENDOUT,&item);

/* Main loop of LZW algorithm */
for (i=0; i<N; i++)
  for (j=0; j<M; j++)
    {
    C=im[i][j];
    /* Check if (P+C) is in T */
    if ((c=inT(T,Tlength,Plength,C,oldc,llinks))>=0)
      {
      /* If true, P=P+C. Increase Plength. */
      Plength++;
      }
    else
      {
      /* If false, write oldc in output file, add P+C in T,
         and make P=C. */
      send2output(fw,(unsigned int)oldc,codesize,
                  &RemainingBits, &BitsRead, ENDOUT,&item);
```

```
            h=AddToTable(T,oldc,(int)C,Tlength,Plength,
                    &tablepointer,&codesize,&maxcode,Nc,llinks);
            if (h==-10) return(-10);
            /* Check if code table has been cleared. */
            if(h==-340)
              {
                send2output(fw,(unsigned int)Ns,codesize,
                    &RemainingBits,&BitsRead,ENDOUT,&item);
                init_lzw(Ns,Nc,&tablepointer,T,Tlength,&Plength,
                        &maxcode,&codesize,llinks);
              }
            c=(int)C;
            Plength=1;
          }
         oldc=c;
      }

   /* Send current prefix position to output. */
   send2output(fw,(unsigned int)oldc,codesize,
       &RemainingBits,&BitsRead,ENDOUT,&item);
   /* Send EOI code to output. */
   send2output(fw,(unsigned int)(Ns+1),codesize,
       &RemainingBits,&BitsRead,ENDOUT,&item);
   /* After the completion of coding, send what has been left
      in 'item' to output (code=0, is irrelevant). */
   ENDOUT=TRUE;
   send2output(fw,0,codesize,&RemainingBits,
       &BitsRead,ENDOUT,&item);

   return(0);
}


int init_lzw(Ns,Nc,tablepointer,T,Tlength,Plength,
             maxcode,codesize,llinks)
int Ns,Nc,*tablepointer, * T, * Tlength,*Plength,
    *maxcode,*codesize;
LPOINTER *llinks;
/* Subroutine that initializes code table and current prefix
   Ns : number of image symbols
   Nc : number of code table symbols
   tablepointer : position of last entry to code table
   T : code table
   Tlength : table of code string lengths
   Plength : length of current prefix P
```

```
        maxcode : maximum code that can be represented
                 with codesize bits
        codesize :current length (in bits) of the code that is read
        llinks :array of the 'children' of each code table string*/
{
    int i,j;
    LPOINTER p;

    /* If subroutine is called for the first time,
       fill first Ns+2 table elements */
    if(*tablepointer==INT_MAX)
      for (i=0;i<=Ns+1;i++)
        {
          T[i]=i; Tlength[i]=1;
        }
    else
      /* If subroutine is called for reinitialization, free all
         table elements after Ns+2 */
      for (i=0;i<Nc;i++)
        for (;llinks[i]!=NULL;)
          { p=llinks[i]->next; free(llinks[i]); llinks[i]=p; }

    for (i=0;i<Nc;i++) llinks[i]=NULL; /* For the first time */
    *Plength=0; *tablepointer=Ns+1;
    *maxcode = Ns<<1; j = 0; i = *maxcode-1;
    while(i) { i = i>>1; j++; }
    *codesize = j;

    return(0);
}


int inT(T,Tlength,Plength,symbol,old,llinks)
int * T, * Tlength, old,Plength;
unsigned char symbol;
LPOINTER *llinks;
/* Subroutine that searches code table to check if
   it includes current prefix appended with 'symbol'

        T:      code table
        symbol: input symbol (C)
        old:    position of current prefix (P) in code table
        llinks: array of the 'children' of each code table string */
{
    int i;
```

```
    LPOINTER p;

    if (old==INT_MAX) return((int)symbol);
    /* Perform serial search of the 'children' of the string
       that corresponds to current prefix (T[old]) */
    if (llinks[old]!=NULL)
     {
       for (p=llinks[old];p!=NULL;p=p->next)
        {
          i=p->n;
          if (Tlength[i]==Plength+1)
            if(T[i]==(int)symbol) return(i);
        }
     }
    return(-310);
}


int AddToTable(T,pos,c,Tlength,Plength,tablepointer,codesize,
               maxcode,Nc,llinks)
int * T, pos, c, * Tlength, Plength, *tablepointer,
    *codesize, *maxcode,Nc;
LPOINTER *llinks;
/* Subroutine that adds T(pos)+c to table at the position
   indicated by tablepointer.
   T:             code table
   pos:           position of string to be written to T
   c:             ASCII code of character to be appended to T(pos)
   tablepointer: position of last entry to T
   codesize:      current length (in bytes) of code that is read
   maxcode:       2^(codesize+1)
   Nc:            number of code table symbols
   llinks : array of the 'children' of each code
            table string */
{
    LPOINTER p;

    (*tablepointer)++;
    if (*tablepointer==*maxcode) {(*maxcode)<<=1; (*codesize)++;}
    T[*tablepointer]=c;
    Tlength[*tablepointer]=Plength+1;
    /* update llinks */
    if (llinks[pos]!=NULL)
     {
       for (p=llinks[pos];p->next!=NULL;p=p->next);
```

```
      if ((p->next=(LPOINTER)
         malloc(sizeof(struct llink)))==NULL) return(-10);
      p=p->next;
   }
   else
   {
      if ((llinks[pos]=(LPOINTER)
         malloc(sizeof(struct llink)))==NULL) return(-10);
      p=llinks[pos];
   }
   p->n=*tablepointer;
   p->next=NULL;
   if (*tablepointer==Nc-1) return(-340);
   return(0);
}


int send2output(fw,outcode,size,RemainingBits,
         BitsRead,ENDOUT,item)
FILE *fw;
unsigned int outcode, *RemainingBits, *BitsRead, *item;
int size,ENDOUT;
/* Subroutine that sends packed codewords in 2 bytes
   to the output file.
   fw:       output file
   outcode:  code table position that must be sent to
             the output file.
   size:     the length of outcode in bits that is
             currently used.
   *RemainingBits: free bits in the 2 byte word that
             must be filled.
   *BitsRead :how many bits in the 2 byte word have
             been filled.
   ENDOUT:   binary flag that becomes TRUE just before
             the end of the compression to output what remains
             in (*item);
   *item:    the 2 byte word that is sent to
             the output file. */
{
   unsigned int StartBit,BitstobeRead,temp;
   if (!ENDOUT)
   {
    if(*RemainingBits==16)
      {
         StartBit=size-1; BitstobeRead=size;
         *item=(outcode>>(StartBit+1-BitstobeRead))
```

```
                 & ~(~0<<BitstobeRead);
        *BitsRead=BitstobeRead; *RemainingBits-=(*BitsRead);
      }
    else
     {
      if (*RemainingBits>size)
        {
         StartBit=size-1; BitstobeRead=size;
        }
      else
        {
         StartBit=(*RemainingBits)-1;
         BitstobeRead= (*RemainingBits);
        }
      temp=(outcode>>(StartBit+1-BitstobeRead))
              & ~(~0<<BitstobeRead);
      temp<<=(*BitsRead); *item+=temp;
      *BitsRead+=BitstobeRead; *RemainingBits-=BitstobeRead;
      if (*BitsRead==16)
        {
         fwrite(item,sizeof(unsigned int),1,fw);
         StartBit=size-1; BitstobeRead=size-BitstobeRead;
         *RemainingBits=16;
         if (BitstobeRead>0)
           {
             *item=(outcode>>(StartBit+1-BitstobeRead))
                    & ~(~0<<BitstobeRead);
             *BitsRead=BitstobeRead;
             *RemainingBits-=BitstobeRead;
           }
        }
     }
   }
  else
   {
    if (*item<256)
     fwrite(item,sizeof(unsigned char),1,fw);
    else
     fwrite(item,sizeof(unsigned int),1,fw);
   }
}
```

Fig. 4.5.2 LZW compression algorithm.

The subroutine init_lzw() initializes the code table by creating the first 258 entries, as shown in Figure 4.5.2. The prefix string is set equal to empty. Subroutine send2output() sends coded information to the output file. This information is packed in the 16 bit buffer item, before it is sent to the output file. The Clear code is sent to indicate the start of a new image (or new image strip). After these initializations are performed, the algorithm works in an iterative manner. At each step, an image byte is read and the current string is created by concatenating the prefix string with the new image byte (S=P+C). If the current string is found in the code table the prefix is updated (P=S) and the procedure is repeated in an effort to find the longest possible string in the code table that is 'similar' to the current string. If the current string is not found in the code table, this means that only the prefix can be found in the code table. Therefore, the codeword of the prefix is sent to the output stream (encoded image). The code table is appended by adding the current string and the prefix is reinitialized to the image pixel just read (P=C). When the end of the image (or image strip) is encountered, the End of information (EOI) code is sent to the output. Each element of T can have a number of 'children', that is, code strings that have the same prefix and have been formed by concatenating an existing code string T[i] with a character (e.g. a pixel value). Thus, a tree-like structure can be superimposed on table T in order to facilitate the search in T. This structure may have more than one root. The search of the current string in the code table T is performed by using subroutine inT() and the array llinks of integer lists. Each element llinks[i] of this array corresponds to code table string T[i] and contains a list of the positions of the children of T[i] in T (i.e. a list of all code strings having the same prefix). The resulting tree structure is exploited in inT() to speed up the search in T. Furthermore, this tree structure allows the storage only of the last character of the corresponding codeword in T[i]. The entire codeword corresponding to T[i] can be constructed by following in llinks the path that leads to T[i]. New elements in T are appended by using AddToTable(). This subroutine appends the new element in T and updates the children of the T elements in table llinks. It is clearly seen that the codebook is built up progressively and can adapt to the image (or image strip) characteristics. If the image is split into strips, each strip has its own codebook. If the code table size becomes 4096, the Clear code is sent to the output and the code table is reinitialized.

LZW decompression follows exactly the opposite procedure. No transmission or storage of the codebook is required; the codebook is retrieved progressively during decompression. An LZW decompression algorithm is shown in Figure 4.5.3 [TIF88]. First, initialization is performed. The initial code length size is determined and the code table T is allocated and initialized by creating its first 258 entries. The routine GetNextCode() reads the next codeword from the encoded image. It is very important to read the correct number of bits. The routine starts by reading 9-bit codewords. As the code table is progressively built up, the number of bits per codeword increases by one at table sizes 511, 1023 and 2047. If a Clear code is encountered, the code table T is initialized. If the string T[code] that corresponds to the new codeword code exists in the code table, the corresponding string is sent to the output and the code table is updated by adding an entry formed by the string T[old] that corresponds to the previous codeword old and the first character of T[code]. If the string that corresponds to the new codeword does not exist in the code table, the output string is formed by concatenating T[old] and the first character of T[old]. This string is sent to the output and also added to the code table. The subroutine output_string() uses code table T to decode image pixels and to send them to the corresponding positions in the output image buffer. This procedure is repeated until an EOI code is encountered. It must be noted that no tree-like structure is needed for LZW decoding. Thus, the code table T is a two-dimensional array of bytes. Several string manipulation routines presented in Figure 4.5.3 are used in the decompression algorithm.

```
#define TRUE  1
#define FALSE 0

unsigned char CC[]  = {"<CC>"};
unsigned char EOI[] = {"<EOI>"};

int lzw_decompress(fp,a,N1,M1,N2,M2,BitsPerSample,Tlength)
FILE *fp;
image a;
int N1,M1,N2,M2,BitsPerSample,Tlength;
/*  Subroutine that decodes an LZW-encoded image.

        fp              : input file containing LZW-encoded image
        a               : decoded image buffer
        N1,M1           : start coordinates
        N2,M2           : end coordinates
        BitsPerSample   : number of bits per pixel in the decoded
                          image (default: 8)
        Tlength         : code table length (default: 4096)*/
{
   int i,ii,j,count,maxcode,code,old, CONT,FIRST_CODE,
       ReadFromInput;
   unsigned int size, isize, temp, item, Nlevels,RemainingBits,
                BitstobeRead,BitsRead;
   unsigned char * aux, * prefix, * prefixQ;
   unsigned char * * T;
   long frem;
   FILE *fp;
```

```
frem=filelength(fileno(fp));
/* Initializations */
/* Initial codesize */
switch(BitsPerSample)
 {
  case 1 : size=3; break;
  default : size=BitsPerSample+1; break;
 }
isize=size; Nlevels=1 << BitsPerSample; maxcode=Nlevels<<1;
/* Allocation of the code table */
T=(unsigned char * *)
  malloc(Tlength*sizeof(unsigned char *));
/* Default code table: ASCII table */
for (i=0;i<Nlevels;i++) T[i]=asciidup(i);
T[Nlevels]=ustrdup(CC); T[Nlevels+1]=ustrdup(EOI);
for (i=Nlevels+2;i<Tlength;i++)
  T[i]=(unsigned char *)NULL;
i=N1; j=M1;
RemainingBits=16; BitsRead=0; BitstobeRead=size;
temp=0; item=0;
ReadFromInput=TRUE; CONT=TRUE;
prefix=(unsigned char *)NULL;
prefixQ=(unsigned char *)NULL;
/* Main decoding loop */
do
 {
   code=GetNextCode(fp,&frem,size, &RemainingBits,&BitsRead,
        &BitstobeRead,&temp,&item,&ReadFromInput);
   if (code==Nlevels)
     { /* RESET */
       /* Reset code length (size), count and maximum code
          (maxcode) to their initial values.
          Set the flag FIRST_CODE TRUE.
          Clear all the table entries above (Nlevels+2). */
       size=isize; count=Nlevels+2; maxcode=Nlevels<<1;
       FIRST_CODE=TRUE;
       for (ii=Nlevels+2;ii<Tlength;ii++)
        if (T[ii]!=(unsigned char *)NULL)
          {
            free((unsigned char *)T[ii]);
            T[ii]=(unsigned char *)NULL;
          }
     }
    else
     if (code==Nlevels+1)
```

```
        { /* EOI. Stop the decoding procedure */
          CONT=FALSE;
        }
      else
        { /* Ordinary code */
          if (FIRST_CODE)
            {
              /* First code after a RESET has occurred */
              output_string(T[code],a,M2,&i,&j);
              /* Output the string for code to the image buffer */
              old=code;
              FIRST_CODE=FALSE;
            }
          else
            {
              if (T[code]!=(unsigned char *)NULL)
                {
                  /* The string T[code] exists in the code table.*/
                  output_string(T[code],a,M2,&i,&j);
                  /* Output the string for code to the
                     image buffer */
                  prefix=ustrdup(T[old]);
                  /* prefix is the translation for old code */
                  aux=asciidup((int)T[code][0]);
                  /* aux is the first character of the
                     string T[code]*/
                  prefixQ=(unsigned char *)malloc(
                      (ustrlen(prefix)+2)*sizeof(unsigned char));
                  ustrcpy(prefixQ,prefix); ustrcat(prefixQ,aux);
                  free((unsigned char *)prefix);
                  /* prefixQ is the concatenation of
                     prefix and aux */
                  prefix=ustrdup(prefixQ);
                  /* prefixQ is copied to prefix */
                  T[count]=ustrdup(prefix);
                  /* Add prefix to the code table */
                  old=code;
                }
              else
                {
                  /* The string T[code] does not exist in the
                     code table. */
                  prefix=ustrdup(T[old]);
                  /* prefix is the translation of
                     the old code */
```

```
                aux=asciidup((int)prefix[0]);
                /* aux is the first character of prefix */
                prefixQ=(unsigned char *)malloc(
                   (ustrlen(prefix)+2)*sizeof(unsigned char));
                ustrcpy(prefixQ,prefix);
                 ustrcat(prefixQ,aux);
                free((unsigned char *)prefix);
                /* prefixQ is the concatenation of
                   prefix and aux */
                prefix=ustrdup(prefixQ);
                /* prefixQ is copied to prefix */
                output_string(prefix,a,M2,&i,&j);
                /* Output the string prefix (i.e. prefixQ)
                   to the image buffer */
                T[code]=ustrdup(prefix);
                /* and add it to code table */
                 old=code;
              }
              /* Clear prefix, aux, and prefixQ */
              free((unsigned char *) prefix);
              free((unsigned char *)aux);
              free((unsigned char *)prefixQ);
              count++;
              if (count==maxcode) { size++; maxcode<<=1; }
           }
         }
    } while(CONT);
    fclose(fp);
    for (ii=0;ii<Tlength;ii++)
      if (T[ii]!=(unsigned char *)NULL)
        {
         free((unsigned char *)T[ii]);
          T[ii]=(unsigned char *)NULL;
        }
    free((unsigned char * *)T);
    return(0);
}

int GetNextCode(fp,frem,size,RemainingBits,BitsRead,
     BitstobeRead,temp,item,ReadFromInput)
FILE *fp;
long *frem;
unsigned int size, *RemainingBits, *BitsRead, *BitstobeRead,
             *temp, *item;
int *ReadFromInput;
```

```
/* Subroutine that retrieves codewords of length size from
   2 byte file records. It returns the current codeword.

   fp:    pointer to the file of input LZW-encoded image
   frem:  how many bytes of the input file that are still unread
   size:  length of the current codeword
   *RemainingBits: unread bits of the 2 byte file record
   *BitsRead: already read bits of the 2 byte file record
   *BitstobeRead:  unread bits of the current codeword
   *temp:    portion of the current codeword that has already
             been read
   *item:          2 byte file record
   *ReadFromInput: binary flag that becomes TRUE if and only
                   if the current codeword is split into two
                   consecutive 2 byte file records. */
{
  unsigned int code;
  unsigned char titem;

  ONCE_MORE:
  if (*ReadFromInput)
    {
    if (*frem==1L)
      {
      fread(&titem,sizeof(unsigned char),1,fp); (*frem)--;
      *item=titem;
      }
    else
      { fread(item,sizeof(unsigned int),1,fp); (*frem)-=2; }
    }
  if (*RemainingBits==16)
    {
    code=(*item) & ~(~0 << (*BitstobeRead));
    code <<= (*BitsRead);
    code += (*temp);
    (*BitsRead) += (*BitstobeRead);
    (*RemainingBits) -= (*BitstobeRead);
    (*ReadFromInput)=FALSE;
    }
  else
    {
    if (*RemainingBits<size)
      {
      if ((*RemainingBits)+(*BitsRead)<=16)
```

```
                *temp= ((*item) >> (*BitsRead))
                   & ~(~0 << (*RemainingBits));
           else
                *temp= ((*item) >>(*BitstobeRead))
                    & ~(~0 << (*RemainingBits));
           *BitsRead=(*RemainingBits);
           (*BitstobeRead)=size-(*BitsRead);
           *RemainingBits=16;
           (*ReadFromInput)=TRUE;
           goto ONCE_MORE;
          }
       else
         {
         code=((*item) >> (*BitstobeRead)) & ~(~0 << size);
         *BitsRead=size+(*BitstobeRead); *RemainingBits-=size;
         (*ReadFromInput)=FALSE;
         }
      }
   return((int)code);
}


int output_string(string,a,M2,row,column)
unsigned char * string;
image a;
int M2, *row, *column;
/* Subroutine that outputs a pixel-stream (i.e. a string of
   unsigned char) to the image buffer.

     string :   pixel-stream that is output to the image buffer
     a      :   image buffer
     M2     :   end image column coordinate
     *row
     *column:   current image coordinates */
{

   int k,i,j;

   i=*row; j=*column;
   k=0;
   while (string[k]!=(unsigned char)'\0')
     {
      a[i][j]=string[k];
      j++;
      if (j==M2) {j=0; i++;}
```

```
     k++;
   }
   *row=i; *column=j;
   return(0);
}

unsigned char * asciidup(i)
int i;
/* Subroutine that produces a string composed of a single
   unsigned character (i-th ASCII character) terminated by '\0'.*/
{
  unsigned char * t;
  t=(unsigned char *)malloc(2*sizeof(unsigned char*));
  t[0]=(unsigned char)i; t[1]=(unsigned char)'\0';
  return(t);
}

unsigned  char * ustrdup(s1)
unsigned char * s1;
/* Subroutine that copies a string of unsigned char's s1 to
     another that is allocated internally.
     It returns the destination string. */
{
  unsigned char * s2;
  int uslen,i;
  uslen=(int)ustrlen(s1); uslen++;
  s2=(unsigned char *)malloc(uslen*sizeof(unsigned char));
  i=0;
  while ( (s2[i] = s1[i])!= (unsigned char)'\0') i++;
  return(s2);
}

int ustrlen(s)
unsigned char * s;
/* Subroutine that evaluates the length of a string of
     unsigned char's. It returns the string length. */
{
  int i; i=0;
  while (s[i] != (unsigned char)'\0') i++;
  return(i);
}

unsigned  char * ustrcpy(s,t)
unsigned char * s, * t;
/* Subroutine that copies the source string of unsigned char's
```

```
    (t) to the destination string (s).
    It returns the destination string s. */
{
  int i; i=0;
  while ( (s[i] = t[i])!= (unsigned char)'\0') i++;
  return(s);
}

unsigned char * ustrcat(s,t)
unsigned char  *s,   *t;
/* Subroutine that appends the strings of unsigned char's
    s with t. It returns the concatenated string s. */
{
  int i,j; i=j=0;
  while ( s[i] != (unsigned char)'\0') i++;
  while ((s[i++]=t[j++])!= (unsigned char)'\0') ;
  return(s);
}
```

Fig. 4.5.3 LZW decompression.

The speed of both LZW compression and decompression depends on the implementation of the code table and on the efficient searching in it. Various such implementations can be devised and usually hashing or tree-based techniques are used for this search [HOR84], [KNU73]. The decompression is usually faster because no search is needed. The output string can be constructed directly by using the corresponding codewords. From a compression point of view, the compression ratio that can be achieved ranges from 1:1.5 to 1:3. Generally speaking, substantial compression can be obtained for binary or bitmap images. LZW encoding of raw grayscale images does not lead to substantial compression in most cases.

4.6 PREDICTIVE CODING

One way to describe information redundancy in digital images is to use predictability in local image neighborhoods. Let $f(n,m)$ be the digital intensity at pixel position (n,m) and $f(n+i, m+j)$, $(i,j) \in A$, the intensities in a local image neighborhood described by the set A. The pixel intensity $f(n,m)$ can be predicted from the intensities of its neighboring pixels, if the image data are correlated within a certain image neighborhood (which is usually the case). The *predictor rule L*:

$$\hat{f}(m,n) = L(f(n-i, m-j), (i,j) \in A, (i,j) \neq (0,0)) \qquad (4.6.1)$$

is usually a linear function of the pixel intensities within the window A. Let us suppose that the window A has a form corresponding to a causal predictor, like the ones shown in Figure 4.6.1. If such a window scans the image plane in a row-wise manner from left to right and from the top to the bottom, it covers only image pixels that belong to the past, that is, the ones that have already been scanned. In this case, we can make a prediction $\hat{f}(m,n)$ based on already reconstructed pixel values $f_r(n-i, m-j)$ in the past:

$$\hat{f}(n,m) = L(f_r(n-i, m-j), \ (i,j) \in A) \qquad (4.6.2)$$

A recursive predictive coding technique can be based on (4.6.2). Let us suppose that an optimal prediction rule L is obtained and that border values are available so that recursion (4.6.2) can be implemented. At each step, it is sufficient to code the error:

$$e(n,m) = f(n,m) - \hat{f}(n,m) \qquad (4.6.3)$$

instead of the pixel intensity $f(n,m)$. If $e_q(n,m)$ is the quantized and coded value of the error $e(n,m)$, the pixel value $f(n,m)$ can be reconstructed as follows:

$$f_r(n,m) = L(f_r(n-i, m-j), \ (i,j) \in A) + e_q(n,m) \qquad (4.6.4)$$

If the prediction $\hat{f}(n,m)$ is good, the error term $e(n,m)$ has a small dynamic range and can be coded by using a relatively small codeword. Thus substantial compression can be achieved. The transmission of the prediction coefficients and of the coded error is needed for the reconstruction of $f(n,m)$ by using (4.6.4), as can be seen in Figure 4.6.2. This coding scheme, called *predictive differential pulse code modulation* (DPCM), is extensively used in telecommunications for speech and image encoding. DPCM is a lossy coding scheme, because of the use of a quantizer in the encoding process. The quantization

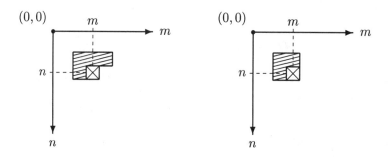

Fig. 4.6.1 Causal windows used in image prediction.

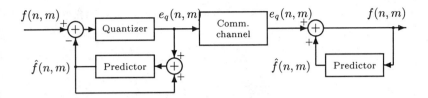

Fig. 4.6.2 Predictive differential pulse code modulation (DPCM).

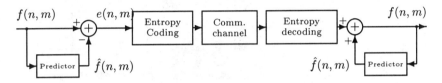

Fig. 4.6.3 DPCM with entropy coding.

of the error signal always creates an irrecoverable amount of distortion. However, distortionless DPCM schemes can be devised as well, as shown in Figure 4.6.3. The digital image is already quantized to a number of intensity levels (typically $0\ldots 255$). The predictor can be forced to produce integer output. In this case the error signal $e(n,m)$ can be encoded by using entropy coding techniques (e.g. Huffman coding) without any quantization. Entropy decoding followed by DPCM decoding results into distortionless image transmission.

The performance of the DPCM depends greatly on the predictor used and on the choice of its coefficients. The simplest approach is to perform one-dimensional DPCM of each image line. Let us suppose that the image line $f(n,m)$ (represented by $f(m)$ for notational simplicity) can be modeled as a stationary AR process [JAY84]:

$$f(m) = \sum_{k=1}^{p} a(k)f(m-k) + \epsilon(m) \qquad E[\epsilon^2(m)] = \sigma^2 \qquad (4.6.5)$$

where $\epsilon(m)$ is a white additive Gaussian noise uncorrelated to $f(m)$. A natural choice for the prediction scheme (4.6.2) is the following:

$$\hat{f}(m) = \sum_{k=1}^{p} a(k) f_r(m-k) \qquad (4.6.6)$$

The predictor coefficients form the vector $\mathbf{a} = [a(1),\ldots,a(p)]^T$. The prediction window A contains the reconstructed pixels $\{f_r(m-1),\ldots,f_r(m-p)\}$. The quantized error signal:

$$e_q(m) = Q[e(m)] = Q[f(m) - \hat{f}(m)] \qquad (4.6.7)$$

is transmitted to the receiver. The image row is reconstructed by using (4.6.4):

$$f_r(m) = \sum_{k=1}^{p} a(k) f_r(m-k) + e_q(m) \qquad (4.6.8)$$

The prediction coefficients can be chosen by solving the set of *normal equations*:

$$\begin{bmatrix} R(0) & R(1) & \cdots & R(p-1) \\ R(1) & R(0) & \cdots & R(p-2) \\ \vdots & \vdots & \vdots & \vdots \\ R(p-1) & R(p-2) & \cdots & R(0) \end{bmatrix} \begin{bmatrix} a(1) \\ a(2) \\ \vdots \\ a(p) \end{bmatrix} = \begin{bmatrix} R(1) \\ R(2) \\ \vdots \\ R(p) \end{bmatrix} \qquad (4.6.9)$$

where $R(k)$ is the autocorrelation function of the signal $f(m)$ [JAY84].

One-dimensional prediction models can be extended to two-dimensional models of the form [JAI89], [LIM90]:

$$\hat{f}(n,m) = \sum_{(i,j) \in A} \sum a(i,j) f(n-i, m-j) \qquad (4.6.10)$$

The 2-d AR prediction model coefficients can be optimally chosen by minimizing the mean square error:

$$E[|\, f(n,m) - \sum_{(i,j) \in A} \sum a(i,j) f(n-i, m-j) \,|^2] \qquad (4.6.11)$$

This minimization leads to the solution of a set of normal equations of the form:

$$R(k,l) = \sum_{(i,j) \in A} \sum a(i,j) R(k-i, l-j) \qquad (4.6.12)$$

where $R(k,l)$ is the 2-d autocorrelation function of the image $f(n,m)$. The solution of the system (4.6.12) leads to the optimal model coefficients $a(i,j)$, $(i,j) \in A$. Once these coefficients are known, the error image:

$$e(n,m) = f(n,m) - \hat{f}(n,m) \qquad (4.6.13)$$

can be obtained and quantized. The digital image can be reconstructed at the receiver by using (4.6.4).

$$f_r(n,m) = \sum_{(i,j) \in A} \sum a(i,j) f_r(n-i, m-j) + e_q(n,m) \qquad (4.6.14)$$

An algorithm will be described in this section that computes the AR coefficients for a two-dimensional non-symmetric half plane (NSHP) prediction window of the form shown in Figure 4.6.4. The size of this window is $(LL + LR + 1) \times K + LL + 1$ pixels. The window shown in Figure 4.6.4

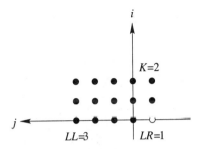

Fig. 4.6.4 Non-symmetric half plane prediction window.

has $LL = 3$, $LR = 1$, $K = 2$ and size 14 pixels. The prediction coefficients $a(i,j)$ are supposed to be stored in a one-dimensional array **A** of size $(LL + LR + 1) \times K + LL + 1$ elements in a row-wise manner from down to top: $\mathbf{A} = [a(0,0), \ldots, a(0, LL), a(1, -LR), \ldots, a(1, LL), \ldots, a(K, -LR), \ldots, a(K, LL)]^T$.

The resulting algorithm arcoefficients() is shown in Figure 4.6.5. This subroutine employs corsample() to calculate the autocorrelation $R(k,l)$ of the image to be modelled. The symmetries of the two-dimensional autocorrelation matrix elements are exploited to minimize the number of calls of corsample(). The normal equations (4.6.12) take the matrix form $\mathbf{RA} = \mathbf{B}$. This system is solved by the subroutine arsolve() and the model coefficients are calculated. The subroutine arcoefficients() also calculates the variance σ^2 of the driving white noise process ϵ.

```
int arcoefficients(ima,N,M,K,LL,LR,A,Pu)
image ima;
int N,M,K,LL,LR;
float A[];
float *Pu;
{
/* Subroutine for 2-d AR model coefficient calculation.
   The model can be non-symmetric half plane (NSHP)
   or quarter plane
   ima: input image buffer
   N,M: size of the image (end coordinates)
   K,LL,LR: size of the support region of the AR model
   A: AR model coefficients (output)
   Pu: variance of the driving white noise process (output) */

   int k,l,i,j;
   int c,c1,c2;
   float *AA;
   float *BB;
```

```
    float * * RR;
    float corsample();

/* Create a temporary matrix and vectors.
   RR: correlation matrix
   AA: coefficient vector
   BB: vector that is used in the Yule-Walker equation
        RR*AA=BB */
    RR=matf2((K+1)*(LL+LR+1),(K+1)*(LL+LR+1));
    AA=(float *)malloc(((K+1)*(LL+LR+1))*sizeof(float));
    BB=(float *)malloc(((K+1)*(LL+LR+1))*sizeof(float));

    /* Calculate the elements of the correlation matrix RR.
       Matrix symmetries and the block Toeplitz form are
       employed. Step 1 */
    j=-1;
    for(k=0; k<=K; k++)
    for(l=0; l<=LL+LR; l++)
        { j++; RR[0][j]=corsample(ima,ima,N,M,k,l); }
    /* Step 2 */
    for(k=0; k<=K; k++)
        { j=k*(LL+LR+1);
          for(l=1; l<=LL+LR; l++)
            RR[1][j]=corsample(ima,ima,N,M,k,-l);
        }
    /* Step 3 */
    for(i=1; i<=LL+LR; i++)
     { j=-1;
       for(k=0; k<=K; k++)
        { j++;
          for(l=1; l<=LL+LR; l++)
            { j++; RR[i][j]=RR[i-1][j-1]; }
        }
     }
    /* Step 4 */
    c=(LL+LR+1);
    for(c1=1; c1<=K; c1++)
     for(c2=c1; c2<=K; c2++)
      for(i=0; i<c; i++)
        for(j=0; j<c; j++)
         RR[c1*c+i][c2*c+j]=RR[(c1-1)*c+i][(c2-1)*c+j];
    /* Step 5 */
    c=(K+1)*(LL+LR+1);
    for(i=1; i<c; i++)
    for(j=0; j<i; j++)
```

```
        RR[i][j]=RR[j][i];
  for(i=LR+1; i<c; i++)
      BB[i]=-RR[i][LR];
/* Solve the system RR*AA=BB. */
  arsolve(RR,BB,AA, LR+1, c);
/* Create routine output. */
  AA[LR]=1.0;
  for(i=LR; i<c; i++)   A[i-LR]=AA[i];
  *Pu=0.0;
  for(i=LR; i<c; i++)
      *Pu += RR[LR][i]*AA[i];

  mfreef2(RR,(K+1)*(LL+LR+1)); free(AA); free(BB);
  return(0);
}


arsolve(sa,sb,ss,s1,sn)
float * * sa;
float *sb;
float *ss;
int s1,sn;
{
  /* Subroutine to solve a system of equations
     (AR model coefficients) A*X=B (sa->A  sb->B  ss->X)
     sa: square matrix of size (sn x sn)
     sb,ss: vectors of size (sn)
     s1: (s1,s1) is the upper left corner in the matrix sa
         from where the contents of the matrix
         will be taken into account. */

  int k,k1,k2,reg;
  float p;

  /* Gauss --> Upper Triangular */
  for(k=s1; k<sn-1; k++)
   { reg=k;
     for(k1=k+1; k1<sn; k1++)
      if(fabs((double)sa[k1][k])>fabs((double)sa[reg][k]))
          reg=k1;
     if( reg != k )
       { for(k2=k; k2<sn; k2++)
          { p=sa[k][k2]; sa[k][k2]=sa[reg][k2];
            sa[reg][k2]=p;
          }
```

```
              p=sb[k]; sb[k]=sb[reg]; sb[reg]=p;
            }
        for(k1=k+1; k1<sn; k1++)
          { for(k2=k+1; k2<sn; k2++)
              sa[k1][k2]=sa[k1][k2]-sa[k][k2]*sa[k1][k]/sa[k][k];
            sb[k1]=sb[k1]-sa[k1][k]/sa[k][k]*sb[k];
          }
        for(k1=k+1; k1<sn; k1++) sa[k1][k]=0.0 ;
     } /* Upper Triangular */

  /* Solve Upper Triangular system */
  ss[sn-1]=sb[sn-1]/sa[sn-1][sn-1];
  for(k=0; k<sn-1-s1; k++)
       { k1=sn-2-k;
         p=sb[k1];
         for(k2=sn-1-k; k2<sn; k2++)  p=p-sa[k1][k2]*ss[k2];
         ss[k1]=p/sa[k1][k1];
       }
}

float corsample(im1,im2,N,M,k,l)
image im1;
image im2;
int N,M;
int k,l;
{
  /* Function that returns the correlation between
     two images im1 and im2 of size (NxM) at the point (k,l).
  */

  int y,x;
  float sum;

  sum=0.0;
  for(y=(k<0) ? -k : 0 ; (k<0 && y<N) ||(k>=0 && y+k<N) ; y++)
  for(x=(l<0) ? -l : 0 ; (l<0 && x<M) ||(l>=0 && x+l<M) ; x++)
    sum +=(unsigned int)im1[y+k][x+l]*(unsigned int)im2[y][x];
  sum /= (float)N*(float)M;
  return(sum);
}
```

Fig. 4.6.5 Algorithm for the calculation of the prediction coefficients.

In many cases, it is sufficient to use only small prediction windows [JAI89]:

$$\hat{f}(n,m) = a_1 f(n-1,m) + a_2 f(n,m-1) + a_3 f(n-1,m-1) + a_4 f(n-1,m+1) \quad (4.6.15)$$

The model coefficients a_i, $i = 1, \ldots, 4$, and the noise variance $\sigma^2 = E[\epsilon^2(n,m)]$ can be obtained by solving the system:

$$\begin{bmatrix} R(0,0) & R(1,-1) & R(0,1) & R(0,1) \\ R(1,-1) & R(0,0) & R(1,0) & R(1,-2) \\ R(0,1) & R(1,0) & R(0,0) & R(0,2) \\ R(0,1) & R(1,-2) & R(0,2) & R(0,0) \end{bmatrix} \begin{bmatrix} a_1 \\ a_2 \\ a_3 \\ a_4 \end{bmatrix} = \begin{bmatrix} r(1,0) \\ r(0,1) \\ r(1,1) \\ r(1,-1) \end{bmatrix} \quad (4.6.16)$$

$$\sigma^2 = R(0,0) - a_1 R(1,0) - a_2 R(0,1) - a_3 R(1,1) - a_4 R(1,-1) \quad (4.6.17)$$

The autocorrelation coefficients $R(i,j)$ can be estimated by using:

$$R(i,j) = \frac{1}{(2N+1)(2M+1)} \sum_{i=-N}^{N} \sum_{j=-M}^{M} f(k,l) f(k+i, l+j) \quad (4.6.18)$$

Predictive DPCM is a simple digital image compression technique and can be easily implemented in both software and hardware. It can be used to achieve moderate compression ratios (in comparison to the transform techniques to be described in the next section). Another disadvantage is its sensitivity to channel noise. Noise bursts propagate in the entire image row, or even in the entire decoded image (if two-dimensional predictive DPCM is used). Finally, optimal AR coefficients are designed by using image statistics obtained over an entire image plane. However, image statistics may change from one image to the other, or even in different image regions. Thus, adaptation of the AR coefficients is recommended [NET88].

4.7 TRANSFORM IMAGE CODING

An approach to digital image coding is to use image transforms in an effort to concentrate the image energy in a few transform coefficients. If energy packing is obtained, a large number of transform coefficients can be discarded and the rest of them can be coded with variable length codewords. Thus, great data compression can be achieved. Image decoding can be easily obtained by using the inverse transform. The transform coding of digital images is illustrated in Figure 4.7.1. Let **f** be a vector representing an image (or image block) of size $L = N \times M$. The transform vector **F** is given by:

$$\mathbf{F} = \mathbf{A}\mathbf{f} \quad (4.7.1)$$

where **A** is the transform matrix. The inverse transform is defined as follows:

$$\mathbf{f} = \mathbf{A}^{-1}\mathbf{F} \quad (4.7.2)$$

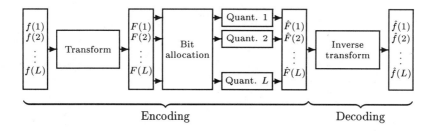

Fig. 4.7.1 Transform encoding/decoding.

A transform is called unitary [JAI89] if it satisfies:

$$\mathbf{AA}^{*T} = \mathbf{A}^T\mathbf{A}^* = \mathbf{I} \qquad (4.7.3)$$

A unitary transform satisfies energy conservation:

$$\|\mathbf{f}\|^2 = \sum_{k=1}^{L} |f(k)|^2 = \sum_{k=1}^{L} |F(k)|^2 = \|\mathbf{F}\|^2 \qquad (4.7.4)$$

In most practical cases, the signal energy is unevenly distributed in the transform coefficients. The DC coefficient and some other 'low-frequency' coefficients $F(k)$, $1 \leq k \leq K \ll L$, tend to concentrate most of the signal energy. Thus, many transform coefficients (e.g. $F(k), k > K$) can be discarded without much loss of information. A varying number of bits n_k, $1 \leq k \leq K$, can be allocated to the remaining coefficients, so that the average number of bits per pixel is equal to a predefined number B. Once the number of bits is allocated, the transform coefficients $F(k)$, $1 \leq k \leq K$, must be quantized. K different quantizers must be designed. If the quantizers are designed, they are applied to the transform coefficients and the result is the encoded image (or image block). The decoding process is very simple. The inverse transform $\mathbf{A}^{-1}$ is applied to the encoded coefficient vector $\hat{\mathbf{F}}$ to produce an estimate $\hat{\mathbf{f}}$ of the original image vector $\mathbf{f}$.

Several problems must be solved in order to obtain a good transform coding algorithm. The first problem is the choice of the transform to be used. Several alternatives exist: discrete Fourier transform, Walsh–Hadamard transform (WHT), discrete cosine transform (DCT), discrete sine transform (DST) etc. [AHM75], [CLA85]. It has been proven by extensive experimentation that DCT gives the best compression performance [CLA85]; it is also used in the JPEG compression standard of CCITT [KON95]. The next problem is the choice of image block size. It is not advisable to apply one transform to the entire image because of the changing image statistics in the various image regions. The image is split into a number of non-overlapping blocks that are coded independently. The typical block size $N \times M$ is 8×8 or 16×16.

Another problem is the determination of bit allocation. Let us suppose that n_k, $k = 1, \ldots, L$, is the number of bits allocated to each transform coefficient $F(k), k = 1, \ldots, L$. If $n_k = 0$, it means that the corresponding coefficient $F(k)$ is discarded. If the average number of bits per pixel is B, the following relation must be satisfied:

$$\frac{1}{L} \sum_{k=1}^{L} n_k = B \qquad (4.7.5)$$

The average distortion due to coefficient quantization is given by [JAI89]:

$$E = \frac{1}{L} \sum_{k=1}^{L} E[|\ F(k) - Q[F(k)]\ |^2] = \frac{1}{N} \sum_{k=1}^{L} \sigma_k^2 q(n_k) \qquad (4.7.6)$$

where $Q[\cdot]$ denotes quantization and σ_k^2 is the variance of the transform coefficient $F(k)$. The quantizer *distortion function* $q(n_k)$ is a monotonically decreasing function having $q(0) = 1$ and $q(\infty) = 0$. The aim of the bit allocation procedure is to minimize the mean square error (4.7.6) under the constraint (4.7.5). A reasonable choice of the bit numbers is given by [CLA85]:

$$n_k = B + \frac{1}{2} \left[\log_2 \sigma_k^2 - \frac{1}{L} \log_2 \left(\prod_{k=1}^{L} \sigma_k^2 \right) \right] \qquad (4.7.7)$$

Rounding is used in (4.7.7) in order to find integer bit lengths. Other algorithms for integer bit allocation are also available [JAI89]. The coefficient variances σ_k^2, $k = 1, \ldots, L$, can be obtained from the transforms of the various image blocks. An example of the bit allocation of a 16×16 block cosine transform is shown in Figure 4.7.2 [LIM90]. The average bit number is 0.5 bits per pixel.

When bit allocation is obtained, the design of optimal quantizers for each transform coefficient must be performed. Ideally, the probability density function of the transform coefficients must be known. A Rayleigh distribution is used to model the pdf of the DC transform coefficients. Gaussian or Laplacian distributions are used to model the rest of the transform coefficients. In practice, uniform quantization is used for the DC coefficients. The maximum bit length is used for these coefficients (usually $n_1 = 8$ bits). All other coefficients $F(k)$ are divided by the corresponding variances σ_k^2 in order to obtain coefficients having unit variance. Optimal quantizers for unit Gaussian distributions are constructed afterwards. One quantizer is constructed for each bit length n, $1 \leq n \leq n_1$. If the maximal bit length is 8, seven quantizers are needed. These quantizers are applied to the corresponding transform coefficients. The quantized coefficients can be encoded by using a Huffman code table, in order to reduce further the bit rate.

Transform decoding is very simple. Huffman decoding is performed first. The transform coefficients are given by:

$$\hat{F}(k) = \begin{cases} F(k) \cdot \sigma_k^2 & 1 \leq k \leq K \\ 0 & K < k \leq L \end{cases} \qquad (4.7.8)$$

8	7	6	5	3	3	2	2	2	1	1	1	1	1	0	0
7	6	5	4	3	3	2	2	1	1	1	1	1	0	0	0
6	5	4	3	3	2	2	2	1	1	1	1	1	0	0	0
5	4	3	3	3	2	2	2	1	1	1	1	1	0	0	0
3	3	3	3	2	2	2	1	1	1	1	1	0	0	0	0
3	3	2	2	2	2	2	1	1	1	1	1	0	0	0	0
2	2	2	2	2	2	1	1	1	1	1	0	0	0	0	0
2	2	2	2	1	1	1	1	1	1	1	0	0	0	0	0
2	1	1	1	1	1	1	1	1	1	0	0	0	0	0	0
1	1	1	1	1	1	1	1	1	0	0	0	0	0	0	0
1	1	1	1	1	1	1	1	0	0	0	0	0	0	0	0
1	1	1	1	1	1	0	0	0	0	0	0	0	0	0	0
1	1	1	1	0	0	0	0	0	0	0	0	0	0	0	0
1	0	0	0	0	0	0	0	0	0	0	0	0	0	0	0
0	0	0	0	0	0	0	0	0	0	0	0	0	0	0	0
0	0	0	0	0	0	0	0	0	0	0	0	0	0	0	0

Fig. 4.7.2 Binary allocation of a 16 × 16 block cosine transform.

and the image block is obtained by using the inverse transform:

$$\hat{\mathbf{f}}(k) = \mathbf{A}^{-1}\hat{\mathbf{F}} \tag{4.7.9}$$

A primary example of the transform coding of still pictures is the baseline of the JPEG international compression standard prepared by the ISO/IEC JTC1/SC2/WG10 photographic image coding committee. The committee existed before 1990 as an *ad hoc* group of ISO/IEC, called the *Joint Photographic Experts Group (JPEG)*. Thus, the compression standard is informally known as JPEG compression. This group collaborated informally with a special raporteur committee of CCITT SGVIII. In this joint work, the ISO/IEC selected, developed and tested coding techniques, whereas CCITT SGVIII provided coding specifications for image communication applications (e.g. fascimile, videotex). Special attention has been paid to other important applications as well, such as medical and scientific imaging and graphic arts.

The JPEG standard can be used for grayscale and color image coding. It is not suitable for binary image coding. It provides specifications for encoding/decoding, specifies the format of the compressed image data and gives information on how to implement the compression/decompression. It covers a broad range of related algorithms, besides its baseline coding process. Its full description could be the subject of a specialized book. This chapter just covers the basics of its operation.

The JPEG standard provides algorithms both for lossless and for lossy image compression. It can be divided into four distinct *modes of operation*: *lossless, sequential DCT-based, progressive DCT-based* and *hierarchical*. The lossless techniques are predictive ones similar to the algorithm described in

TRANSFORM IMAGE CODING   233

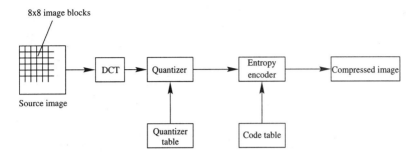

Fig. 4.7.3 Simplified diagram of a JPEG baseline encoder.

Figure 4.6.3. The simplest lossy technique is sequential DCT-based coding and is referred to as the *baseline sequential process*. It is sufficient for a variety of applications and must be implemented in any JPEG codec. Therefore, it will be described here in more detail. The flow diagram of the encoder is described in Figure 4.7.3. The first preprocessing step is a *level shift* applied to the entire source image. Let B be the number of bits used for image representation. In most cases 256 grayscales are used and $B = 8$. All image pixels are transformed to a signed representation by subtracting 2^{B-1}. The level shift is 128 for $B = 8$. JPEG compression uses blocks of 8×8 pixels. If the image dimensions are not multiples of 8, the image is appended with the necessary image blocks. The appended data are discarded at decompression. DCT is applied to each image block. The following transform definition is used [KON95]:

$$C(k_1, k_2) = \frac{1}{4} v_1(k_1) v_2(k_2) \sum_{n_1=0}^{7} \sum_{n_2=0}^{7} x(n_1, n_2)$$
$$\cdot \cos \frac{(2n_1 + 1)k_1 \pi}{16} \cos \frac{(2n_2 + 1)k_2 \pi}{16} \qquad (4.7.10)$$

where $v_1(k_1), v_2(k_2) = 1/\sqrt{2}$ for $k_1, k_2 = 0$ and $v_1(k_1), v_2(k_2) = 0$ otherwise. Fast algorithms described in Chapter 2 can be used for the implementation of both forward and inverse DCT transforms.

After the forward DCT computation, the 64 DCT coefficients are quantized by a uniform quantizer:

$$C_q(k_1, k_2) = round(C(k_1, k_2)/Q[k_1][K_2]) \qquad (4.7.11)$$

Rounding to the nearest integer is performed. Thus, the higher the quantization step, the larger the quantization error produced. The quantizer steps are included in the quantization table $Q[k_1][k_2]$. No default quantization tables are specified in JPEG but typical quantization tables are included as examples. Such a table is shown in Figure 4.7.4. Quantization tables possess circular symmetry and quantize severely high-frequency coefficients. The quantized

16	11	10	16	24	40	51	61
12	12	14	19	26	58	60	55
14	13	16	24	40	57	69	56
14	17	22	29	51	87	80	62
18	22	37	56	68	109	103	77
24	35	55	64	81	104	113	92
49	64	78	87	103	121	120	101
72	92	95	98	112	100	103	99

Fig. 4.7.4 An 8 × 8 quantization table for image luminance.

0	1	5	6	14	15	27	28
2	4	7	13	16	26	29	42
3	8	12	17	25	30	41	43
9	11	18	24	31	40	44	53
10	19	23	32	39	45	52	54
20	22	33	38	46	51	55	60
21	34	37	47	50	56	59	61
35	36	48	49	57	58	62	63

Fig. 4.7.5 Zig-zag scanning of an 8 × 8 DCT coefficient table.

DCT values are signed 2's complement integers with 11 bit precision (for 8 bit source pixel precision).

After quantization, entropy or arithmetic encoding is performed. The quantized DC coefficient $C(0,0)$ carries a lot of energy and is encoded differentially prior to entropy or arithmetic encoding. Let $C_i(0,0)$, $C_{i-1}(0,0)$ denote the DC coefficients of two adjacent image blocks. At the beginning of a scan, the predictor $C_{i-1}(0,0)$ is initialized to zero. The difference $d = C_i(0,0) - C_{i-1}(0,0)$ is used for entropy encoding. The rest of the 63 AC quantized DCT coefficients are converted to a one-dimensional *zig-zag* sequence $ZZ(1), \ldots, ZZ(63)$ by scanning the 8 × 8 DCT block as shown in Figure 4.7.5. Many quantized AC coefficients are zero, especially those corresponding to high frequencies. Runs of zeros are identified and coded efficiently. If the last part of the ZZ string is zero, this fact is encoded by an *end of block* (EOB) codeword.

There are two options in coding the preprocessed DCT coefficients: Huffman coding and arithmetic coding. The baseline sequential DCT compression uses Huffman coding, whereas extended DCT-based schemes may use either Huffman or arithmetic coding. Different Huffman code tables are used for AC and DC DCT coefficient coding respectively. Their syntax and creation is described in the standard. In the case of arithmetic coding, *conditioning table* specifications are provided.

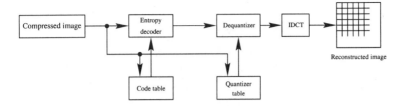

Fig. 4.7.6 Simplified diagram of JPEG baseline decoder.

The JPEG baseline decoding process is described in Figure 4.7.6. The first step is Huffman (or arithmetic) decoding. AC DCT coefficient decoding produces the elements $ZZ(1), \ldots, ZZ(63)$ of the zig-zag string. The DC coefficient difference d is decoded and used to produce the DC term $C_i(0,0) = C_{i-1}(0,0) + d$. Dequantization is performed by multiplying the decoded coefficients by the quantization table entries $C(k_1, k_2) = C_q(k_1, k_2) Q[k_1][K_2]$. It is understandable that the quantization–dequantization procedure introduces losses in the quality of the reconstructed image. The final step in the JPEG baseline decoding algorithm is inverse DCT [KON95]:

$$x(n_1, n_2) = \frac{1}{4} \sum_{k_1=0}^{7} \sum_{k_2=0}^{7} v_1(k_1) v_2(k_2) C(k_1, k_2)$$
$$\cdot \cos \frac{(2n_1 + 1) k_1 \pi}{16} \cos \frac{(2n_2 + 1) k_2 \pi}{16} \qquad (4.7.12)$$

and inverse level shift. Level shifting for 256 grayscale images is obtained by adding 128 to all image pixels obtained by inverse DCT.

An example of DCT coding is illustrated in Figure 4.7.7. Blocks of size 8×8 have been used for coding. The reconstructed black and white image which has been compressed to 0.25 bits per pixel is shown in Figure 4.7.7b. The severe compression ratio results in blocking effects that are evident in the reconstructed image. Much better quality is obtained by using compression to 1 bit/pixel. The decoded image is shown in Figure 4.7.7c. Its subjective quality is very good.

4.8 JPEG2000 COMPRESSION STANDARD

In order to overcome the limitations and disadvantages of the JPEG compression algorithm or other existing image coding solutions, efforts are currently made by ISO/IEC JTC1/SC29/WG1 to design a newer standard, called JPEG 2000 [ISO99, CHA99], which will address the evolving needs of multimedia technologies in the field of still image compression. JPEG 2000 is envisioned to be a coding system for different types of still images (bi-level, gray-level, color, multi-component) with different characteristics (natural images, scien-

Fig. 4.7.7 (a) Original image; (b) image encoded at 0.25 bits per pixel; (c) image encoded at 1 bit per pixel.

tific, medical, etc) allowing different imaging models (client/server, real-time transmission, image library archival, limited buffer and bandwidth resources, etc) [ISO99]. It is intended to provide superior performance with respect to both rate-distortion and subjective image quality against existing image coding solutions. To its current form, JPEG 2000 achieves better compression rates and improved subjective quality, especially at low bit-rate cases, than its predecessor JPEG. The most annoying effect of blocking artifacts in highly compressed JPEG images is now eliminated. JPEG 2000 will further address functionalities that current solutions either cannot address efficiently or they do not address at all. Both lossless and lossy compression will be provided in a single codestream, a feature that no other existing solution addresses, to serve different application requirements (e.g. lossless compression for image archiving). Encoding of very large images will be enabled (the JPEG algorithm allows encoding of up to 64K by 64K images without tiling). Robustness to the presence of bit errors will be considered. Different types of progressive image encoding will be offered, such as progressive transmission by resolution (PBR) or by pixel accuracy (PBA). In the former type of progressiveness, the image size increases up to its original size when more bits are received. In the latter type of progressiveness, the image quality is enhanced to its original one as more bits are received. Progressiveness is very useful when the transmission channel is of limited bandwidth. The user will be able to stop the transmission as soon as he is satisfied with the resolution or quality of the received image (e.g. in Internet-based environments). A very important feature of the JPEG 2000 standard will be Region Of Interest (ROI) coding and random codestream access and processing. ROI coding means that a region of the image is coded with better quality than the background. Furthermore, the user will be able to randomly access image regions with less distortion and process them independently. JPEG 2000 will also attain increased capacity for color information. Furthermore, it will address image security aspects by means of one or a combination of watermarking, labeling, encryption or scrambling methods. Content or object-based functionalities and metadata descriptions will be incorporated in its structure to efficiently manage the content of JPEG 2000 coded images. The enhanced capabilities of the JPEG 2000 standard will serve different markets and application fields like Internet, facsimile (including color facsimile), printing, scanning, digital photography, remote sensing, mobile applications, medical imagery, digital libraries or e-commerce.

The JPEG 2000 standard is still in progress and is expected to be finalized by the end of the year 2000. However, certain aspects of the standard have already been decided upon. It is very likely, for example, that JPEG 2000 will be based on the Discrete Wavelet Transform [ISO98, BAL98]. Finally, JPEG 2000 is intended to be an open standard allowing new functionalities to be incorporated in its structure without rewriting the standard, while additionally attaining backwards compatibility with JPEG.

References

AHM75. N. Ahmed, K.R. Rao, *Orthogonal transforms for digital signal processing*, Springer Verlag, 1975.

BAL98. I. Balasingham, M. Adams, T. Ramstad, F. Kossentini, H. Coward, A. Perkis, G. Oien, "Performance Evaluation of Different Filter Banks in the JPEG-2000 Baseline System," *IEEE Int. Conf. on Image Processing (ICIP'98)*, vol. 2, pp. 569-573, 4-7 Oct. 1998.

CHA99. M. Charrier, D.S. Cruz, M. Larsson, "JPEG2000, the Next Millennium Compression Standard for Still Images," *IEEE Int. Conf. on Multimedia Computing and Systems (ICMCS'99)*, vol. 1, pp. 131-132, 7-11 June 1999.

CLA85. R.J. Clark, *Transform coding of images*, Academic Press, 1985.

GAL68. R.G. Gallagher, *Information theory and reliable communications*, Wiley, 1968.

HOR84. E. Horowitz, S. Sahni, *Fundamentals of computer algorithms*, Computer Science Press, 1984.

ISO98. I. Balasingham, T. Ramstad, A. Perkis, G. Oien, "Performance of Different Filter Banks and Wavelet Transforms," *ISO/IEC JTC1/SC29/WG1 N740*, March 1998.

ISO99. T. Ebrahimi (editor), "JPEG2000 Requirements and Profiles, version 6.0," *ISO/IEC JTC1/SC29/WG1 N1385*, July 1999.

JAIN89. A.K. Jain, *Fundamentals of digital image processing*, Prentice-Hall, 1989.

JAY84. N.S. Jayant, P. Noll, *Digital coding of waveforms*, Prentice-Hall, 1984.

KNU73. D.E. Knuth, *The art of computer programming*, vol. 3, Addison-Wesley, 1973.

KON95. K. Konstantinides, V. Bhaskaran, *Image and video compression standards*, Kluwer Academic, 1995.

LIM90. J.S. Lim, *Two-dimensional signal and image processing*, Prentice-Hall, 1990.

NET88. A.N. Netravali, B.G. Haskell, *Digital pictures: Representation and compression*, Plenum Press, 1988.

RAO90. K.R. Rao, P. Yip, *Discrete cosine transform: Algorithms, advantages, applications*, Academic Press, 1990.

RIM90. S. Rimmer, *Bit-mapped graphics*, Windcrest, 1990.

TIF88. *TIFF5.0: Aldus/Microsoft technical memorandum*, 1988.

WEL84. T.A. Welch, "A technique for high performance data compression," *IEEE Computer*, vol. 17, no. 6, June 1984.

5
Edge Detection Algorithms

5.1 INTRODUCTION

Edges are basic image features. They carry useful information about object boundaries which can be used for image analysis, object identification and for image filtering applications as well. Despite their fundamental importance in digital image processing and analysis, no precise and widely accepted mathematical definition of an edge is yet available. This fact is explained by the complexity of the image content and by the interference of high-level vision mechanisms in the human perception of the boundary of an object. In the following, we shall consider as an edge the border between two homogeneous image regions having different illumination intensities. This definition implies that an edge is a local variation of illumination (but not necessarily vice versa).

Several edge detectors have been proposed in the literature [GON87], [BAL82], [JAI89]. They can be grouped into two classes: (a) *local techniques*, which use operators on local image neighborhoods, and (b) *global techniques*, which use global information and filtering methods to extract edge information. In the following, we shall present algorithms for edge, line and isolated point detection. Furthermore, we shall describe algorithms for edge linking, boundary (or line) following and for the detection of parametric curves.

5.2 EDGE DETECTION

Image edges have already been defined as local variations of image intensity. Therefore, local image differentiation techniques can produce edge detector operators. The image *gradient* $\nabla f(x, y)$:

$$\nabla f(x,y) = [\partial f/\partial x \quad \partial f/\partial y]^T \triangleq [f_x \quad f_y]^T \quad (5.2.1)$$

provides useful information about local intensity variations. Its magnitude:

$$e(x,y) = \sqrt{f_x^2(x,y) + f_y^2(x,y)} \quad (5.2.2)$$

can be used as an edge detector. Alternatively, the sum of the absolute values of partial derivatives f_x, f_y can be employed:

$$e(x,y) = |f_x(x,y)| + |f_y(x,y)| \quad (5.2.3)$$

for computational simplicity. Local edge direction can be described by the direction angle:

$$\phi(x,y) = \arctan(f_y/f_x) \quad (5.2.4)$$

Gradient estimates can be obtained by using gradient operators of the form:

$$\hat{f}_x = \mathbf{w}_1^T \mathbf{x} \quad (5.2.5)$$

$$\hat{f}_y = \mathbf{w}_2^T \mathbf{x} \quad (5.2.6)$$

where $\mathbf{x}$ is the vector containing image pixels in a local image neighborhood. Weight vectors $\mathbf{w}_1, \mathbf{w}_2$ are described by gradient masks. Such masks are shown in Figure 5.2.1 for the *Prewitt* and *Sobel* edge detectors, respectively. Relations (5.2.5–5.2.6) are essentially two-dimensional linear convolutions with the 3 × 3 kernels shown in Figure 5.2.1. They can be easily implemented in the spatial domain. The Sobel edge detector algorithm is described in Figure 5.2.2.

```
int sobel(a,b, N1,M1,N2,M2)
image a,b;
int N1,M1,N2,M2;

/* Subroutine for Sobel edge detector
   a,b: input and output buffers
   N1,M1: start coordinates
   N2,M2: end   coordinates              */

{ int i,j,c,cc;

  for(i=N1+1; i<N2-1; i++)
```

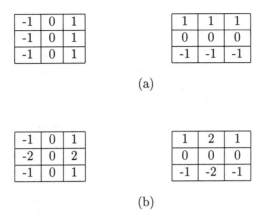

Fig. 5.2.1 (a) Prewitt edge detector masks; (b) Sobel edge detector masks.

```
for(j=M1+1; j<M2-1; j++)
   { cc=-(int)a[i-1][j-1]-2*(int)a[i-1][j]-(int)a[i-1][j+1];
     cc+=(int)a[i+1][j-1]+2*(int)a[i+1][j]+(int)a[i+1][j+1];
     c=abs(cc);
     cc=-(int)a[i-1][j-1]-2*(int)a[i][j-1]-(int)a[i+1][j-1];
     cc+=(int)a[i-1][j+1]+2*(int)a[i][j+1]+(int)a[i+1][j+1];
     c += abs(cc) ;
     if(c>255) c=255;
     b[i][j]=(unsigned char)c;
   }
  return(0);
}
```

Fig. 5.2.2 Sobel edge detector algorithm.

The Sobel edge detector provides good performance and is relatively insensitive to noise. Better noise characteristics can be achieved by using larger neighborhoods at the expense of computational effort. However, large neighborhoods tend to produce thick edges.

Edge templates are masks that can be used to detect edges along different directions. Such masks of size 3×3 are shown in Figure 5.2.3. They can detect edges only at four different directions (0, 45, 90 and 135 degrees). Templates of higher size can be more sensitive to edge orientation. Let $\mathbf{w}_i$, $i = 1, \ldots, n$, be the weight vector associated with each template. Edge detection can be performed in the following way. All templates are applied to each image pixel.

244 EDGE DETECTION ALGORITHMS

-1	0	1
-1	0	1
-1	0	1

1	1	1
0	0	0
-1	-1	-1

0	1	1
-1	0	1
-1	-1	0

1	1	0
1	0	-1
0	-1	-1

Fig. 5.2.3 Kirsch templates of size 3 × 3.

The template that produces the maximal output is the winner:

$$e(k,l) = |\mathbf{w}_i^T \mathbf{x}| \text{ if } |\mathbf{w}_i^T \mathbf{x}| \geq |\mathbf{w}_j^T \mathbf{x}|, \quad j = 1,\ldots,n \qquad (5.2.7)$$

The corresponding output $|\mathbf{w}_i^T \mathbf{x}|$ is a measure of confidence of the edge detector output. If it is close to zero, no edge is present at this pixel location. If all templates give approximately equal output magnitudes, no reliable information can be extracted for edge direction. An algorithm for edge detection based on templates is shown in Figure 5.2.4.

```
int compass(a,b, th, N1,M1,N2,M2)
image a,b;
int th;
int N1,M1,N2,M2;

/* Subroutine to detect edges
   a,b: input and output buffers
   th: line direction (0, 45, 90, 135)
   N1,M1: start coordinates
   N2,M2: end   coordinates          */
{ int i,j,c;
  if(th==0)
    { for(i=N1+1; i<N2-1; i++)
      for(j=M1+1; j<M2-1; j++)
       { c=(int)a[i+1][j-1]+(int)a[i+1][j]+(int)a[i+1][j+1];
         c=c-(int)a[i-1][j-1]-(int)a[i-1][j]-(int)a[i-1][j+1];
         c=abs(c);
         if(c>255) c=255;
         b[i][j]=(unsigned char)c;
       }
    }
  else if(th==45)
    { for(i=N1+1; i<N2-1; i++)
      for(j=M1+1; j<M2-1; j++)
       { c=(int)a[i+1][j+1]+(int)a[i+1][j]+(int)a[i][j+1];
```

```
          c=c-(int)a[i-1][j]-(int)a[i][j-1]-(int)a[i-1][j-1];
          c=abs(c);
          if(c>255) c=255;
          b[i][j]=(unsigned char)c;
        }
    }
  else if(th==90)
    { for(i=N1+1; i<N2-1; i++)
        for(j=M1+1; j<M2-1; j++)
        { c=(int)a[i-1][j+1]+(int)a[i][j+1]+(int)a[i+1][j+1];
          c=c-(int)a[i-1][j-1]-(int)a[i][j-1]-(int)a[i+1][j-1];
          c=abs(c);
          if(c>255) c=255;
          b[i][j]=(unsigned char)c;
        }
    }
  else if(th==135)
    { for(i=N1+1; i<N2-1; i++)
        for(j=M1+1; j<M2-1; j++)
        { c=(int)a[i][j-1]+(int)a[i+1][j-1]+(int)a[i+1][j];
          c=c-(int)a[i][j+1]-(int)a[i-1][j+1]-(int)a[i-1][j];
          c=abs(c);
          if(c>255) c=255;
          b[i][j]=(unsigned char)c;
        }
    }
  return(0);
}
```

Fig. 5.2.4 Edge detection based on template matching.

Some of the above-mentioned algorithms have been applied to the image shown in Figure 5.2.5a. The result of the Sobel edge detector is shown in Figure 5.2.5b. The results of the template edge detector are shown in Figures 5.2.5c and 5.2.5d for horizontal and vertical edges respectively. Templates can also be used for line and point detection. The corresponding templates are shown in Figure 5.2.6. Lines at four different directions (0, 45, 90 and 135 degrees) can be detected by using the templates of Figure 5.2.6b.

Another approach to edge detection is the use of the *Laplace operator*. It is defined in terms of the second-order partial derivatives of $f(x,y)$ with respect to x, y respectively:

$$\nabla^2 f(x,y) = \frac{\partial^2 f}{\partial x^2} + \frac{\partial^2 f}{\partial y^2} \tag{5.2.8}$$

246 EDGE DETECTION ALGORITHMS

Fig. 5.2.5 (a) Original image; (b) Sobel edge detector output; (c) horizontal edge detection; (d) vertical edge detection.

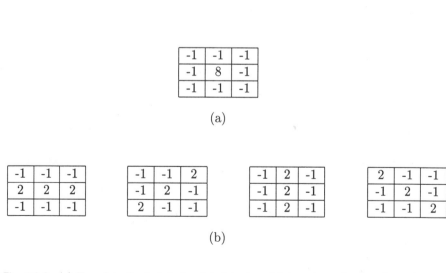

Fig. 5.2.6 (a) Template for isolated point detector; (b) templates for line detection.

The first-order derivatives have local maxima or minima at image edges due to large local intensity changes. Therefore, the second-order derivatives have zero-crossings (e.g. transitions from positive to negative values and vice versa) at edge locations. Thus, an approach to edge detection is to estimate the Laplace operator output and to find zero-crossing positions. An estimator of the Laplace operator (5.2.8) is given by:

$$\nabla^2 f(x,y) \simeq f(x,y) - \frac{1}{4}[f(x,y+1)+f(x,y-1)+f(x+1,y)+f(x-1,y)] \quad (5.2.9)$$

The corresponding edge detector algorithm is described in Figure 5.2.7.

```
int laplace(a,b, N1,M1,N2,M2)
image a,b;
int N1,M1,N2,M2;

/* Subroutine for Laplace edge detector
   a,b: input and output buffers
   N1,M1: start coordinates
   N2,M2: end   coordinates            */

{ int i,j,c;

  for(i=N1+1; i<N2-1; i++)
  for(j=M1+1; j<M2-1; j++)
     { c=-4*(int)a[i][j]+(int)a[i-1][j]+(int)a[i+1][j];
       c=c+(int)a[i][j-1]+(int)a[i][j+1];
       if(c<0) c=0;
       else if(c>255) c=255;
       b[i][j]=(unsigned char)c;
     }
  return(0);
}
```

Fig. 5.2.7 Laplacian edge detector algorithm.

Differentiation is a high-pass operation. Thus, second-order differentiation tends to enhance image noise. The Laplace operator creates several false edges, especially in areas where the image variance is small, because small intensity perturbations (noise) tend to produce false zero-crossings only. One method to reduce its noise sensitivity is to consider zero-crossings only in the areas where the local variance $\sigma^2(i,j)$ given by:

$$\sigma^2(i,j) = \left(\frac{1}{2M+1}\right)^2 \sum_{k=i-M}^{i+M} \sum_{l=j-M}^{j+M} (f(k,l) - \bar{f}(i,j))^2 \quad (5.2.10)$$

$$\bar{f}(i,j) = \left(\frac{1}{2M+1}\right)^2 \sum_{k=i-M}^{i+M} \sum_{l=j-M}^{j+M} f(k,l) \qquad (5.2.11)$$

is large, because actual image edges occur in areas having large local image dispersion. Based on this observation, local data dispersion measures can be used as edge detectors themselves. Local variance (5.2.10–5.2.11) is such a measure. The local image *range* [DAV81], [PIT90]:

$$w(k,l) = \max_A\{f(k,l)\} - \min_A\{f(k,l)\} \qquad (5.2.12)$$

can be used as a local dispersion measure and as an edge detector. The max and min operators in (5.2.12) describe local maxima and minima in an image neighborhood A. The resulting range edge detector [PIT86] is described in Figure 5.2.8.

```
int range(a,b, NW,MW, N1,M1,N2,M2)
image a,b;
int NW,MW;
int N1,M1,N2,M2;
/* Subroutine for range edge detection
   a,b: input and output buffers
   NW,MW: window size
   N1,M1: start coordinates
   N2,M2: end   coordinates       */

{ int k,l, i,j, NW2,MW2, smax,smin;
  NW2=NW/2; MW2=MW/2;
  for(k=N1+NW2; k<N2-NW2; k++)
  for(l=M1+MW2; l<M2-MW2; l++)
     { smax=(int)a[k][l];
       smin=smax;
       for(i=0; i<NW; i++)
       for(j=0; j<MW; j++)
       {
         if(smin>a[k+i-NW2][l+j-MW2]) smin=a[k+i-NW2][l+j-MW2];
         if(smax<a[k+i-NW2][l+j-MW2]) smax=a[k+i-NW2][l+j-MW2];
       }
       b[k][l]=(unsigned char)(smax-smin);
     }
  return(0);
}
```

Fig. 5.2.8 Range edge detector algorithm.

The range edge detector is very fast since it requires only local maximum and minimum calculation. Running max/min algorithms described in Chapter 3 can be used to speed up the calculation still further.

Edge detection has been recently extended for angular data defined on a circle [NIK98]. The hue information in the HLS and HIS color systems forms such angular images. Classical edge detectors cannot be used on such images due to hue periodicity in $[\pi, \pi]$. Angular edge detectors must be used in this case.

5.3 EDGE THRESHOLDING

Most edge detectors described in the previous section produce a grayscale output image $e(k,l)$. This image carries information about the edge magnitude. If the edge detector output is large, a local edge is present. Otherwise, this pixel location corresponds to background. Therefore, a thresholding operation is required after edge detection:

$$E(k,l) = \begin{cases} 1 & \text{if } e(k,l) \geq T \\ 0 & \text{otherwise} \end{cases} \quad (5.3.1)$$

The threshold T can be chosen by inspecting the edge detector output histogram, so that only a small percentage of the pixels $e(k,l)$ above it. Thresholding (5.3.1) is global, because T is chosen based on global information and (5.3.1) is applied to the entire image. In many applications, the edge detector output has regions possessing different statistical properties. Therefore, global thresholding may produce thick edges in one region and thin or broken edges in another region. Thus, locally adapted thresholding is desirable. The thresholding operation is still described by (5.3.1), with threshold $T(k,l)$ adapted locally. Several heuristic adaptation techniques can be used. One method is to calculate the local arithmetic mean of the edge detector output:

$$\bar{e}(k,l) = \left(\frac{1}{2M+1}\right)^2 \sum_{i=k-M}^{i+M} \sum_{l=j-M}^{j+M} e(i,j) \quad (5.3.2)$$

and to use it in the threshold calculation [HAR73], [HAR75], [LEV85]:

$$T(k,l) = \bar{e}(k,l)(1+p) \quad (5.3.3)$$

where p is a percentage indicating the level of the threshold above the local arithmetic mean.

5.4 HOUGH TRANSFORM

Edge detector output thresholding produces a binary image having 1s at edge locations. This image must be further processed to produce more useful in-

EDGE DETECTION ALGORITHMS

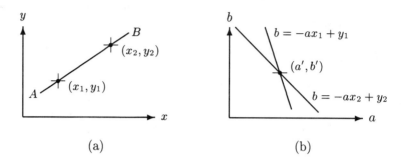

Fig. 5.4.1 (a) Straight line on the image plane; (b) its representation in the parameter space.

formation that can be used in the detection of simple shapes (e.g. straight lines, circles) or arbitrarily shaped objects. In the following, we shall describe a method for the detection of parametric curves in general and for the detection of straight lines in particular. Let us suppose that we search for straight lines on a binary image having size $N = N_1 \times N_2$ pixels. The simplest approach is to find all possible lines determined by pairs of pixels and to check if subsets of the binary image pixels belong to any of these lines. The maximal number of possible lines is $N(N-1)/2$. In the worst case every image pixel must be checked if it belongs to such a line. The computational complexity of this simple method is of the order $O(N^3)$ in the worst case, thus prohibiting its use in practical applications.

The *Hough transform* uses a parametric description of simple geometrical shapes (curves) in order to reduce the computational complexity of their search in a binary image. We shall start with the presentation of this method for the search of straight lines. Their parametric description is a linear equation:

$$y = ax + b \quad (5.4.1)$$

Each line is represented by a single point (a', b') in the parameter space (a, b). Let us suppose that a straight line passes from two image points (x_1, y_1) and (x_2, y_2) on the image plane (x, y), as seen in Figure 5.4.1. Any line passing through the point (x_1, y_1) corresponds to the line $b = -ax_1 + y_1$ in the parameter space. Similarly, any line passing through (x_2, y_2) corresponds to the line $b = -ax_2 + y_2$ in the space (a, b). The intersection (a', b') of these two lines determines uniquely the straight line passing through (x_i, y_i), $i = 1, 2$. Therefore, a simple procedure for straight line detection is the following. The parameter space is discretized and a parameter matrix $P(a, b)$, $a_1 \leq a \leq a_K$, $b_1 \leq b \leq b_L$, is formed. For every pixel (x_i, y_i) that possesses value 1 at the binary edge detector output, the equation $b = -ax_i + y_i$ is formed.

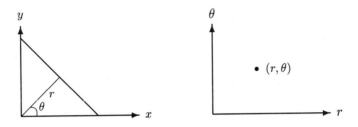

Fig. 5.4.2 Polar representation of a straight line.

For every parameter value a, $a_1 \leq a \leq a_K$, the corresponding parameter b is calculated and the appropriate parameter matrix element $P(a, b)$ is increased by 1:
$$P(a, b) = P(a, b) + 1 \qquad (5.4.2)$$
This process is repeated until the entire binary image is scanned. At the end of the procedure, each parameter matrix element $P(a, b)$ shows the number of binary edge detector output pixels that satisfy (5.4.1). If this number is above a certain threshold a line of the form (5.4.1) is declared. The accuracy of the method depends of course on the quantization interval of the parameters a, b, and, thus, on the size $K \times L$ of matrix $P(a, b)$. The computational complexity of the Hough transform is of the order $O(KN)$ in the worst case, where K is the number of subdivisions of the parameter a and N is the number of pixels in the binary edge detector output. Therefore, the Hough transform is much faster than the simple straight line detection algorithm described in the beginning of this section.

The parametric model (5.4.1) has some difficulties in the representation of vertical straight lines, because the parameter a must tend to infinity. A polar representation of a straight line given by:
$$r = x \cos \theta + y \sin \theta \qquad (5.4.3)$$
can be used instead. It describes a line having orientation θ at distance r from the origin, as can be seen in Figure 5.4.2. A line passing through the point (x_1, y_1) represents a sinusoidal curve $r = x_1 \cos \theta + y_1 \sin \theta$ in the parameter space (r, θ). Collinear points (x_i, y_i) on the binary image space correspond to intersections of sinusoids on the parameter space. Therefore, a similar algorithm to the one described in Figure 5.4.2 can be used by employing the model (5.4.3) instead of (5.4.1). The range of the parameters (r, θ) to be used is given by:
$$-\sqrt{N_1^2 + N_2^2} \leq r \leq \sqrt{N_1^2 + N_2^2} \qquad (5.4.4)$$
$$-\pi/2 \leq \theta < \pi/2 \qquad (5.4.5)$$

for a binary image of size $N_1 \times N_2$. The description of the Hough transform algorithm is shown in Figure 5.4.3.

```
int hough(xin,p,n,m,N1,N2,COS,SIN)
int N1,N2,m,n;
image xin;
imatrix p;
float *COS,*SIN;
{  /* Subroutine to perform Hough transform. The result is
      a two-dimensional matrix p in the
      r,theta parameter space.
      xin: input image
      p: parameter matrix  in the r,theta space
      N1: number of rows of image
      N2: number of columns of image
      n: number of rows of matrix p
      m: number of columns of matrix p
      COS: look-up table for cos function
      SIN: look-up table for sin function */
   int k,l,i,j,kk,ll;
   float r,b;
   float SQRTD=sqrt((float)N1*(float)N1+(float)N2*(float)N2);
   /* Initialize matrix p */
   for(kk=0; kk<n; kk++)
     for(ll=0; ll<m; ll++)
       p[kk][ll]=0;
   /* For any pixel having value 1 in the input image
      increment corresponding elements of matrix p,
      according to r, theta Hough transform.
      Theta ranges from -pi/2 to pi/2.
      r ranges from -sqrt(N1*N1+N2*N2) to sqrt(N1*N1+N2*N2).*/
   for(k=0; k<N1; k++)
    for(l=0; l<N2; l++)
      {
        if (xin[k][l] == 1 )
        {
          for(i=0; i<n; i++)
            {
              r = k * COS[i] + l * SIN[i]; b = SQRTD;
              r+= b ; r/= (SQRTD*2.0); r*= (m-1); r+= 0.5;
              j=floor(r); p[i][j]++;
            }
        }
      }
```

```
      }
  return(0);
}
int ihough(xin,xout,p,N1,N2,n,m,COS,SIN)
image xin,xout;
imatrix p;
int N1,N2,n,m;
float *COS,*SIN;

{ /* Function to produce the inverse Hough transform
     from the r,theta space to the x,y space.
     xin: input image
     xout: output image
     p: parameter matrix  in the r,theta space
     N1: number of rows of image
     N2: number of columns of image
     n: number of rows of matrix p
     m: number of columns of matrix p
     COS: look-up table for cos function
     SIN: look-up table for sin function              */

   int k,l,i,j;
   float r,y;
   float SQRTD=sqrt((float)N1*(float)N1+(float)N2*(float)N2);
   /* Initialize output image */
   for(i=0; i<N1; i++)
     for(j=0; j<N2; j++)
       xout[i][j] = 0;
   /* Parameter matrix p has already been thresholded.
      For every element of p having value 1, the corresponding
      straight line (consisting of 1s) is drawn in the output
      matrix xout. And operation with the input image xin
      is performed!!!! */
   for(k=0; k<n; k++)
     for(l=0; l<m; l++)
       {
          y=(float)0.0;
          if(p[k][l]==1)
            {
               for(i=0; i<N1; i++)
                 {
                    r=(float) l * 2.0 * SQRTD/(m-1) - SQRTD;
                    if ( SIN[k]==(float)0.0 )   y=y+1 ;
                    else  y=( r - (float) i * COS[k] ) / SIN[k];
                    j = floor(0.5+y) ;
```

```
                if ( j >= 0  &&  j < N2 )
                  if (xin[i][j]==1) xout[i][j] = 1;
             }
           }
        }
   return(0);
}
int thres_par(p,n,m,t)
imatrix p;
int n,m,t;
{ /* Subroutine to threshold parameter matrix (type imatrix).
     Each element of matrix p which is above the
     given threshold t declares a line.
     p: parameter matrix  in the r,theta space
     n: number of rows of matrix p
     m: number of columns of matrix p
     t: threshold                                              */
  int a,b;
  for(a=0; a<n; a++)
   for(b=0; b<m; b++)
    if (p[a][b]>t) p[a][b]=1; else p[a][b]=0;
   return(0);
}

int look_up_table(COS,SIN,m)
int m;
float *COS,*SIN;
{ /* Subroutine to produce look-up tables for
     COS and SIN functions.
     COS: look-up table for cos function.
     SIN: look-up table for sin function.
     m:   number of rows for COS, SIN look-up tables           */
  int i;
  float th;
  float R_TO_D=0.017453;
  for(i=0; i<m; i++)
   { th = (float) i * 180.0/(m-1) - 90.0; th = th * R_TO_D;
     COS[i] = (double) cos((double) th);
     SIN[i] = (double) sin((double) th);
   }
  return(0);
}
```

Fig. 5.4.3 Hough transform algorithm.

The parameter matrix is denoted by p. It is of type imatrix, because its elements may take relatively large values that cannot be accommodated within the range provided by the unsigned characters used in the type image. Its elements are updated by using (5.4.3). A threshold value is used to declare elements of the parameter matrix as actual straight line parameters. Parameter matrix thresholding is performed by subroutine thres_par(). The image pixels that belong to straight lines detected by the Hough transform are produced by subroutine ihough(). For every element of matrix p having value 1, the corresponding straight line (consisting of 1s) is drawn in the output matrix xout. An and operation with the input image xin is performed. The application of the Hough transform in the detection of straight line segments is shown in Figure 5.4.4. Image 5.4.4b shows the sinusoids in the parameter space that correspond to (5.4.3). Several accumulation points can be easily identified. The straight lines corresponding to such accumulation points are shown in Figure 5.4.4c after an and operation with the thresholded edge detector output. The overlay of the detected line segments on the original image is shown in Figure 5.4.4d. The threshold that has been applied on the parameter matrix was relatively small. Therefore, even small straight vertical line segments have been detected. In certain cases, the concatenation of such small segments creates the illusion that curves have been detected.

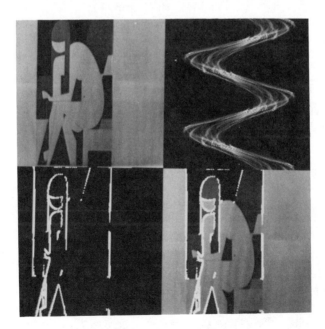

Fig. 5.4.4 (a) Original image; (b) sinusoids in the parameter space representing straight lines in the image plane; (c) detected straight lines; (d) detected line segments in the image plane.

Certain edge detectors (e.g. the Sobel edge detector) can produce not only edge magnitude but edge direction $\phi(x, y)$ as well. The directional information can be used to facilitate Hough transform calculation by reducing the two-dimensional search to a one-dimensional search. If both sides of (5.4.3) are differentiated with respect to x, the following equation gives the line gradient:

$$\frac{dy}{dx} = -\cot\theta = \tan(\pi/2 + \theta) \qquad (5.4.6)$$

The derivative dy/dx is also equal to $\tan\phi$. The line orientation θ can be calculated from the edge detector output:

$$\theta = \frac{\pi}{2} - \phi \qquad (5.4.7)$$

Therefore, only one parameter matrix element given by (5.4.7) and (5.4.3) must be updated for each edge pixel (x_i, y_i) of the binary edge detector output. The use of the edge gradient reduces the computational complexity of the Hough transform to the order $O(N)$. However, care must be taken to avoid using noisy edge direction estimates that will reduce the performance of the algorithm.

The Hough transform can be generalized to detect any parametric curves of the form $f(\mathbf{x}, \mathbf{a}) = 0$, where $\mathbf{a}$ is the parameter vector [BAL82], [SCH89]. The memory required for the parametric matrix $P(\mathbf{a})$ increases as K^p where p is the parameter number. Thus, this method is practical only for curves having a small number of parameters. Such a curve is the circle:

$$(x - a)^2 + (y - b)^2 = r^2 \qquad (5.4.8)$$

Its parameters are the radius r and the center coordinates (a, b). A three-dimensional parameter matrix $P(r, a, b)$ is needed. The Hough transform is implemented in the following way. Let (x_i, y_i) be a candidate binary edge image pixel. The center coordinates (a, b) of a circle having radius $r = R$ and passing through (x_i, y_i) lie on a circle of the form:

$$x_i = a + R\cos\theta \qquad (5.4.9)$$

$$y_i = b + R\sin\theta \qquad (5.4.10)$$

as can be seen in Figure 5.4.5a. For any radius r, $0 < r \leq r_{max}$, the coordinates (a, b) given by (5.4.9–5.4.10) are calculated and the corresponding elements of matrix $P(a, b, r)$ are increased by one. These points belong to a cone surface. This process is repeated for any eligible pixel of the binary edge detector output. The elements of the matrix $P(a, b, r)$ having a final value that is larger than a certain threshold denote the circles that are present in the edge detector output. If the edge direction $\phi(x_i, y_i)$ is available, Hough transform computation is greatly facilitated. Ideally, the local edge element is a tangent to any circle of radius R passing through (x_i, y_i), as can be seen

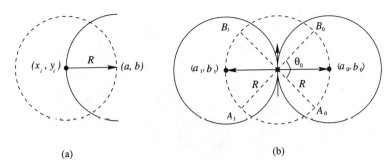

Fig. 5.4.5 (a) Locus of circle centers passing through (x_i, y_i); (b) centers of circles that pass through (x_i, y_i) and are tangent to the corresponding edge element.

in Figure 5.4.5b. Thus, only two such circle centers (a_0, b_0), (a_1, b_1) are allowable. Only the corresponding locations in the matrix $P(a, b, r)$ must be updated. Since edge direction $\phi(x_i, y_i)$ is usually noisy, a certain tolerance θ_0 is permitted for (5.4.9–5.4.10) and only the circle centers corresponding to the arcs $A_0 B_0, A_1 B_1$ shown in Figure 5.4.5b are allowed to be updated in the matrix $P(a, b, r)$.

The disadvantages of classical Hough Transform are its large storage and computational time requirements and its rather poor performance in noisy images, where it frequently leads to false detections. Fuzzy versions of Hough transform have been developed to solve such robustness problems. Fuzzy Cell Hough Transform [CHA97] uses fuzzy cells in the parameter space that result to a fuzzy voting algorithm. The array that is created after the fuzzy voting process is smoother than in the classical approach. Local maxima that correspond to the effect of noise or any kind of uncertainty are reduced and subsequently the number of correct detections of curves is increased. Moreover, curves are detected with better accuracy in comparison to classical Hough Transform. The computation time is slightly increased by this method but it can be traded with accuracy, if needed. An example is shown in Figure 5.4.6. Figure 5.4.6a shows the original image which was used as an input to an edge detector to produce the thresholded image shown in Figure 5.4.6b. By using this image, classical Hough Transform and Fuzzy Cell Hough Transform were applied to detect two circles. The results are presented in Figures 5.4.6c and 5.4.6d, respectively. Fuzzy Cell Hough Transform produces correct results, whereas the classical HT fails.

5.5 EDGE-FOLLOWING ALGORITHMS

In many cases it is desirable to follow object boundaries for object recognition applications. One approach to boundary following is to perform edge detection and to follow the local edge elements. Let $e(\mathbf{x}) = e(x, y)$ and $\phi(\mathbf{x}) = \phi(x, y)$

Fig. 5.4.6 (a) The original image; (b) edge detector output; (c) two circles detected by using classical HT; (d) two circles detected by using FCHT.

be the edge magnitude and direction produced by the edge detector output at the location $\mathbf{x} = (x, y)$. Edge detector output tends to have approximately constant intensity along object boundaries. The difference $|e(\mathbf{x}_i) - e(\mathbf{x}_j)|$ can be used as a similarity measure for neighboring edge elements at neighbor positions $\mathbf{x}_i$, $\mathbf{x}_j$. Another frequently used similarity measure is the direction difference $|\phi(\mathbf{x}_i) - \phi(\mathbf{x}_j)|$. Its use stems from the fact that image edges and lines are smooth and tend to have low curvature. Small local edge direction differences ensure smooth object boundaries. Two neighboring edge elements can be linked if the magnitude and direction differences are below certain thresholds and their magnitudes are relatively large:

$$|e(\mathbf{x}_i) - e(\mathbf{x}_j)| \leq T_1 \qquad (5.5.1)$$

$$|\phi(\mathbf{x}_i) - \phi(\mathbf{x}_j)| \bmod 2\pi \leq T_2 \qquad (5.5.2)$$

$$|e(\mathbf{x}_i)| \geq T \qquad |e(\mathbf{x}_j)| \geq T \qquad (5.5.3)$$

The inequalities (5.5.3) ensure that small perturbations will not be mistaken as edge elements.

A simple algorithm for edge following can be based on (5.5.1–5.5.3). Let us suppose that we start from a point $\mathbf{x}_A$ satisfying $|e(\mathbf{x}_A)| \geq T$. If there is no neighboring edge element satisfying (5.5.1–5.5.3), the algorithm stops. If more than one neighbor satisfy (5.5.1–5.5.3), the element x_N that possesses the minimal differences $|e(x_N) - e(x_A)|$, $|\phi(x_N) - \phi(x_A)|$ is chosen. The procedure continues recursively, with the new edge element x_N as a starting element. A modification of this algorithm is shown in Figure 5.5.1.

```
int edge_follow(x,e,ka,la,T1,T,a)
image x,e;
int ka,la,T1,T,a;

{ /* Subroutine to perform edge-following starting from
     pixel ka,la. Search is based on the intensity
     difference of neighboring edge elements.
     The procedure continues recursively.
     x: output image
     e: input image
     ka: row of input pixel
     la: column of input pixel
     T1: threshold of difference
     T: threshold for following pixels
     a: factor which determines the
        convergence of the algorithm                         */

  int diff,mindiff,i,j,cont,l,k,ABS,label,lastx,lasty;
  unsigned char val;
  diff=0; mindiff=0; cont=0; lastx=ka; lasty=la;

  /* Search for edge element successor is performed on the
     entire 8-point neighborhood of the current edge pixel.
     Search starts from the NW neighbor and continues in row-
     wise manner until it finds the first valid successor.
     The successor must not be equal to the current pixel
     predecessor.
     Subroutine stops if no valid successor has been found. */
  for( ; ; )
    {
    label:
    k=ka; l=la; x[ka][la]=1;
    val=e[ka][la];
    e[ka][la]=0;
    mindiff=300;
    for(i=-1; i<2; i++)
      for(j=-1; j<2; j++)
        {
        if(!(i==0 && j==0) && k+i>=0 && k+i<NMAX && l+j>=0
```

```
              && l+j<MMAX && !(k+i==lastx && l+j==lasty))
              {
                ABS=abs((int)(val-e[k+i][l+j]));
                if(e[k+i][l+j]>T && ABS<T1)
                {
                  diff=a*ABS;
                  if(diff<mindiff)
                    {
                      mindiff=diff; lastx=ka; lasty=la;
                      ka=k+i; la=l+j;
                      if(e[ka][la]==0) return(0);
                      goto label;
                    }
                  cont=1;
                }
              }
            }
        if(cont==0)  break;
      }
    return(0);
  }
```

Fig. 5.5.1 Simple edge-following algorithm.

The subroutine `edge_follow()` performs edge following starting from pixel (`ka,la`). The search is based on the intensity difference of neighboring edge elements. The search for the edge element successor is performed on the entire eight-point neighborhood of the current edge pixel. The search starts from the NW neighbor and continues in a row-wise manner until it finds the first valid successor. Valid successors are those satisfying (5.5.1), (5.5.3). Orientation information is not used in this subroutine. The successor must not be equal to the current pixel predecessor. The subroutine stops if no valid successor is found.

A more comprehensive approach to edge following is based on graph searching. Edge elements at position x_i can be considered as graph nodes. The nodes are connected to each other if local edge linking rules (e.g. the one described by (5.5.1–5.5.3)) are satisfied. Such a graph is shown in Figure 5.5.2. The generation of a contour (if any) from the pixel x_A to the pixel x_B is equivalent to the generation of a path in the directed graph. Several graph search algorithms can be used for this purpose. In the following we shall describe the heuristic search algorithm [NIL80], which has been proposed for edge following in [MAR72]. Let us suppose that we form a cost function for a path

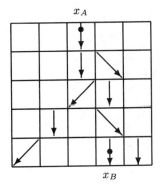 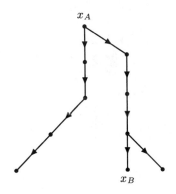

Fig. 5.5.2 Transformation of an edge image to a directed graph.

connecting nodes $\mathbf{x}_1 = \mathbf{x}_A$ to $\mathbf{x}_N = \mathbf{x}_B$:

$$C(\mathbf{x}_1, \mathbf{x}_2, \ldots, \mathbf{x}_N) \stackrel{\triangle}{=} -\sum_{k=1}^{N} |e(\mathbf{x}_k)|$$

$$+ a \sum_{k=2}^{N} |\theta(\mathbf{x}_k) - \theta(\mathbf{x}_{k-1})| + b \sum_{k=2}^{N} |e(\mathbf{x}_k) - e(\mathbf{x}_{k-1})| \qquad (5.5.4)$$

For any intermediate node $\mathbf{x}_i$, the cost function can be split into two components: $C(\mathbf{x}_1, \ldots, \mathbf{x}_i)$, which is the cost of the path from the start node to node $\mathbf{x}_i$, and $C(\mathbf{x}_i, \ldots, \mathbf{x}_N)$, which is the cost from the node $\mathbf{x}_i$ to node $\mathbf{x}_N$. The heuristic search algorithm tries to produce a minimum cost path from $\mathbf{x}_A$ to $\mathbf{x}_B$. The algorithm is based on the cost function (5.5.4) and on the choice of the successors of a node $\mathbf{x}_i$ by using edge linking criteria of the form (5.5.1–5.5.3). The cost function in the algorithm may be replaced by $C(\mathbf{x}_A, \mathbf{x}_2, \ldots, \mathbf{x}_i)$, that is, by the cost from the start node to the current node. This variation of the algorithm is a minimum cost search. Another variation of this algorithm is shown in Figure 5.5.3.

```
int modified_heuristic_edge_search(xA,xB,xin,
                                   T,index,last,outstack)
int T;
image xin;
CPIXEL *index,last,xA,xB;
NODE ** outstack;

{ /* Function to produce a path from starting pixel xA to
     ending pixel xB using a heuristic search method.
```

262 EDGE DETECTION ALGORITHMS

```
   Two stacks are used. Stack 1 contains the pixels of the
   path (the winner of every step of the algorithm).
   Stack p contains the probable winners of every step
   (in the 8-neighborhood).
   xA: starting pixel
   xB: goal pixel
   xin: input image
   T: threshold for pixels in the path
   index: array used to find winner
   last: variable to hold the row, column and cost of the
   last pixel processed
   Return: -92 stack is empty */

int i,d,flag; int NOS=9;
CPIXEL xi,xo,xt;
NODE * tempstack=NULL;
if(xA.row==xB.row && xA.column==xB.column) return(0);
last=xA;
/* Push xA into stack. */
push(outstack,xA);
/* Mark pixel to avoid second pass. */
xin[xA.row][xA.column]=0;
/* Find possible edge element successors.
   Sort the successors according to their local cost.
   Put possible successors in list p.
   Choose as most possible successor the one with the
   minimal local cost. */
find_successors(xin,xA,T,index,last);
sort_pixel(index);
for(i=NOS-1;i>=0;i--) if(index[i].cost!=-1)
                           push(&tempstack,index[i]);
for(i=0; i<NOS; i++)
 if (index[i].cost!=-1 && xin[index[i].row][index[i].column])
  { xi=index[i]; break; }
initialize_index(index);

/* Main edge follow loop. */
for( ; ; )
 {
  flag=0;
  push(outstack,xi);
  xin[xi.row][xi.column]=0;
  /* If end pixel (goal) has been reached, return. */
  if (xi.row==xB.row && xi.column==xB.column) return(0);
  /* Find possible edge element successors. Find the
```

         possible edge successor that is in the direction
         defined by previous and current edge pixel. If this is a
         valid direction and the corresponding edge pixel
         has not been marked, we choose this pixel as the current
         winner. Otherwise,
         sort the successors according to their local cost.
         Put possible successors in list p.
         Choose as most possible successor the one with the
         minimal local cost. If such successor does not exist, we
         backtrace by using pixels from stack p.
         If stack p is empty, -1 is returned. */
      find_successors(xin,xi,T,index,last);
      d = find_direction(xi,last);
      last=xi;
      if((d>=0) && (index[d].cost!=-1) &&
         (xin[index[d].row][index[d].column]!=0))
       {
        xi=index[d];
        sort_pixel(index);
        for(i=NOS-1; i>=0; i--)
          if(index[i].cost!=-1) push(&tempstack,index[i]);
       }
      else
       {
        sort_pixel(index);
        for(i=NOS-1; i>=0; i--)
          if(index[i].cost!=-1) push(&tempstack,index[i]);

        for(i=0; i<NOS; i++)
        if(index[i].cost!=-1 &&
            xin[index[i].row][index[i].column]!=0)
          { xi=index[i]; flag=1; break; }
        if(flag==0)
         {
          while(tempstack != NULL && xin[xi.row][xi.column]==0)
            xi = pop(&tempstack);
          if(tempstack == NULL) return(-92);
          while(*outstack!=NULL && xo.row!=xi.row &&
                xo.column!=xi.column) xo = pop(outstack);
         }
       }
      initialize_index(index);
    }
}

264 EDGE DETECTION ALGORITHMS

```
int find_successors(e,x,TT,index,last)
image e;
int TT;
CPIXEL *index,last,x;

{ /* Subroutine to find successors in the 8-neighborhood
     of the current edge pixel.
     e: image
     x: current pixel
     TT: threshold
     index: array used to store possible successors temporarily
     last: variable to hold the row, column and cost of the
     previous pixel being processed
     (predecessor of the current pixel)              */

  /* Successor pixel must lie within the image limits,
     and must be different from the current pixel and its
     predecessor.
     The corresponding edge image intensity must be greater
     than threshold TT. All successors are stored temporarily
     on array index. */
  int i,j; int k=-1;
  for(i=-1; i<2; i++)
   {
     for(j=-1; j<2; j++)
      if ((i!=0 || j!=0) && (x.row+i!=last.row ||
          x.column+j!=last.column) && x.row+i>=0 &&
          x.row+i<NMAX && x.column+j>=0 && x.column+j<MMAX)
       {
         k++;
         if(e[x.row+i][x.column+j]>TT)
          {
            index[k].row = x.row+i;
            index[k].column = x.column+j;
            index[k].cost =
              abs(last.cost-e[x.row+i][x.column+j]);
          }
       }
   }
  return(0);
}

int initialize_index(index)
CPIXEL *index;
{ /* Function to initialize array index of size NOS
```

      (Number Of Successors) of CPIXEL structure which holds the
      row, column and cost of every pixel in the 8-neighborhood
      of the pixel being processed.
      index: array of size NOS of CPIXEL   */
   int i; int NOS=9;
   for(i=0; i<NOS; i++)
     { index[i].row=-1; index[i].column=-1; index[i].cost=-1; }
   return(0);
}

int sort_pixel(index)
CPIXEL *index;
{ /* Subroutine to sort array index using the cost of every
      array element. The elements are put in ascending cost
      order.
      index: array of CPIXEL of size NOS.   */
   int i,j,k; int NOS=9;
   CPIXEL min;

   for(i=0; i<NOS-1; i++)
     { k=i; min=index[k];
       for(j=i+1; j<NOS; j++)
         if (index[j].cost < min.cost) {k=j; min=index[k];}
       index[k] = index[i];
       index[i] = min;
     }
   return(0);
}

int find_direction(xi,last)
CPIXEL xi,last;
{ /* Subroutine to find the direction defined by the current
      pixel and its predecessor.
      xi: current pixel.
      last: current pixel predecessor.
      The subroutine returns the code of this direction
      in the range [0...8].*/
if(xi.row-last.row==1 && xi.column-last.column==1) return(8);
if(xi.row-last.row==1 && xi.column-last.column==0) return(7);
if(xi.row-last.row==1 && xi.column-last.column==-1) return(6);
if(xi.row-last.row==0 && xi.column-last.column==1) return(5);
if(xi.row-last.row==0 && xi.column-last.column==-1) return(3);
if(xi.row-last.row==-1 && xi.column-last.column==1) return(2);
if(xi.row-last.row==-1 && xi.column-last.column==0) return(1);
if(xi.row-last.row==-1 && xi.column-last.column==-1) return(0);

```
    else return(-1);
}


int stack2image(outstack,xout)
image xout;
NODE ** outstack;
{
  /*Subroutine to create an image using the contents of stack 1.
      mxout: output image   */
  int i,j;
  CPIXEL x;
  for(i=0; i<NMAX; i++)
    for(j=0; j<MMAX; j++)
      xout[i][j] = 0;
  while(*outstack != NULL)
   { x=pop(outstack); xout[x.row][x.column] = 1; }
  return(0);
}
```

Fig. 5.5.3 Heuristic graph search algorithm.

This algorithm employs stacks of pixels. Each stack element is of the type:

```
struct pixel {int row; int column; int cost;}
typedef struct pixel CPIXEL;
struct node { CPIXEL info; struct node * link;}
typedef struct node NODE;
```

Thus, stack elements contain the coordinates of each pixel plus a value related to local path cost. Push and pop operations can be performed on these stacks [AMM87]. Subroutine modified_heuristic_edge_search() produces a path from a starting pixel xA to goal pixel xB using a modification of the heuristic search method. Two stacks are used. Stack outstack contains the winner pixels of the path from the starting pixel to the current edge pixel. Stack tempstack contains all probable winners at every step of the algorithm (until current edge pixel). Possible edge element successors are found for the current edge pixel and lie in its eight-pixel neighborhood. A valid candidate must satisfy (5.5.1, 5.5.3). The possible edge successor that is in the direction defined by the previous and current edge pixel is examined first. If it is rejected, all possible successors are sorted according to their local cost calculated by subroutine find_successors(). The successor having the minimal local cost is chosen. If such a successor does not exist, the algorithm backtraces to other possible paths by using pixels from stack tempstack until the goal pixel is reached or stack tempstack becomes empty.

The basic disadvantage of the heuristic search algorithm is the need to keep track of all current best paths. The number of these paths may increase dramatically during algorithm execution. Another problem is that short paths (close to the origin) may have smaller cost than longer paths that are more likely to be the final winners. These problems may be solved by pruning certain paths (e.g. the short ones, or the paths having high cost per unit length). Furthermore, other graph search techniques (e.g. depth-first search, least maximum cost) can be employed for edge following [BAL82].

Another useful approach to edge following is based on *dynamic programming* [BEL62]. The fundamental observation underlying this method is that any optimal path between two nodes of a graph has optimal subpaths for any node lying on it. Thus the optimal path between two nodes x_A, x_B can be split into two optimal subpaths $x_A x_i$ and $x_i x_B$ for any x_i lying on the optimal path $x_A x_B$. Let us use the following objective function to be maximized:

$$F(x_1, x_2, \ldots, x_N) = \sum_{k=1}^{N} |e(x_k)| - a \sum_{k=2}^{N} |\theta(x_k) - \theta(x_{k-1})| \quad (5.5.5)$$

The start and goal nodes are $x_1 = x_A$ and $x_N = x_B$, respectively. The goal function F can be written in a recursive form:

$$F(x_1, x_2, \ldots, x_k) = F(x_1, x_2, \ldots, x_{k-1}) + f(x_{k-1}, x_k) \quad (5.5.6)$$

where

$$f(x_{k-1}, x_k) = |e(x_k)| - a|\theta(x_k) - \theta(x_{k-1})| \quad (5.5.7)$$

The optimal path $\hat{x}_1 \hat{x}_k$ can be split into two optimal paths $\hat{x}_1 \hat{x}_{k-1}$ and $\hat{x}_{k-1} \hat{x}_k$ satisfying the following recursive relation:

$$\begin{aligned}
\hat{F}(\hat{x}_1, \ldots, \hat{x}_k) &= \max_{x_i, i=1,\ldots,k} F(x_1, \ldots, x_k) \\
&= \max_{x_i, i=1,\ldots,k} \{F(x_1, \ldots, x_{k-1}) + f(x_{k-1}, x_k)\} \\
&= \max_{x_k} \{\hat{F}(\hat{x}_1, \ldots, \hat{x}_{k-1}) + f(\hat{x}_{k-1}, x_k)\} \quad (5.5.8)
\end{aligned}$$

The initial value for $\hat{F}(\hat{x}_1)$ is given by:

$$\hat{F}(\hat{x}_1) = |e(x_1)| \quad (5.5.9)$$

By using the recursive equation (5.5.8), the dynamic programming approach splits the optimization of (5.5.5) into N independent optimization steps. At each step, we search for nodes x_k such that the objective function $\hat{F}(\hat{x}_1, \ldots, \hat{x}_k)$ is maximized. A modification of this algorithm is shown in Figure 5.5.4.

```
int edge_dynamic_prog(xA,xB,xin,T,index,last, outstack)
int T;
image xin;
CPIXEL *index,last,xA,xB;
```

NODE ** outstack;
{ /* Subroutine to produce a path from starting pixel xA
    to goal pixel xB by using a dynamic programming method.
    Two stacks are used. Stack outstack
      contains the pixels of the path (the winner of every step
      of the algorithm). Stack tempstack contains the probable
      winners of every step (in the
      8-neighborhood).
    xA: starting pixel
    xB: goal pixel
    xin: input image
    T: threshold
    index: array of CPIXEL of size NOS
    last: predecessor of current pixel.
    outstack: pointer to the start pointer of output stack */

  int i,d; int NOS=9;
  CPIXEL xi,xo;
  NODE * tempstack=NULL;
  if(xA.row==xB.row && xA.column==xB.column) return(0);
  last=xA;
  /* Push xi into stack. */
  push(outstack,xA);
  /* Mark pixel to avoid second pass. */
  xin[xA.row][xA.column]=0;
  /* Find possible edge element successors.
     Sort the successors according to their local cost.
     Put possible successors in list p.
     Return: -91 current pixel has no successors. */
  initialize_index(index);
  calculate_Fx(xin,xA,T,index,last);
  sort_pixel(index);
  if (index[NOS-1].cost == -1) return(-91);
  for(i=0; i<NOS; i++)
    if(index[i].cost!=-1) push(&tempstack,index[i]);
  if (xin[index[NOS-1].row][index[NOS-1].column]!=0)
    xi = index[NOS-1];
  initialize_index(index);

/* Main search loop. */
  for(; ;)
    {
      /* Put current edge pixel at stack 1 and mark this
      pixel to avoid second pass. */
      push(outstack,xi);

```
    xin[xi.row][xi.column]=0;
    /* If end pixel (goal) has been reached, return */
    if (xi.row==xB.row && xi.column==xB.column) return(0);
    /* Find possible edge element successors.
     Find the possible edge successor that is in the direction
     defined by previous and current edge pixel.
     If this is a valid direction and the corresponding edge
     pixel is a valid candidate, we choose this pixel as the
     current winner. Otherwise,
     sort the successors according to their cost.
     Put possible successors in list p.
     Choose as most possible successor the one with the maximal
     cost. If such successor does not exist, we backtrace
     by using pixels from stack p. If stack p is empty, -1
     is returned. */
    calculate_Fx(xin,xi,T,index,last);
    d = find_direction(xi,last);
    last=xi;
    if((d>=0) && (index[d].cost!=-1) &&
       (xin[index[d].row][index[d].column]!=0)) xi = index[d];
    sort_pixel(index);
    for(i=0; i<NOS; i++)
     if(index[i].cost!=-1) push(&tempstack,index[i]);
     else
       {
         sort_pixel(index);
         for(i=0; i<NOS; i++)
            if(index[i].cost!=-1) push(&tempstack,index[i]);
         if(index[NOS-1].cost!=-1 &&
               xin[index[NOS-1].row][index[NOS-1].column]!=0)
            xi = index[NOS-1];
         else
          {
            while(tempstack != NULL&&xin[xi.row][xi.column]==0)
              xi = pop(&tempstack);
            if(tempstack == NULL) return(-92);
            while(*outstack!=NULL && xo.row!=xi.row &&
                  xo.column!=xi.column)
              xo = pop(outstack);
          }
       }
    initialize_index(index);
   }
}
```

```
int calculate_Fx(e,x,TT,index,last)
image e;
int TT;
CPIXEL *index,last,x;
{ /* Subroutine to produce probable successors of the current
    edge pixel.
    e: image
    x: pixel
    TT: threshold
    index: array of CPIXEL of size NOS
    last: variable to hold the row, column and cost of the
    last pixel processed                                      */

  int i,j,c,t;int k=-1;
  /* Successor pixel must lie within the image limits,
     and must be different from the current pixel and its
     predecessor. The corresponding edge image intensity must
     be greater than threshold TT. Successor cost is
     incremental. All successors are stored temporarily on
     array index. */
  for(i=-1; i<2; i++)
   for(j=-1; j<2; j++)
    { k++;
      if ((i!=0 || j!=0) && (x.row+i!=last.row ||
                             x.column+j!=last.column) &&
          x.row+i>=0 && x.row+i<NMAX && x.column+j>=0 &&
          x.column+j<MMAX)
       {
         if(e[x.row+i][x.column+j]>TT)
          {
            index[k].row=x.row+i; index[k].column=x.column+j;
            index[k].cost=last.cost+e[x.row+i][x.column+j];
          }
       }
    }
  return(0);
}
```

Fig. 5.5.4 Dynamic programming algorithm for edge following.

Subroutine edge_dynamic_prog() produces a path from a starting pixel xA to goal pixel xB using a modification of the dynamic programming method. Two stacks are used, as has already been described for the case of the modified heuristic search algorithm. Possible edge element successors are found

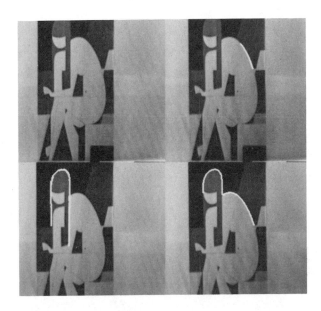

Fig. 5.5.5 (a) Original image; (b) output of the simple edge-following algorithm; (c) output of the heuristic graph search algorithm; (d) output of the dynamic programming algorithm.

for the current edge pixel and lie in its eight-pixel neighborhood. A valid candidate must satisfy (5.5.3). A cost function of the form (5.5.8) is used having $f(\mathbf{x}_{k-1}, \mathbf{x}_k) = |e(\mathbf{x}_k)|$. No orientation differences are employed in this subroutine. The possible edge successor that lies in the direction defined by the previous and current edge pixel is examined first. If it is rejected, all possible successors are sorted according to their cost calculated by subroutine calculate_Fx(). The successor having the maximal cost is chosen. Again, if such a successor does not exist, the algorithm backtraces to other possible paths by using pixels from stack tempstack until the goal pixel is reached or stack tempstack becomes empty.

The application of the algorithms described in this section is illustrated in Figure 5.5.5. It can be easily seen that the simple edge-following algorithm produces relatively small edge segments, because it breaks when the edges have small gaps (e.g. 1–2 pixels long). On the contrary, the modified heuristic search algorithm and the dynamic programming algorithm can follow the edge elements that connect two image pixels lying far apart, as can be seen in Figure 5.5.5.

References

AMM87. L. Ammerdaal, *Programs and data structures in C*, Wiley, 1987.

BAL82. D.H. Ballard, C.M. Brown, *Computer vision*, Prentice-Hall, 1982.

BEL62. R.E. Bellman, S. Dreyfus, *Applied dynamic programming*, Princeton University Press, 1962.

CHA97. V. Chatzis, I. Pitas, "Fuzzy Cell Hough Transform for Curve Detection," *Patern Recognition, Elsevier*, vol. 30, pp. 2031-2042, 1997.

DAV81. H.A. David, *Order statistics*, Wiley, 1981.

GON87. R.C. Gonzalez, P. Wintz, *Digital image processing*, Addison-Wesley, 1987.

HAR73. R.M. Haralick, K. Shanmugan, "Computer classification of reservoir sandstones," *IEEE Transactions on Geoscience Electronics*, vol. GE-11, no. 4, pp. 171-177, Oct. 1973.

HAR75. R.M. Haralick, I. Dinstein, "A spatial clustering procedure for multiimage data," *IEEE Transactions on Circuits and Systems*, vol. CAS-22, no. 5, pp. 440-450, May 1975.

JAI89. A.K. Jain, *Fundamentals of digital image processing*, Prentice-Hall, 1989.

LEV85. M.D. Levine, *Vision in man and machine*, McGraw-Hill, 1985.

MAR72. A. Martelli, "Edge detection using heuristic search methods," *Computer Graphics and Image Processing*, vol. 1, no. 2, pp. 169-182, Aug. 1982.

NIK98. N. Nikolaidis and I. Pitas, "Nonlinear processing and analysis of angular signals," *IEEE Trans. on Signal Processing*, vol. 46, no. 12, pp. 3181-3194, December 1998.

NIL80. N.J. Nilson, *Principles of artificial intelligence*, Tioga, 1980.

PIT86. I. Pitas, A.N. Venetsanopoulos, "Edge detectors based on order statistics," *IEEE Transactions on Pattern Analysis and Machine Intelligence*, vol. PAMI-8, no. 4, pp. 538-550, July 1986.

PIT90. I. Pitas, A.N. Venetsanopoulos, *Nonlinear digital filters: Principles and applications*, Kluwer Academic, 1990.

SCH89. R.J. Schalkof, *Digital image processing and computer vision*, Wiley, 1989.

6
Image Segmentation Algorithms

6.1 INTRODUCTION

The shape of an object can be described either in terms of its boundary or in terms of the region it occupies. Shape representation based on boundary information requires image edge detection and following. Region-based shape representation requires image segmentation in several homogeneous regions. Thus, edge detection and region segmentation are dual approaches in image analysis. Image regions are expected to have homogeneous characteristics (e.g. intensity, texture) that are different in each region. These characteristics form the feature vectors that are used to discriminate one region from the other. The features are employed during the segmentation procedure in the rules that check region homogeneity.

Let us suppose that an image domain X must be segmented in N different regions $R_1, \ldots, R_N$ and that the segmentation rule is a logical predicate of the form $P(R)$. Both image domain X and its regions $R_1, \ldots, R_N$ can be conveniently described by subsets of the image plane $\mathbf{Z}^2$. Image segmentation partitions the set X into the subsets R_i, $i = 1, \ldots, N$, having the following properties:

$$X = \bigcup_{i=1}^{N} R_i \qquad (6.1.1)$$

$$R_i \cap R_j = 0 \text{ for } i \neq j \qquad (6.1.2)$$

$$P(R_i) = TRUE \text{ for } i = 1, 2, \ldots, N \qquad (6.1.3)$$

$$P(R_i \cup R_j) = FALSE \text{ for } i \neq j \tag{6.1.4}$$

The regions $R_i, i = 1, \ldots, N$, must cover the entire image, as it is expressed in property (6.1.1). Property (6.1.2) ensures that two different regions are disjoint sets. The logical predicate $P(R_i)$ must be $TRUE$ over each region R_i to ensure region homogeneity, as can be seen in (6.1.3). Finally, the predicate $P(R)$ must fail in the union of two image regions, since this union is expected to correspond to an inhomogeneous region.

The form of the segmentation predicate P and the features it uses play an important role in the segmentation result. P is usually a logical predicate of the form $P(R, \mathbf{x}, \mathbf{t})$, where $\mathbf{x}$ is a feature vector associated with an image pixel and $\mathbf{t}$ is a set of parameters (usually thresholds). Textural characteristics (to be described in a subsequent section) are usually included in the feature vector. In the simplest case, the feature vector $\mathbf{x}$ contains only image intensity $f(k, l)$ and the threshold vector contains only one threshold T. A simple segmentation rule has the form:

$$P(R): \quad f(k, l) < T \tag{6.1.5}$$

In the case of color images the feature vector $\mathbf{x}$ can be the three RGB image components $[f_R(k, l), f_G(k, l), f_B(k, l)]^T$. A simple segmentation rule may have the form:

$$P(R, \mathbf{x}, \mathbf{t}): (f_R(k, l) < T_R) \&\& (f_G(k, l) < T_G) \&\& (f_B(k, l) < T_B) \tag{6.1.6}$$

In many applications, region connectedness plays an important role in image segmentation. A region R is called *connected* if any two pixels (x_A, y_A), (x_B, y_B) belonging to R can be connected by a path $(x_A, y_A), \ldots, (x_{i-1}, y_{i-1})$, $(x_i, y_i), (x_{i+1}, y_{i+1}), \ldots, (x_B, y_B)$, whose pixels (x_i, y_i) belong to R and any pixel (x_i, y_i) is adjacent to the previous pixel (x_{i-1}, y_{i-1}) and the next pixel (x_{i+1}, y_{i+1}) in the path. A pixel (x_k, y_k) is said to be *adjacent* to the pixel (x_l, y_l) if it belongs to its immediate neighborhood. We can define two types of neighborhoods. The *4-neighborhood* of a pixel (x, y) is the set that includes its horizontal and vertical neighbors:

$$N_4((x, y)) = \{(x + 1, y), (x - 1, y), (x, y + 1), (x, y - 1)\} \tag{6.1.7}$$

The *8-neighborhood* of (x, y) is a superset of the 4-neighborhood and contains the horizontal, vertical and diagonal neighbors:

$$\begin{aligned} N_8((x, y)) &= N_4((x, y)) \cup \{(x + 1, y + 1), (x - 1, y - 1), \\ &\quad (x + 1, y - 1), (x - 1, y + 1)\} \end{aligned} \tag{6.1.8}$$

The paths that are defined by using the 4-neighborhood consist of horizontal and vertical streaks of length $\Delta x = \Delta y = 1$. The paths which use the 8-neighborhood consist of horizontal and vertical streaks of length 1 and of diagonal streaks having length $\sqrt{2}$.

Region segmentation techniques can be grouped in three different classes. *Local techniques* are based on the local properties of the pixels and their neighborhoods. *Global techniques* segment an image on the basis of information obtained globally (e.g. by using the image histogram). *Split, merge* and *growing* techniques use both the notions of homogeneity and geometrical *proximity*. Two regions can be merged if they are similar, i.e. if $P(R_i \cup R_j) = TRUE$, and are adjacent to each other. An inhomogeneous region can be split into smaller subregions. A region can grow by appending pixels in such a way that it remains homogeneous, $P(R_i) = TRUE$. In the following, we shall first describe some global techniques based on thresholding and we shall continue with split and merge techniques.

6.2 IMAGE SEGMENTATION BY THRESHOLDING

The simplest image segmentation problem occurs when an image contains an object having homogeneous intensity and a background with a different intensity level. Such an image can be segmented in two regions by simple thresholding:

$$g(x,y) = \begin{cases} 1 & \text{if } f(x,y) > T \\ 0 & \text{otherwise} \end{cases} \quad (6.2.1)$$

A simple thresholding algorithm is shown in Figure 6.2.1.

```
int thres(a,b, t, N1,M1,N2,M2)
image a,b;
int t;
int N1,M1,N2,M2;
/* Subroutine to threshold an image
   a,b: buffers
   t: threshold (integer)
   N1,M1: start coordinates
   N2,M2: end   coordinates    */
{ int i,j;
  for(i=N1; i<N2; i++)
    for(j=M1; j<M2; j++)
      if(a[i][j]<t) b[i][j]=0; else b[i][j]=1;
  return(0);
}
```

Fig. 6.2.1 Image thresholding algorithm.

The choice of threshold T can be based on the image histogram. If the image contains one object and a background having homogeneous intensity, it usually possesses a bimodal histogram like the one shown in Figure 6.2.2. The

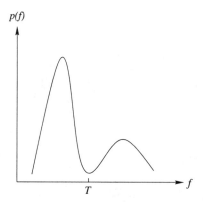

Fig. 6.2.2 Bimodal image histogram.

threshold is chosen to be at the local histogram minimum lying between the two histogram peaks. The local histogram maxima can be easily detected by using the top-hat transformation described in Chapter 7 [PIT90]. The calculation of the local histogram minimum is difficult if the histogram is noisy. Therefore, histogram smoothing (e.g. by using one-dimensional low-pass filtering) is recommended in this case. In certain applications, the intensity of the object or of the background is slowly varying. The image histogram may not contain two clearly distinguished lobes in this case, so a spatially varying thresholding can be applied [CHO72]. The image is divided into square blocks and the histogram and the corresponding threshold are calculated at each block. If the local histogram is not bimodal, the threshold is computed by interpolating the thresholds of the neighboring image blocks. When a local threshold is obtained, the thresholding operation (6.2.1) is applied to this block.

In certain cases the region boundary is desired, rather than the region itself. If the segmented image $g(x,y)$ is available, the boundary can be easily obtained by finding the transitions from one region to the other:

$$b(x,y) = \begin{cases} 1 & \text{if } ((g(x,y) \in R_i \text{ and } g(x, y-1) \in R_j, i \neq j) \\ & \text{or } (g(x,y) \in R_i \text{ and } g(x-1, y) \in R_j, i \neq j)) \\ 0 & \text{otherwise} \end{cases} \quad (6.2.2)$$

Exactly the same procedure can be used to obtain the boundaries, if more than two regions exist in the segmented image. Multiple thresholding can be used for segmenting images containing N objects, provided that each object R_i occupies a distinct intensity range which is defined by two thresholds T_{i-1}, T_i. The thresholding operation takes the following form:

$$g(x,y) = R_i \quad \text{if } T_{i-1} \leq f(x,y) \leq T_i, \quad i = 1, \ldots, N \quad (6.2.3)$$

The thresholds can be obtained from the image histogram by finding the $N-1$ minima between N consecutive histogram peaks. In many cases, the various histogram lobes are not clearly distinguished owing to the contribution of pixels lying in the transitions between two regions. If this contribution is eliminated or moderated, the resulting histogram has lobes that are clearly distinguished. The simplest method for histogram modification is to perform edge detection and to exclude all pixels belonging to edges from the histogram calculation. Another approach is to define a modified histogram of the form [NAG73], [WAH87]:

$$h(i) = \sum_{k=0}^{N_1-1} \sum_{l=0}^{N_2-1} t(e(k,l)) \delta(f(k,l) - i) \qquad (6.2.4)$$

where $e(k,l)$ is an edge detector output and $\delta(i)$ is the delta function. The modified histogram (6.2.4) differs from the classical histogram definition:

$$h(i) = \frac{1}{N_1 N_2} \sum_{k=0}^{N_1-1} \sum_{l=0}^{N_2-1} \delta(f(k,l) - i) \qquad (6.2.5)$$

because the contribution of each pixel $f(k,l)$ depends on a function of the local edge detector output. A monotonically decreasing function t can be chosen for histogram modification:

$$t(e(k,l)) = \frac{1}{1 + |e(k,l)|} \qquad (6.2.6)$$

Pixels close to edges (i.e. in transitions between two image regions) have a moderate contribution to the modified histogram. Histogram (6.2.4) tends to have well-defined peaks and valleys.

Finally, in certain applications (e.g. halftoning and pseudocoloring) the image must be segmented in N different regions each possessing an equal intensity range of the original image. The thresholding operation is described by the following equation:

$$g(k,l) = \begin{cases} R_i & \text{if } i[L/N] \leq f(k,l) < (i+1)[L/N], \; i = 0, 1, \ldots, N-2 \\ R_{N-1} & \text{if } (N-1)[L/N] \leq f(k,l) < L \end{cases} \qquad (6.2.7)$$

The corresponding algorithm is shown in Figure 6.2.3.

```
int segm(a,b, nr, N1,M1,N2,M2)
image a,b;
int nr;
int N1,M1,N2,M2;

/* Subroutine to segment an image in n regions
   a,b: buffers
```

```
       nr: number of regions (integer)
       N1,M1: start coordinates
       N2,M2: end    coordinates     */

{ int i,j;
  nr=256/nr;
  for(i=N1; i<N2; i++)
    for(j=M1; j<M2; j++)
      b[i][j]=((int)a[i][j]/nr)*nr;
  return(0);
}
```

Fig. 6.2.3 Image segmentation in N equirange regions.

The application of this algorithm to the segmentation of an image in four equirange regions is shown in Figure 6.2.4. If the image histogram is concentrated in a small intensity range, uniform thresholding (6.2.7) does not give good results, because a small number of regions may contain most of the image pixels, whereas the rest of the regions are almost non-existent. Non-uniform thresholding creates much better results in this case. A non-uniform thresholding technique is based on the histogram equalization technique, described in Chapter 3. Let $G(f(k,l))$ be the transformation function used in histogram equalization. The segmentation procedure is described by the equation:

$$g(k,l) = \begin{cases} R_i & \text{if } i[L/N] \leq G(f(k,l)) < (i+1)[L/N] \\ R_{N-1} & \text{if } (N-1)[L/N] \leq G(f(k,l)) < L \end{cases} \quad (6.2.8)$$

Non-uniform thresholding (6.2.8) generally results in image segmentations that are better than those given by (6.2.7).

Thresholding techniques described in this section guarantee region homogeneity but no region connectedness, because thresholding is performed on a

Fig. 6.2.4 (a) Original image; (b) image segmentation in four equirange regions.

pixel basis and no neighborhood information is taken into account. Proximity information is combined with homogeneity rules to produce connected regions in the techniques described in the next section.

Thresholding is essentially a special case of a pattern recognition approach applied to single feature cases (one dimensional data). Such a feature is the pixel luminance. The pattern recognition approach to image segmentation can be extended to the multi-dimensional case. The multidinensional data could be the pixels lying in an image neighborhood. The data dimensionality is equal to the number of pixels, that is, a 3 × 3 neighborhood results to a 9-dimensional feature space. Multidimensional data can be created by using texture features to be described in a subsequent section. Any pattern recognition approach can be used for performing image segmentation. Neural network algorithms (multilayer perceptrons, LVQs, RBFs) have been extensively used as pattern classifiers for image segmentation [SIM90], [KOH90]. Self-organizing feature maps (SOFMs, LVQs) and their variants have been extensively used for image classification/segmentation [KOT98IJCM]. They cluster feature vectors in classes. Each class can represent an image region. The number of classes can be predefined or adapted dynamically [KOT93CAIP]. LVQs that employ nonlinear operators (e.g. median) have been proven more robust to image noise than their linear counterparts [PIT96TIP]. The radial basis function networks (RBFs) have been used for image segmentation due to their functional approximation and localization properties. RBF networks have been suggested to be used for jointly modeling and segmenting image objects [BOR99]. We consider the gray level and image coordinates as input features. Gaussian functions, which geometrically model ellipses in 2-D, are used as RBF hidden unit activation functions. Gaussian functions are characterized by two parameters. The center corresponds to the center of the ellipse in the coordinate space and to the most likely value in the gray level domain, while the variance represents the width of the ellipse and the gray level variance respective. The cross correlation in coordinate domain means the orientation of the ellipse. The cross correlation between coordinates and gray level is set to 0. Various ellipses are joined together by means of the output weights in order to describe objects. Alpha-trimmed mean RBF algorithm has been proposed for modeling and segmenting objects [BOR99]. Alpha-trimmed mean is a robust statistics based algorithm which does not take into account the outliers. This algorithm can estimate quite accurately the shape of ellipses that are corrupted by noise. In the case of complex shapes the number of RBFs increases accordingly. An example of applying alpha-trimmed mean RBF networks in segmenting microscopy images of tooth pulp is presented in Figure 6.2.5a. Such images are very noisy due to the tissue structure. The segmentation of the blood vessels is provided in Figure 6.2.5b.

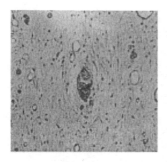
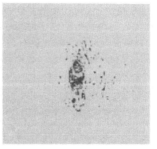

Fig. 6.2.5 (a) Microscopy images of tooth pulp; (b) Blood vessel segmented in tooth pulp. (*Courtesy of Dr. K. Lyroudia, Department of Dentistry, Aristotle University of Thessaloniki, Greece.*)

6.3 SPLIT/MERGE AND REGION GROWING ALGORITHMS

Geometrical proximity plays an important role in image segmentation. Pixels lying in the same neighborhood tend to have similar statistical properties and belong in the same image regions. Therefore, segmentation algorithms must incorporate, if possible, both proximity and homogeneity to produce connected image regions.

A simple approach to image segmentation is to start from some pixels (*seeds*) representing distinct image regions and to *grow* them, until they cover the entire image. The pixel seeds are usually chosen by the user in a supervised mode. At least one seed s_i, $i = 1, \ldots, N$, per image region R_i is chosen. In order to implement region growing, we need a rule describing a growth mechanism and a rule checking the homogeneity of the regions after each growth step. The growth mechanism is simple: at each stage (k) and for each region $R_i^{(k)}$, $i = 1, \ldots, N$, we check if there are unclassified pixels in the 8-neighborhood of each pixel of the region border. Before assigning such a pixel **x** to a region $R_i^{(k)}$, we check if the region homogeneity:

$$P(R_i^{(k)} \cup \{\mathbf{x}\}) = TRUE \qquad (6.3.1)$$

is still valid. The algorithm describing region growing is shown in Figure 6.3.1.

```
int region_grow(a,b,N1,M1,N2,M2,N,T)
image a;
image b;
int N1,N2,M1,M2,N,T;
/* Subroutine for region growing. Seed pixels labeled by their
   region number [1,...,N] are initially contained in array b.
   a: input image buffer
   b: output image buffer (initially it contains seed points)
```

```
   N1,M1: start coordinates
   N2,M2: end   coordinates
   N: number of regions
   T: threshold */
{
  image XC; image BR;
  int * NP; unsigned char * M;
  long * S;
  int x,y,dy,dx,k,l; int m,c,ap,i,ima,j,stop,times;
  long allc;
  /* Local image buffers.
     XC: Pixels that have already been allocated to a region
         have flag 1.
     BR: Buffer containing border information. Border pixels
         are labeled by their region number. */
  XC=build_image(NMAX,MMAX);
  BR=build_image(NMAX,MMAX);
  /* Local buffers.
     NP: number of points of each region.
     M: mean value of each region.
     S: sum of the intensities of the pixels of each region.*/

  for (y=M1; y<M2; y++)
  for (x=N1; x<N2; x++)
  if (b[y][x]>N) return(-97);

  NP=(int *)malloc((unsigned int)N*sizeof(int));
  M=(unsigned char *)malloc((unsigned int)N*
                                  sizeof(unsigned char));
  S=(long *)malloc((unsigned int)N*sizeof(long));
  /* Initialize arrays BR,XC and arrays NP,M,S. */
  for (y=M1;y < M2 ;y++)
    for (x=N1; x< N2;x++)
      { BR[y][x]=0; XC[y][x]=0; }
  for (x=0;x<N;x++)
   { NP[x]=0; M[x]=0; S[x]=0; }
  /* Use seeds from image b to initialize BR,XC,NP,M,S */
  i=0;
  for (y=M1;y<M2;y++)
    for (x=N1;x<N2;x++)
     if(b[y][x]!=0)
      { BR[y][x] = b[y][x]; XC[y][x] = 1;
        M[i] = a[y][x]; S[i] = a[y][x];
        NP[i]++ ; i++;
      }
```

```c
/* Variable allc contains the number of pixels that
   have not been labeled yet. */
m=0;c=0;allc=0;times=0; stop=0;

/* Main segmentation loop. */
while ( stop!=1 )
 { times++;
   /* Scan the entire image */
   for(y=M1+1; y<M2-1; y++)
    for(x=N1+1; x<N2-1; x++)
     { m=BR[y][x];
       /* For any current border point check if there is any
          pixel in its neighborhood that can be merged to the
          region to which the border belongs. */
       if ( m != 0)
       {
         for (dy=-1;dy<=1;dy++)
          for (dx=-1;dx<=1;dx++)
           { /* Candidate pixel must not be assigned to a
                region and its intensity must be close to the
                mean value of the region under consideration
                (their difference must be less than threshold
                T). Update NP,S,M,XC,BR. */
             c=XC[y+dy][x+dx];
             if((c!=1) && (abs(a[y+dy][x+dx]-M[m-1])<T))
              { b[y+dy][x+dx]=m; NP[m-1]++;
                S[m-1]=S[m-1]+a[y+dy][x+dx];
                XC[y+dy][x+dx] = 1; XC[y][x]=1;
                BR[y+dy][x+dx] = m;   BR[y][x] = 0;
              }
           }
         BR[y][x]=0;
       }
     }

   /* Count pixels that have not been assigned to any region.
      If this number did not change in this iteration, stop.*/
   ap=allc; allc=0;
   for(y=M1+1; y<M2-1; y++)
    for(x=N1+1; x<N2-1; x++)
     allc += abs(XC[y][x]-1);
   if (abs(allc-ap) == 0 ) stop=1;
 }

/*Calculate the mean of each region */
```

```
for (i=1; i<=N; i++) M[i-1]=(unsigned char)(S[i-1]/NP[i-1]);

/* Scan output image and label the assigned pixels by the
   mean value of the corresponding region. */
for (y=M1;y<M2;y++)
  for (x=N1;x<N2;x++)
    if(i=b[y][x]) b[y][x]=M[i-1];
mfreeuc2(XC,MMAX); mfreeuc2(BR,MMAX);
return(0);
}
```

Fig. 6.3.1 Region growing algorithm.

The inputs of the subroutine `region_grow()` are the image to be segmented and an image containing the seeds. These seeds grow iteratively. Each region has border pixels stored on image BR. For any current border pixel, we check if there is any pixel in its 3 × 3 neighborhood that can be merged with the region to which the border belongs. The candidate pixel **x** must not have been assigned to a region and its intensity must be close to the mean value of the region R under consideration, i.e. $|f(\mathbf{x}) - m_R|$ must be less than a threshold. The algorithm stops when no more region growing is possible. The performance of this algorithm depends heavily on the choice of the initial seeds. Ideally, one seed pixel per image region must be provided by the user. This process can be automated by using the global image histogram. Image pixels, whose intensity corresponds to histogram peaks, can be used as seeds. Usually there is more than one seed per image region. A merging procedure must be devised in order to merge adjacent regions that have similar statistical properties. The arithmetic mean m_i and the standard deviation s_i of a class R_i having n pixels:

$$m_i = \frac{1}{n} \sum_{(k,l) \in R_i} f(k,l) \qquad (6.3.2)$$

$$\sigma_i = \sqrt{\frac{1}{n} \sum_{(k,l) \in R_i} (f(k,l) - m_i)^2} \qquad (6.3.3)$$

can be used to decide if the merging of two regions R_1, R_2 is allowed. If their arithmetic means are close to each other:

$$|m_1 - m_2| < k\sigma_i \quad i = 1,2 \qquad (6.3.4)$$

the two regions are allowed to merge.

Region merging can be incorporated in the growing mechanism, in order to construct a combined merging/growing segmentation algorithm [LEV81],

[LEV85]. Let us suppose that no a priori information is available about the image. The image can be scanned in a row-wise manner. Each pixel is assigned to a region during the scanning procedure. Let us suppose that we are currently at the pixel (k,l). All pixels $(k-1,l)$ of the previous image row and pixels (k,j), $0 \leq j < l$, of the current image row have been assigned to image regions during image scanning in a row-wise manner. Therefore, pixel (k,l) has some adjacent regions R_i in its neighborhood, unless pixel (k,l) is the starting pixel $(0,0)$. First, we try to merge pixel (k,l) with one of its adjacent regions R_i. If this merge fails, or if no adjacent region exists (e.g. for the starting pixel $(0,0)$), this pixel is assigned to a new region. The performance of this algorithm is based on the choice of the merging rule $P(R_i \cup (k,l))$ of a pixel (k,l) to a class R_i. The merging rule can be based on the region mean m_i and standard deviation s_i described by (6.3.2, 6.3.3). If merging $R \cup (k,l)$ was allowed, the updated mean and standard deviation would be given by:

$$m'_i = \frac{1}{n+1}(f(k,l) + nm_i) \qquad (6.3.5)$$

$$\sigma'_i = \sqrt{\frac{1}{n+1}\left(n\sigma_i^2 + \frac{n}{n+1}[f(k,l) - m_i]^2\right)} \qquad (6.3.6)$$

Merging is allowed if the pixel intensity $f(k,l)$ is close to the class mean m_i:

$$|f(k,l) - m_i| \leq T_i(k,l) \qquad (6.3.7)$$

The threshold T_i may vary with the class R_i and the pixel intensity $f(k,l)$. It can be chosen in the following way:

$$T_i(k,l) = \left(1 - \frac{\sigma'_i}{m'_i}\right) T \qquad (6.3.8)$$

where m'_i, σ'_i are given by (6.3.5, 6.3.6). If no region in the pixel neighborhood satisfies expression (6.3.7), this pixel is assigned to a new region. If more than one region in the pixel neighborhood satisfy (6.3.7), the pixel is merged to the region R_i having minimal difference $|f(k,l) - m_i|$. The growth of the region depends heavily on the parameter T used in the threshold calculation. If it is small, the threshold $T_i(k,l)$ is small for all regions and the merging is difficult. Thus, many small regions are created. If it is large, it may create regions having less homogeneity and large standard deviation. Thresholds $T_i(k,l)$ also depend on the ratio σ'_i/m'_i. If the regions are homogeneous, this ratio tends to zero and the threshold $T_i(k,l)$ tends to T. Thus, T denotes the maximal allowed difference of $|f(k,l) - m_i|$ for the merging of a pixel (k,l) with a highly homogeneous region R_i. Less homogeneous regions have higher σ'_i/m'_i ratio. The corresponding threshold $T_i(k,l)$ becomes smaller and the growth of such regions is restricted. A region merging subroutine, called region_merge(), is described in Figure 6.3.2.

```
int region_merge(a,b,N1,M1,N2,M2,T,REGMAX)
image a;
image b;
int N1,N2,M1,M2,T,REGMAX;
/* Subroutine for region segmentation by merging.
   a: input image buffer
   b: output image buffer
   N1,M1: start coordinates
   N2,M2: end   coordinates
   T: threshold */
{
  int x,y,dy,dx,merge,nubs; int m,i,j,c,dmin,min,cmin,ml;
  image XC;
  int * NP;
  unsigned char * ME; long * S;

  /* Local image buffers. XC: Pixels that have already been
     allocated to a region have flag 1.*/
  XC=build_image(NMAX,MMAX);
  /* Local buffers.
     NP: number of points of each region.
     ME: mean value of each region.
     S: sum of the intensities of the pixels of each region. */
  NP=(int *)malloc((unsigned int)REGMAX*sizeof(int));
  ME=(unsigned char *)malloc((unsigned int)REGMAX
                                    *sizeof(unsigned char));
  S=(long *)malloc((unsigned int)REGMAX*sizeof(long));
  /* Initializations */
  for (y=M1; y<M2; y++)
   for (x=N1; x<N2; x++)
       { b[y][x]=0 ; XC[y][x]=0; }
   for (i=0;i<REGMAX;i++)
     { NP[i] = 0; S[i]=0; ME[i] = 0; }
  /* nubs: number of regions.
     merge: flag to test if merge will be performed. */
  nubs=1; merge=0;

  /* Main segmentation loop */
  for(y=M1+1; y<M2-1; y++)
   for(x=N1+1; x<N2-1; x++)
     {
       /* Create first region for the pixel at (1,1).
          Update XC, b, NP, S, ME.*/
       if ( (x==1) && (y==1) )
        { XC[y][x] = 1; b[y][x] = nubs;
```

```
         NP[nubs-1] = 1;  S[nubs-1] =(long)a[y][x];
         ME[nubs-1] = a[y][x];
  }
  else
  { /* Rest of the image pixels.
       For each pixel, we examine its NSHP 4-neighborhood
       and we test if this pixel can be merged with any
       region in this neighborhood. Merging is allowed if
       the pixel intensity is close to the region mean value.
       If more than one merge are possible, the region
       with the closest mean value is chosen. */
    dmin = abs ( a[y][x] -ME [0] ) ; min = 1;
    for (dy = -1;dy<=0;dy++)
     for (dx=-1;dx<=1;dx++)
      if  ( ( (x+dx != x) || (y+dy!= y) ) &&
            ( (y+dy!=y) || (x+dx!= x+1) )  )
        if(XC[y+dy][x+dx]!=0)
         { c=abs( a[y][x] - ME[XC[y+dy][x+dx]-1]);
           if(c<T)
            { merge = 1;
              if(c<dmin) { dmin=c; min=XC[y+dy][x+dx]; }
            }
         }
    /* If merge is allowed,update XC,b,NP,S,ME */
    if(merge==1)
     { m = min;  XC[y][x] = m;  b[y][x] =   m;
       NP[m-1]++;  S[m-1]   += a[y][x];
       ME[m-1] = (unsigned char)(S[m-1] / NP[m-1]) ;
     }
    /* If merge is not allowed with the immediately
       neighboring region, try to merge it with
       any other possible region. */
    if (merge==0)
    {
      cmin=255; ml=1;
      for (m=1;m<=nubs;m++)
       { c=abs( a[y][x] - ME[m-1]  );
         if (c<cmin){cmin=c; ml=m;}
         /* If merge is allowed, update XC,b,NP,S,ME. */
         if ( c <  T )
          { merge=1 ; XC[y][x] = ml; b[y][x] =   ml;
            NP[ml-1]++ ; S[ml-1]   += a[y][x];
            ME[ml-1]=(unsigned char)(S[ml-1]/NP[ml-1]);
          }
       }
```

```
        /* If no merge is allowed and the max number
           of regions is not exceeded, create a new
           region and update XC,b,NP,S,ME. */
        if (merge == 0 && nubs<REGMAX)
          { nubs++; XC[y][x]=nubs; b[y][x]= nubs;
            NP[nubs-1]=1; S[nubs-1]=a[y][x];
            ME[nubs-1]=a[y][x];
          }
        else merge=0 ;
      }
    else merge = 0;
   }  /*else*/
 }    /*for loop*/

/* Scan output image and label the assigned pixels by
   the mean value of the corresponding region. */
for (y=M1;y<M2;y++)
 for (x=N1;x<N2;x++)
  for (i=1;i<=REGMAX;i++)
    if (i=b[y][x]) b[y][x]=ME[i-1] ;
mfreeuc2(XC,MMAX);

/* Return the number of regions created. */

 return(nubs);
}
```

Fig. 6.3.2 Merging algorithm for image segmentation.

The candidate pixel (k,l) for merging to a region R_i must have an intensity close to the mean value m_i of this region, i.e. $|f(k,l) - m_i|$ must be less than a threshold. The subroutine region_merge() tries first to merge the current pixel (k,l) with neighboring regions. For this task, it employs the non-symmetric half plane neighborhood $(k-1,l-1)$, $(k-1,l)$, $(k-1,l+1)$, $(k,l-1)$. If such merging is not possible, it tries to merge it with any of the known regions. If this merging fails and the maximum number of allowable regions is not exceeded, a new region is created.

The opposite approach to region merging is *region splitting*. It is a top-down approach and it starts with the assumption that the entire image is homogeneous. If this is not true, the image is split into four subimages. This splitting procedure is repeated recursively until homogeneous image regions are encountered. The splitting algorithm is described in Figure 6.3.3.

290   IMAGE SEGMENTATION ALGORITHMS

```c
int region_split(a,b,N1,M1,N2,M2,N,T)
image a;
image b;
int N1,N2,M1,M2,T;
int *N;

/* Subroutine for region splitting.
   a: input image buffer
   b: output image buffer
   N1,M1: start coordinates
   N2,M2: end   coordinates
   N: number of regions (originally 0);
   T: threshold */
{
   int i,j,test; long sum;

   /* Test homogeneity of an image region. */
   test=test_homogeneity(a,N1,M1,N2,M2,T);
   /* If it is inhomogeneous, split it into four regions. */
   if (  (test==0) && (N2-N1 > 1) && (M2-M1 > 1) )
     {
        region_split(a,b,N1,M1,N1+(N2-N1)/2,M1+(M2-M1)/2,N,T);
        region_split(a,b,N1+(N2-N1)/2,M1,N2,M1+(M2-M1)/2,N,T);
        region_split(a,b,N1,M1+(M2-M1)/2,N1+(N2-N1)/2,M2,N,T);
        region_split(a,b,N1+(N2-N1)/2,M1+(M2-M1)/2,N2,M2,N,T);
     }
   /* If the region is homogeneous, calculate its mean and
      update output image buffer. */
   else
     {
       (*N)++; sum=0;
       for (j=M1;j<M2;j++)
        for (i=N1;i<N2;i++)
         sum +=a[j][i];
       sum /=((N2-N1)*(M2-M1));
       for (j=M1;j<M2;j++)
        for (i=N1;i<N2;i++)
         b[j][i]=(unsigned char)sum;
     }
   return(0);
}

int test_homogeneity(a,N1,M1,N2,M2,T)
image a;
int N1,N2,M1,M2,T;
```

```
/* Subroutine to test region homogeneity.
   a: input image buffer
   N1,M1: start coordinates
   N2,M2: end   coordinates
   T: threshold */
{
  int i,j,max,min;
  max=0; min=255;
  for (i=N1;i<N2;i++)
   for (j=M1;j<M2;j++)
    { if(a[i][j]<min) min=a[i][j];
      if(a[i][j]>max) max=a[i][j];
    }
  if(abs(max-min)<T) return(1); else return(0);
}
```

Fig. 6.3.3 Region splitting algorithm.

Subroutine `region_split()` is recursive. If the region under consideration is inhomogeneous, it calls itself four times, one for each quarter of the region. Region homogeneity is tested by the subroutine `test_homogeneity()`. This subroutine calculates the range of image intensities in the region (maximal intensity minus minimal intensity) and checks if the range is smaller than a certain threshold. Other segmentation rules can be easily employed by modifying this subroutine. Region splitting has some interesting properties. If the original image is square $N \times N$ having dimensions that are powers of 2 ($N = 2^n$), all regions produced by the splitting algorithm are squares having dimensions $M \times M$, where M is a power of 2 as well: $M = 2^m$, $m \leq n$. Since the procedure is recursive, it produces an image representation that can be described by a tree whose nodes have four sons each. Such a tree is called a *quadtree* and is a very convenient region representation scheme [BAL82], [SAM90]. A quadtree is shown in Figure 6.3.4. Pure splitting techniques have a major disadvantage. They create regions R_i, R_j that may be adjacent and homogeneous: $P(R_i \cup R_j) = TRUE$. Ideally, these regions should be merged. This observation leads to the so-called *split and merge* algorithm [HOR74]. It is an iterative algorithm that includes both splitting and merging at each iteration:

1. If a region R is inhomogeneous ($P(R) = FALSE$) it is split into four subregions.

2. If two adjacent regions R_i, R_j are homogeneous ($P(R_i \cup R_j) = TRUE$), they are merged.

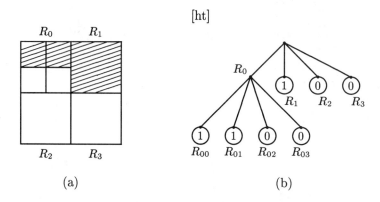

Fig. 6.3.4 (a) Original image; (b) quadtree representation.

The algorithm stops when no further splitting or merging is possible. A description of a split and merge algorithm is given in Figure 6.3.5.

```
int region_split_merge_c(a,b,MARRAY,LABEL,N1,M1,N2,M2,N,T,REGMAX)
image a;
image b;
unsigned char * MARRAY;
int ** LABEL;
int N1,N2,M1,M2,T,REGMAX;
int *N;

/* Subroutine for split and merge algorithm.
   This algorithms updates the region intensity means during
   merging.
   a: input image buffer
   b: output image buffer
   MARRAY: vector containing the mean values of each region.
   LABEL:  int array containing region labels.
   N1,M1: start coordinates
   N2,M2: end   coordinates
   N: number of regions (originally 0);
   T: threshold
   REGMAX: maximal allowable number of image regions */

{
  int i,j,test;
  int ret=0,ret1,ret2,ret3,ret4;
  long sum;
```

```
   /* Test homogeneity of an image region */
   test=test_homogeneity(a,N1,M1,N2,M2,T);
   /* If the region is inhomogeneous and its size is greater
      than 1 pixel, split it into four regions */
   if ( (test==0) && (N2-N1 > 1) && (M2-M1 > 1) )
    {
      ret1=region_split_merge_c(a,b,MARRAY,LABEL,
                  N1,M1,N1+(N2-N1)/2,M1+(M2-M1)/2,N,T,REGMAX);
      ret2=region_split_merge_c(a,b,MARRAY,LABEL,
                  N1+(N2-N1)/2,M1,N2,M1+(M2-M1)/2,N,T,REGMAX);
      ret3=region_split_merge_c(a,b,MARRAY,LABEL,
                  N1,M1+(M2-M1)/2,N1+(N2-N1)/2,M2,N,T,REGMAX);
      ret4=region_split_merge_c(a,b,MARRAY,LABEL,
                  N1+(N2-N1)/2,M1+(M2-M1)/2,N2,M2,N,T,REGMAX);
      if(ret1==-1||ret2==-1||ret3==-1||ret4==-1) ret=-1;
    }
   else
    {
     /* If the region is homogeneous, calculate its mean value
        and assign a label and the mean value to the region.
        Labels take values 1,...,N. */
     sum=0;
     (*N)++;
     if ( (*N) > REGMAX) return(-92);
     for (j=M1;j<M2;j++)
      for (i=N1;i<N2;i++)
        { sum +=a[j][i];
          LABEL[j][i]=(*N);
        }
     sum /=((N2-N1)*(M2-M1));
     for (j=M1;j<M2;j++)
      for (i=N1;i<N2;i++)
        b[j][i]=(unsigned char)sum;
     MARRAY[ (*N) ]=(unsigned char)sum ;
     /* perform merging */
     if ( (*N) > 1 ) rmerge_c(a,b,MARRAY,LABEL,N1,M1,N2,M2,N,T);
    }
      return(ret);
}


int rmerge_c(a,b,MARRAY,LABEL,N1,M1,N2,M2,N,T)
image a;
image b;
unsigned char * MARRAY;
```

```
int **LABEL;
int N1,N2,M1,M2;
int *N;
int T;

/* Subroutine for merging adjacent regions.
   Region intensity means are updated after merging.
   a: input image buffer
   b: output image buffer
   MARRAY: vector containing the mean values of each region.
   LABEL:  int array containing the region labels.
   N1,M1: start coordinates
   N2,M2: end   coordinates
   N: number of regions
   T: threshold */
{
  int i,x,xd,xu,yd,yu,y,cmin,c,merging_label,merge=0;
  long sum,count;
  sum=0; count=0;

  /* Calculate the region having the closest mean,
     along the upper boundary of current region */
  if (N1-1>=0) xd=N1-1;else xd=0;
  if (N2+1<NMAX) xu=N2+1;else xu=N2;
  if (M1-1>=0) yd=M1-1;else yd=0;
  if (M2+1<MMAX) yu=M2+1;else yu=M2;

  /* Initialize min variables to a big value */
  cmin=255;
  if(M1-1>=0)
   {
    y=M1-1;
    for (x=xd;x<xu;x++)
     { if (b[y][x]!=0)
        {
          c=(int)abs((float)b[y][x]-(float)MARRAY[ (*N) ]);
             if (c<cmin)
             { cmin=c ;
               merging_label=LABEL[y][x];
             }
        }
     }
   }
  /* Calculate the region having the closest mean,
     along the lower boundary of current region */
```

```
if(M2+1<MMAX)
 {
  y=M2+1;
  for (x=xd;x<xu;x++)
   { if (b[y][x]!=0)
       {
          c=(int)abs((float)b[y][x]-(float)MARRAY[ (*N) ]) ;
             if (c<cmin)
                { cmin=c ;
                  merging_label=LABEL[y][x];
                }
       }
   }
 }
/* Calculate the region having the closest mean,
   along the left boundary of current region */
if(N1-1>=0)
 {
  x=N1-1;
  for (y=yd;y<yu;y++)
   { if (b[y][x]!=0)
       {
          c=(int)abs((float)b[y][x]-(float)MARRAY[ (*N) ]) ;
             if (c<cmin)
               { cmin=c ;
                 merging_label=LABEL[y][x];
               }
       }
   }
 }
/* Calculate the region having the closest mean,
   along the right boundary of current region */
if(N2+1<NMAX)
 {
  x=N2+1;
  for (y=yd;y<yu;y++)
   { if (b[y][x]!=0)
       {
          c=(int)abs((float)b[y][x]-(float)MARRAY[ (*N) ]) ;
             if (c<cmin)
               { cmin=c ;
                 merging_label=LABEL[y][x];
               }
       }
    }
```

```
    }
/* Perform merging */
if(cmin<T)
{
   /* Calculate the mean intensity of the new region */
   sum=0;count=0;
    for (y=0;y<MMAX;y++)
     for (x=0;x<NMAX;x++)
      if(LABEL[y][x]==*N || LABEL[y][x]==merging_label)
      { sum +=a[y][x];
        count++;
      }
   /* Update the mean value of the new region */
   if(count!=0)
    { sum /= count;
      for (y=0;y<MMAX;y++)
       for (x=0;x<NMAX;x++)
        if(LABEL[y][x]==*N || LABEL[y][x]==merging_label)
        { b[y][x]=(unsigned char) sum;
          LABEL[y][x]=merging_label;
        }
    }
   /* Update the vector of the mean values */
   MARRAY[merging_label]=(unsigned char)sum;
   (*N)--;
 }
}
```

Fig. 6.3.5 Split and merge algorithm.

Subroutine `region_split_merge()` is recursive. If a region is inhomogeneous, it is split into four subregions by calling this subroutine four times. If the region is homogeneous, its mean is calculated and compared to the means of the neighboring regions. Possible mergings are checked by subroutine `merge()`. If more than one merge is possible, the best merge is chosen, i.e. with the neighboring region that has the closest mean value to the mean of the region under consideration. The split and merge algorithm produces more compact regions than the pure splitting algorithm. Its major disadvantage is that it does not produce quadtree region descriptions. Several modifications of the basic split and merge algorithm have been proposed to solve this problem. The most straightforward procedure is to use the splitting algorithm and to postpone merging until no further splitting is possible. This procedure produces a quadtree representation at the end of the splitting stage. The adjacent quadtree regions are merged during the final merging step.

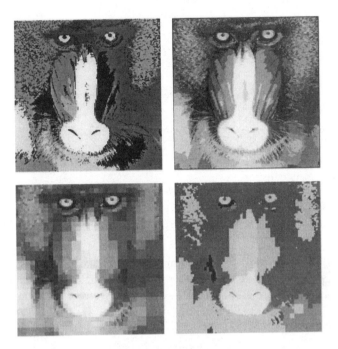

Fig. 6.3.6 (a) Result of the region growing algorithm; (b) output of the region merge algorithm; (c) result of the region split subroutine; (d) output of the split and merge algorithm.

The results of all the methods described in this section are shown in Figure 6.3.6. The test image of Figure 6.2.4a has been employed. Figure 6.3.6a shows the result of a region growing algorithm when eight seeds corresponding to eight different regions are employed. The output of the region merge algorithm is shown in Figure 6.3.6b. It can be seen that the results are very good. The region split algorithm produces good results as well, as can be seen in Figure 6.3.6c. However, it fails to preserve small regions and region boundaries. Furthermore, the square shape of the resulting regions is evident. Finally, the number of regions produced is large, because no merging of the adjacent regions is allowed. The result of region split and merge is shown in Figure 6.3.6d. The number of resulting regions is smaller and their shape is smoother.

6.4 RELAXATION ALGORITHMS IN REGION ANALYSIS

All region segmentation methods described in the previous sections are deterministic, in the sense that they assign each image pixel to just one region R_i, $i = 1, \ldots, N$. Although such a segmentation is ultimately desirable, it is

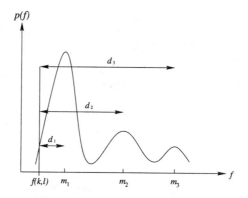

Fig. 6.4.1 Distances on an image histogram.

not always useful to employ such segmentation methods, because they treat ambiguous cases (e.g. pixels lying in transition regions) in a rather inflexible way. Instead, it is more useful to produce confidence vectors $\mathbf{p}_k$ for each pixel $\mathbf{x}_k$ that contain the probabilities $p_k(i)$ that a pixel $\mathbf{x}_k$ belongs to a class R_i, $i = 1, \ldots, N$:

$$\mathbf{p}_k = [p_k(1), \ldots, p_k(N)]^T \tag{6.4.1}$$

Confidence vectors can easily be used to produce deterministic segmentation: pixel $\mathbf{x}_k$ is assigned to the region R_l having the maximal probability $p_k(l)$. Probabilities $p_k(l)$, also called *confidence weights*, must satisfy the following relations:

$$0 \le p_k(i) \le 1 \tag{6.4.2}$$

$$\sum_{i=1}^{N} p_k(i) = 1 \tag{6.4.3}$$

There are several ways to obtain an initial estimate of the confidence vectors. Image histograms can be used for this purpose [NAG80]. Let m_i, $i = 1, \ldots, N$, be the arithmetic means of the intensity of each region that usually correspond to histogram peaks and $f(\mathbf{x}_k) = f(n, l)$ the pixel intensity at location $\mathbf{x}_k = (n, l)$. The initial estimate of the confidence weight $p_k^{(0)}(i)$, $i = 1, \ldots, N$, is given by:

$$p_k^{(0)}(i) = \frac{1/|f(n,l) - m_i|}{\sum_{i=1}^{N} 1/|f(n,l) - m_i|} \qquad i = 1, \ldots, N \tag{6.4.4}$$

It is inversely proportional to the distance $d_i = |f(n,l) - m_i|$ of the pixel intensity from the region arithmetic mean m_i, $i = 1, \ldots, N$. Such distances are shown in Figure 6.4.1. In many cases, it is highly probable that two adjacent image pixels belong to two specific classes R_i, R_j. For example, it is probable that adjacent pixels belong to the regions 'sea' and 'boat' in a natural scene. Such classes are called compatible. Incompatible regions are those that are not expected to be found in adjacent image locations. For example, the regions 'water' and 'fire' can be highly incompatible. The compatibility between two regions R_i, R_j is described in terms of a *compatibility function* $r(i, j)$, whose range is $-1 \leq r(i, j) \leq 1$. The value of the compatibility function has the following meaning:

$$r(i,j) \begin{cases} < 0 & \text{Regions } R_i, R_j \text{ are incompatible} \\ = 0 & \text{Regions } R_i, R_j \text{ are independent} \\ > 0 & \text{Regions } R_i, R_j \text{ are compatible} \end{cases} \quad (6.4.5)$$

Compatibility functions are known a priori or can be estimated from an initial image segmentation. Once they are known, they can be used to change the confidence vectors of an image in the following way. Incompatible regions tend to compete in adjacent image pixels, whereas compatible regions tend to cooperate. Competition and cooperation can be continued in an iterative way until a steady state is reached. Each pixel $\mathbf{x}_k$ receives confidence contributions from any pixel $\mathbf{x}_e$ lying in its neighborhood. The resulting change in confidence weight $p_k(i)$ of the pixel $\mathbf{x}_k$ at step (n) is the following:

$$\Delta p_k^{(n)}(i) = \sum_l d_{kl} \left[\sum_{j=1}^N r_{kl}(i,j) p_l^{(n)}(j) \right] \quad (6.4.6)$$

where the summation is performed over all confidence weights $p_l(j)$ of all pixels $\mathbf{x}_l$ lying in the neighborhood of pixel $\mathbf{x}_k$. The parameters d_{kl} weigh the contributions to pixel $\mathbf{x}_k$ coming from neighboring pixels $\mathbf{x}_l$. The sum is chosen to be equal to 1:

$$\sum_l d_{kl} = 1 \quad (6.4.7)$$

The updated probabilities $p_k^{n+1}(i)$ for the pixel x_k are given by [ROS76]:

$$p_k^{(n+1)}(i) = \frac{p_k^{(n)}(i)[1 + \Delta p_k^{(n)}(i)]}{\sum_{i=1}^N p_k^{(n)}(i)[1 + \Delta p_k^{(n)}(i)]} \quad (6.4.8)$$

The summation in the denominator ensures the validity of (6.4.2, 6.4.3) at each iteration. The iterative equations (6.4.6, 6.4.8) form the *relaxation labeling* algorithm for region segmentation. The iterations stop when convergence is achieved. The algorithm is expected to produce relatively large connected homogeneous image regions by removing small spurious noisy regions within larger regions. When the algorithm stops, a deterministic segmentation can be easily obtained by choosing the maximal confidence weights $p_k(i)$ and the corresponding classes at each image pixel $\mathbf{x}_k$. The performance of relaxation techniques depends greatly on the choice of the compatibility functions $r_{kl}(i, j)$. Wrong compatibility functions may result in algorithmic instabilities. A method to estimate compatibility functions from the initial probabilities $p_k^{(0)}(i)$ (6.4.4) is described in [PEL78]. Let $\mathbf{x}_l$ be the pixels belonging to the 4-connected neighborhood of $\mathbf{x}_k$. The mutual information concept leads to the following estimates of the compatibility functions:

$$r_{kl}(i,j) = \ln \frac{N^2 \sum_{k=1}^{N^2} p_k^{(0)}(i) p_l^{(0)}(i)}{\sum_{k=1}^{N^2} p_k^0(i) \sum_{k=1}^{N^2} p_l^{(0)}(i)} \qquad (6.4.9)$$

where N^2 is the number of pixels in the original image. The compatibility functions (6.4.9) must be scaled to ensure that they take values in the range $[-1, 1]$.

6.5 CONNECTED COMPONENT LABELING

Digital image segmentation produces either a binary or a multivalued image output $g(k,l)$. Each image region is labeled by a region number. However, each region may consist of several disconnected subregions. An important task is to identify and label the various connected components of each region. In the following we shall limit our discussion to binary images $g(k,l)$. In this case, *connected component labeling* assigns a unique number to each blob of 1s. If each blob corresponds to a single object, connected component labeling counts the objects that exist in a binary image. Labeling algorithms can be divided into two large classes: (a) local neighborhood algorithms and (b) divide-and-conquer algorithms. The algorithms belonging to the first class perform iterative local operations. Such a labeling algorithm is shown in Figure 6.5.1.

```
int grass_label(a,b,N,h,NS,N1,M1,N2,M2)
image a, b;
int *N;
vector h;
int NS;
int N1,M1,N2,M2;

/* Subroutine for counting objects in a binary image
    a: input image buffer
    b: output labeled image buffer
    *N: number of objects
    h: buffer to store object areas
    NS: maximal allowable number of objects
        (NS must be less than 255!)
    N1,M1: start coordinates
    N2,M2: end   coordinates      */

{ int col, row, i, j, number;
  for(i=0;i<NS;i++) h[i]=0.0;
```

```
    number=0;
    for(i=N1; i<N2; i++)
      for(j=M1; j<M2; j++)
        if(a[i][j]==1)
          {
            grass(a,b,i,j,number+1,h+number,N1,M1,N2,M2);
            number++;
            if(number>NS) goto RET;
          }
    RET:
    *N=number;
    return(0);
}

int grass(a,b,row,col,number,area,N1,M1,N2,M2)
image a,b;
int row,col,number;
float * area;
int N1,M1,N2,M2;

/* Subroutine to clear a binary object that starts at
   (row, col) and calculates its area
   a: input image buffer
   b: labeled image buffer
   row,col: pixel coordinates
   number: current object number
   *area: object area (in pixels)
   N1,M1: start coordinates
   N2,M2: end   coordinates        */

{ int i,j;
  a[row][col]=0;
  b[row][col]=number;
  (*area)++;
  for(i=-1; i<=1; i++)
  for(j=-1; j<=1; j++)
    if( (row+i)>=N1 && (row+i)<N2 && (col+j)>=M1 && (col+j)<M2)
      if(a[row+i][j+col]==1)
          grass(a,b,row+i,col+j,number,area,N1,M1,N2,M2);
  return(0);
}
```

Fig. 6.5.1 Connected component labeling algorithm.

302 IMAGE SEGMENTATION ALGORITHMS

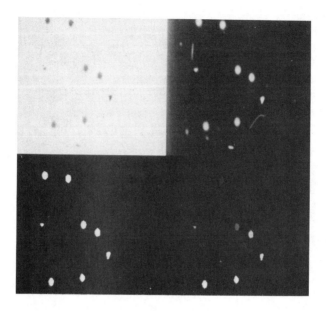

Fig. 6.5.2 (a) Test image; (b) negative image; (c) thresholded image; (d) labeled connected regions.

The algorithm uses the grassfire concept implemented in the recursive subroutine `grass()`. The image is scanned in a row-wise manner until the first pixel at the boundary of a binary image object is hit. A 'fire' is set at this pixel that propagates to all pixels belonging to the 8-neighborhood of the current pixel. This operation is continued recursively until all pixels of the image object are 'burnt' and the fire is extinguished. After the end of this operation, all pixels belonging to this object have value 0 and cannot be distinguished from the background. A byproduct of the subroutine `grass()` is the area of the object. This procedure is repeated until all objects in the image are counted. Subroutine `grass()` is recursive. Therefore, a relatively large stack is needed when it is used. The application of this algorithm is shown in Figure 6.5.2. A digital image containing particle traces and shown in Figure 6.5.2a is used as a test image. Its negative image is shown in Figure 6.5.2b. Thresholding can be applied to the negative image to produce the image of Figure 6.5.2c. The labeling algorithm identifies nine objects. They are shown in Figure 6.5.2d by using different grayscales (some of them are depicted by using a low intensity that cannot be easily distinguished from the background). Another local neighborhood labeling algorithm is the following. Each pixel $f(n, l)$ having value 1 is labeled by the concatenation of its n, l coordinates [CYP89]. Subsequently, we scan the labeled image and assign to each pixel the minimum of the labels in its 4-connected or 8-connected neighborhood. This process is repeated until no more label changes are made. This algorithm converges relatively fast for small convex objects. The convergence is slower when long and thin objects exist in the image. Parallel versions of this algorithm and similar algorithms are presented in [CYP89], [ROS83]. Another local neighborhood algorithm is the *blob coloring* algorithm described in [BAL82]. This algorithm has two passes. In the first pass, colors are assigned to image pixels by using a three-pixel L-shaped mask. Color equivalences are established and stored, when needed. In the second pass, the pixels of each connected region are labeled with a unique color by using the color equivalences obtained in the first pass. Another algorithm of this class is based on the *shrinking* operation proposed by Levialdi [LEV72]. If a pixel $f(n,l)$ has value 1, it retains this value after local shrinking if and only if at least one of its three East, South or South-East neighbors has value 1. This local operation is described by the following recursive relation:

$$f(n,l) = h(h(f(n,l-1) + f(n,l) + f(n+1,l) - 1) \\ + h(f(n,l) + f(n+1,l-1) - 1)) \quad (6.5.1)$$

where function $h(t)$ is given by:

$$h(t) = \begin{cases} 0 & \text{for } t \leq 0 \\ 1 & \text{for } t > 0 \end{cases} \quad (6.5.2)$$

After repeated scanning of the binary image with (6.5.1), each connected component shrinks to the North-West corner of its bounding box, before it vanishes at the next shrinking operation.

A simple divide-and-conquer algorithm for connected component labeling can work along similar lines to the split and merge algorithm described in a previous section. Inhomogeneous regions consisting of 0s and 1s are split recursively until we reach homogeneous regions consisting only of 1s. These regions are assigned a unique label. This is the split step. Label equivalences can be established by checking the borders of all homogeneous regions. Those regions having equivalent labels are merged to a single connected component. Several variations as well as parallel implementations of this basic algorithm have been proposed [NAS80].

6.6 TEXTURE DESCRIPTION

An important characteristic of an image is its *texture*. Texture features are extensively used in region segmentation. Despite this broad use, there is no unique and universally agreed definition of image texture. Generally speaking, it is a measure of image coarseness, smoothness and regularity. Texture description techniques can be grouped into three large classes: *statistical, spectral* and *structural*. Statistical descriptors are based on region histograms, their extensions and their moments; they measure contrast, granularity and coarseness. Spectral techniques are based on the autocorrelation function of a region or on the power distribution in the Fourier transform domain in order to detect texture periodicities. Finally, structural techniques describe the texture by using pattern primitives accompanied by certain placement rules. In the following, we shall present some statistical and spectral textural descriptors.

The simplest texture descriptors are based on the image histogram $p_f(f)$. Let f_k, $k = 1, \ldots, N$, be the various image intensity levels. The first four central moments are given by [LEV85]:

1. Mean

$$\mu = \sum_{k=1}^{N} f_k p_f(f_k) \tag{6.6.1}$$

2. Variance

$$\sigma^2 = \sum_{k=1}^{N} (f_k - \mu)^2 p_f(f_k) \tag{6.6.2}$$

3. Skewness

$$\mu_3 = \frac{1}{\sigma^3} \sum_{k=1}^{N} (f_k - \mu)^3 p_f(f_k) \tag{6.6.3}$$

4. Kurtosis

$$\mu_4 = \frac{1}{\sigma^4} \sum_{k=1}^{N} (f_k - \mu)^4 p_f(f_k) - 3 \tag{6.6.4}$$

Image *entropy* is defined in terms of the histogram as well:

$$H = -\sum_{k=1}^{N} p_f(f_k) \ln p_f(f_k) \tag{6.6.5}$$

Let us suppose that the histogram was calculated within an image region. The mean μ gives an estimate of the average intensity level in this region and the variance σ^2 is a measure of the dispersion of region intensity. Histogram skewness is a measure of histogram symmetry. It shows the percentage of the region's pixels that favor intensities on either side of the mean. Kurtosis is a measure of the tail of the histogram; long-tailed histograms correspond to spiky regions. The subtraction in (6.6.4) ensures that the kurtosis of a Gaussian distribution is normalized to zero. The big advantage of the texture descriptors (6.6.1–6.6.5) is their computational simplicity. Subroutines for their calculation are shown in Figure 6.6.1.

```
float meanvalu(hv)
    vector hv;
/* Subroutine to calculate histogram mean value
    hv: histogram buffer*/
```

```
{
 int i;
 float meanvalue;
 meanvalue=0.0F;
 for (i=0; i<NSIG; i++) meanvalue+=i*hv[i];
 return(meanvalue);
}

float varianc(hv)
vector hv;
/* Subroutine to calculate histogram variance
   hv: histogram buffer*/
{
 float variance,meanvalue,temp;
 int i;
 float meanvalu(vector);
 meanvalue=meanvalu(hv);
 variance=0.0F;
 for (i=0; i<NSIG; i++)
    { temp=i-meanvalue;  variance+=temp*temp*hv[i]; }
 return(variance);
}

float skewnes(hv)
vector hv;
/* Subroutine to calculate histogram skewness
   hv: histogram buffer*/
{
 int i;
 float temp,meanvalue,variance,skewness;
 float meanvalu(vector);
 float varianc(vector);
 meanvalue=meanvalu(hv);
 variance=varianc(hv);
 skewness=0.0F;
 for (i=0; i<NSIG; i++)
    { temp=i-meanvalue; skewness+=temp*temp*temp*hv[i]; }
 temp=(float)sqrt((double)variance);
 temp=temp*temp*temp;
 skewness=skewness/temp;
 return(skewness);
}

float kurtosi(hv)
  vector hv;
```

```
/* Subroutine to calculate histogram kurtosis
    hv: histogram buffer*/
{
 int i;
 float temp,meanvalue,kurtosis;
 float meanvalu(vector);
 meanvalue=meanvalu(hv);
 kurtosis=0.0F;
 for (i=0; i<NSIG; i++)
   { temp=i-meanvalue; kurtosis+=temp*temp*temp*temp*hv[i]; }
 kurtosis=kurtosis/4;
 kurtosis=kurtosis-3.0;
 return(kurtosis);
}

float entrop(hv)
 vector hv;
/* Subroutine to calculate histogram entropy
    hv: histogram buffer*/
{
 int i;
 float entropy;
 entropy=0.0F;
 for (i=0; i<NSIG; i++)
  if(hv[i]>(float)1e-5)
    entropy-=hv[i]*(float)log((double)hv[i]);
 return(entropy);
}
```

Fig. 6.6.1 Subroutines for the calculation of the central moments of a histogram.

Their major limitation is that they cannot express spatial texture characteristics. Spatial information can be described by using the histograms of *gray-level differences*. Let $\mathbf{d} = (d_1, d_2)$ be the displacement vector between two image pixels, and $g(\mathbf{d})$ the gray-level difference at distance $\mathbf{d}$:

$$g(\mathbf{d}) = |f(k,l) - f(k+d_1, l+d_2)| \qquad (6.6.6)$$

We denote by $p_g(g, \mathbf{d})$ the histogram of the gray-level differences at the specific distance $\mathbf{d}$ [WES76], [LEV85]. One distinct histogram $p_g(g, \mathbf{d})$ exists for each distance $\mathbf{d}$. It contains valuable information about the spatial organization of image intensities. If an image region has coarse texture, the histogram $p_g(g, \mathbf{d})$ tends to concentrate around $g = 0$ for small displacements $\mathbf{d}$. If the region texture is thin, $p_g(g, \mathbf{d})$ tends to spread, even for small displacement vectors **d** that exceed the texture grain size. Several texture measures can be extracted from the histogram of gray-level differences:

1. **Mean**

$$\mu_{\mathbf{d}} = \sum_{k=1}^{N} g_k p_g(g_k, \mathbf{d}) \qquad (6.6.7)$$

Small mean values $\mu_{\mathbf{d}}$ indicate coarse texture having a grain size equal to or larger than the magnitude of the displacement vector **d**. If the distance d is expressed in polar coordinates $\mathbf{d} = (r, \theta)$, the mean value μ_θ can give directional information about image texture.

2. **Variance and contrast**

$$\sigma_{\mathbf{d}}^2 = \sum_{k=1}^{N} (g_k - \mu_{\mathbf{d}})^2 p_g(g_k, \mathbf{d}) \qquad (6.6.8)$$

$$c_{\mathbf{d}} = \sum_{k=1}^{N} g_k^2 p_g(g_k, \mathbf{d}) \qquad (6.6.9)$$

The variance $\sigma_{\mathbf{d}}^2$ is a measure of the dispersion of the gray-level differences at a certain distance **d**. 'Deterministic' textures tend to have relatively small variance $\sigma_{\mathbf{d}}^2$. The contrast $c_{\mathbf{d}}$ is the moment of inertia of the histogram $p_g(g, \mathbf{d})$ about $g = 0$. It measures the contrast in gray-level differences.

3. **Entropy**

$$H_{\mathbf{d}} = -\sum_{k=1}^{N} p_g(g_k, \mathbf{d}) \ln p_g(g_k, \mathbf{d}) \qquad (6.6.10)$$

This is a measure of the homogeneity of the histogram $p_g(g, \mathbf{d})$. It is maximized for uniform histograms. The major advantages of the histograms of gray-level differences are their computational simplicity and their capability to give information about the spatial texture organization. An algorithm for their computation is shown in Figure 6.6.2.

```
int difhist(a,hv,dx,dy,N1,M1,N2,M2)
image a;
vector hv;
int dx,dy;
int N1,M1,N2,M2;
/* Subroutine to calculate the histogram of
   gray-level differences.
   a: input image buffer
   hv: output histogram buffer
   dx,dy: displacement vector
   N1,M1: start coordinates
```

```
         N2,M2: end coordinates */
{
  int xbegin,ybegin,xend,yend,difference,y,x;
  int i,j;
  vector lh;
  lh=(float *)malloc((unsigned int)NSIG*sizeof(float));
  /* Initialize histogram. */
  for (i=0;i<NSIG;i++) lh[i]=0;
  /* Calculate valid search area. */
  if (dx<0) xbegin=M1-dx; else xbegin=M1;
  if (dy<0) ybegin=N1-dy; else ybegin=N1;
  if (dx>0) xend=M2-dx; else xend=M2;
  if (dy>0) yend=N2-dy; else yend=N2;
  /* Perform histogram calculation. */
  for (y=ybegin; y<yend; y++)
    for (x=xbegin; x<xend; x++)
      {
        difference=abs( (int)a[y][x] -(int)a[y+dy][x+dx] );
        lh[difference]++;
      }
  for (i=0; i<NSIG; i++)
    hv[i]=(float)lh[i]/((float)(M2-M1-abs(dx))*
          (float)(N2-N1-abs(dy)));
  free(lh);
  return(0);
}
```

Fig. 6.6.2 Algorithm for the histogram of gray-level differences.

Spatial texture organization can also be revealed in terms of the run-length statistics [LEV85], [GAL75]. A *run length l* of pixels having equal intensity f in a direction θ is an event denoted by (l, f, θ). The run lengths reveal both directionality and coarseness of image texture. Coarse textures tend to produce long gray-level runs whereas directional texture tends to produce long runs at specific directions θ. Gray-level runs can be computed very easily. An algorithm for computing run lengths at direction $\theta = 0$ degrees is shown in Figure 6.6.3. Similar algorithms can be constructed for run-length calculations at orientations of 45, 90 and 135 degrees.

```
int runlen(a,rc,theta,N1,M1,N2,M2)
image a;
imatrix rc;
int theta;
int N1,M1,N2,M2;
```

```c
/* Subroutine for the calculation of the run-length matrix
   a: input image buffer
   rc: output run-length matrix (integers)
   theta: run orientation  (0, 45, 90, 135 degrees)
   N1,M1: start coordinates
   N2,M2: end coordinates */
{
  int y,x,tx,ty,value,valuetemp,sum;
  int cont,direction,flag;
  long int sum1;
  sum1=(long int)((long int)(N2-N1)*(N2-N1)+
       (long int)(M2-M1)*(M2-M1));
  for(y=0;y<NSIG;y++)
     for (x=0;x<(int)fabs(sqrt((double)sum1))+1;x++)
        rc[y][x]=0;
  /* Horizontal run-length calculation */
  switch(theta)
  {
    case 0 :
      for (y=N1;y<N2;y++)
        {
          x=M1;
          while (x<M2-1)
            {
              sum=0;flag=1;
              while(flag && x<M2-1)
                {
                  if (a[y][x]==a[y][x+1] ) sum++;
                  else { flag=0; rc[a[y][x]][sum]++; }
                  x++;
                }
            }
          if (a[y][x]==a[y][x-1]) rc[a[y][x]][sum]++;
          else rc[a[y][x]][0]++;
        }
      break;

  }
  return(0);
}
```

Fig. 6.6.3 Algorithm for the calculation of horizontal gray-level run lengths.

310   IMAGE SEGMENTATION ALGORITHMS

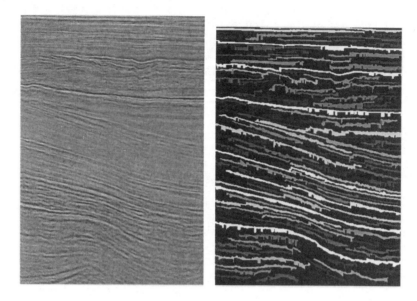

Fig. 6.6.4 (a) Original image; (b) binary run-length image.

This algorithm has been applied to the segmentation of the seismic image shown in Figure 6.6.4a [PIT92PR],[PIT94CDS]. The original image is segmented to a binary image and the binary run lengths are calculated. The resulting run-length image is shown in Figure 6.6.4b. Let $N(l, f, \theta)$ denote the number of events (l, f, θ) in an image having dimensions $N_1 \times N_2$. Let N_R be the total number of existing runs and let T_R denote the double sum:

$$T_R = \sum_{k=1}^{N} \sum_{l=1}^{N_R} N(l, f_k, \theta) \qquad (6.6.11)$$

The ratio $N(l, f, \theta)/T_R$ is the histogram of the gray-level runs at a specific direction θ. The following texture features can be calculated from the gray-level run lengths [LEV85], [GAL75]:

1. Short-run emphasis

$$A_1 = \frac{1}{T_R} \sum_{k=1}^{N} \sum_{l=1}^{N_R} \frac{1}{k^2} N(l, f_k, \theta) \qquad (6.6.12)$$

This is a measure that emphasizes the short run lengths existing in the image.

2. Long-run emphasis

$$A_2 = \frac{1}{T_R} \sum_{k=1}^{N} \sum_{l=1}^{N_R} k^2 N(l, f_k, \theta) \qquad (6.6.13)$$

This is a measure emphasizing the long runs of a gray-level image.

3. Gray-level distribution

$$A_3 = \frac{1}{T_R} \sum_{k=1}^{N} \left[\sum_{l=1}^{N_R} N(l, f_k, \theta) \right]^2 \qquad (6.6.14)$$

The sum $\sum_{l=1}^{N_R} N(l, f_k, \theta)$ gives the total number of runs for a certain gray-level value f_k at direction θ. Therefore, A_3 is a measure of the gray-level distribution.

4. Run-length distribution

$$A_4 = \frac{1}{T_R} \sum_{l=1}^{N_R} \left[\sum_{k=1}^{N} N(l, f_k, \theta) \right]^2 \qquad (6.6.15)$$

The sum $\sum_{k=1}^{N} N(l, f_k, \theta)$ gives the total number of occurrences of a certain run length l for any gray level at direction θ. Thus, A_4 is a measure of run-length distribution.

5. Run percentages

$$A_5 = \frac{1}{N_1 N_2} \sum_{k=1}^{N} \sum_{l=1}^{N_R} N(l, f_k, \theta) \qquad (6.6.16)$$

Another set of texture descriptors is based on the *gray-level co-occurrence matrices* [HAR79], [GON87], [LEV85]. Let $p(f_k, f_l, \mathbf{d})$ denote the joint probability of two pixels f_k, f_l lying at distance $\mathbf{d}$ in the image. This probability can be easily estimated from an image by counting the number n_{kl} of occurrences of the pixel values (f_k, f_l) lying at distance $\mathbf{d}$ in the image. Let n be the total number of any possible joint pairs lying at distance $\mathbf{d}$ in the image. The co-occurrence matrix elements c_{kl} are given by:

$$c_{kl} = \hat{p}(f_k, f_l, \mathbf{d}) = \frac{n_{kl}}{n} \qquad (6.6.17)$$

The co-occurrence matrix $\mathbf{C_d}$ has dimension $N \times N$, where N is the number of gray levels in the image. An algorithm for its calculation for various distance vectors is given in Figure 6.6.5.

```
int glcmarr(a,pdf,dx,dy,N1,M1,N2,M2)
image a;
matrix pdf;
int dx,dy;
int N1,M1,N2,M2;
```

```
/* Subroutine to calculate gray-level co-occurrence matrices.
   a: input image buffer
   pdf: output GLCM
   dx,dy: displacement vector
   N1,M1: start coordinates
   N2,M2: end coordinates*/
{
 int y,x,yr,ys;
 float sum=0.0F;
 int xbegin,xend,ybegin,yend;
 xbegin=M1;ybegin=N1;xend=M2;yend=N2;

 if (dx<0) xbegin=M1-dx; else xbegin=M1;
 if (dy<0) ybegin=N1-dy; else ybegin=N1;
 if (dx>0) xend=M2-dx; else xend=M2;
 if (dy>0) yend=N2-dy; else yend=N2;

 for (y=0; y<NSIG; y++)
    for (x=0; x<NSIG; x++)
       pdf[y][x]=0.0F;
 for (y=ybegin; y<yend; y++)
    for (x=xbegin; x<xend; x++)
     { yr=(int)(a[y][x]);
       ys=(int)(a[y+dy][x+dx]);
       pdf[yr][ys]++;
     }
 for (y=0; y<NSIG; y++)
    for (x=0; x<NSIG; x++)
       sum +=pdf[y][x];
 for (y=0; y<NSIG; y++)
    for (x=0; x<NSIG; x++)
       pdf[y][x] /=sum;
 return(0);
}
```

Fig. 6.6.5 Co-occurrence matrix calculation.

Co-occurrence matrices carry very useful information about spatial texture organization. If the texture is coarse, their mass tends to be concentrated around the main diagonal, i.e. at the elements c_{kl}, $\mid k-l \mid < t$. If the texture is fine, co-occurrence matrix values are much more spread. If texture carries strong directional information, the co-occurrence matrices $\mathbf{C_d}$ tend to have their mass in the main diagonal, for displacement vectors $\mathbf{d}$ corresponding to the texture direction. Several texture descriptors have been proposed to characterize the co-occurrence matrix content [GON87]:

1. Maximum probability

$$\mathbf{p_d} = \max_{k,l} c_{kl} \qquad (6.6.18)$$

This descriptor indicates the most probable gray level couple (f_k, f_l) lying at a distance **d** in the image.

2. Entropy

$$H_\mathbf{d} = -\sum_{k=1}^{N}\sum_{l=1}^{N} c_{kl} \ln c_{kl} \qquad (6.6.19)$$

Entropy is maximized when the joint probability $p(f_k, f_l, \mathbf{d})$ is uniformly distributed.

3. Element-difference moment of order m

$$I_\mathbf{d} = \sum_{k=1}^{N}\sum_{l=1}^{N} |k-l|^m c_{kl} \qquad (6.6.20)$$

This descriptor is minimized when the co-occurrence matrix mass is concentrated around the main diagonal, because the differences $|k-l|^m$ are very small. It attains small values in the case of coarse textures. This texture descriptor is equivalent to the *inertia* of the co-occurrence matrix for $m = 2$ [JAI89].

The spectral characterization of the image texture is based either on the autocorrelation function of a two-dimensional image or on its power spectrum [LIM90], [JAI89]. The autocorrelation function $R_{ff}(k,l)$ of an image $f(i,j)$ is given by the following relation:

$$R_{ff}(k,l) = \frac{1}{(2N_1+1)(2N_2+1)} \sum_{i=-N_1}^{N_1}\sum_{j=-N_2}^{N_2} f(i,j)f(i+k,j+l) \qquad (6.6.21)$$

It can be calculated both for positive and negative lags (k,l). It usually attains a maximum for zero lag $(0,0)$ and drops exponentially with (k,l) (positive or negative). The drop-off rate is an indication of the texture coarseness. If the texture is fine the drop-off rate is large, whereas small drop-off rates indicate large texture elements. The circular symmetry of the autocorrelation function indicates spatial isotropy. Finally, the periodicity in the peaks of $R_{ff}(k,l)$ gives an indication of the spatial periodicity of the texture patterns. Direct computation of the autocorrelation function is preferred for a small number of lags (k,l) by using (6.6.21). If the calculation of $R_{ff}(k,l)$ for a large number of lags is needed, the following property of the autocorrelation function can be very useful:

$$R_{ff}(k,l) = IDFT[F(\omega_1,\omega_2)F^*(\omega_1,\omega_2)] \qquad (6.6.22)$$

where $F(\omega_1, \omega_2)$ is the discrete time Fourier transform of the image $f(n,m)$. Therefore, the autocorrelation function can be calculated by employing the discrete Fourier transform:

$$F(k,l) = \sum_{n=0}^{N-1} \sum_{m=0}^{M-1} f(n,m) \exp\left[-i\left(\frac{2\pi nk}{N} + \frac{2\pi ml}{M}\right)\right] \quad (6.6.23)$$

and the inverse DFT:

$$R_{ff}(k,l) = \frac{1}{NM} \sum_{u=0}^{N-1} \sum_{v=0}^{M-1} F(u,v) F^*(u,v) \exp\left[i\left(\frac{2\pi ku}{N} + \frac{2\pi lv}{M}\right)\right] \quad (6.6.24)$$

More details on the calculation of the autocorrelation function can be found in Chapter 2.

The autocorrelation function $R_{ff}(k,l)$ is closely related to the power spectrum $P_{ff}(\omega_1, \omega_2)$ of an image, as has already been described in Chapter 2. An estimate of the power spectrum is given by the two-dimensional *periodogram*:

$$P_{ff}(\omega_1, \omega_2) = |F(\omega_1, \omega_2)|^2 = F(\omega_1, \omega_2) F^*(\omega_1, \omega_2) \quad (6.6.25)$$

and can be easily calculated by employing (6.6.23). Premultiplication of the image $f(m,n)$ by a two-dimensional window $w(m,n)$ produces a relatively smooth power spectrum estimate. Other power spectrum estimation methods have already been presented in Chapter 2. The power spectrum carries important information about image texture. Its peaks give information about the fundamental spatial period of the texture patterns. Texture that has strong directional characteristics produces a power spectrum concentrated along lines perpendicular to the texture direction. Let us suppose that we use the polar coordinates:

$$r = \sqrt{\omega_1^2 + \omega_2^2} \quad (6.6.26)$$

$$\phi = \arctan\left(\frac{\omega_2}{\omega_1}\right) \quad (6.6.27)$$

for power spectrum description $P_{ff}(r, \phi)$. The average $P_\phi(\phi)$

$$P_\phi(\phi) = \int_0^{r_{max}} P_{ff}(r, \phi) dr \quad (6.6.28)$$

is a very good descriptor of texture directionality. A texture that has strong directional characteristics along direction θ creates a maximum of $P_\phi(\phi)$ at angle $\phi = \theta + \frac{\pi}{2}$. The integral (6.6.28) can be approximated by a summation within a wedge $\phi_1 \leq \phi < \phi_2$ in the spectral domain:

$$P_\phi(\phi) \simeq \sum_{\sqrt{\omega_1^2 + \omega_2^2} < r_{max},\ \phi_1 \leq \phi < \phi_2} |F(\omega_1, \omega_2)|^2 \quad (6.6.29)$$

Spatial texture isotropy produces circular symmetry in the frequency domain. In this case, the radial distribution of the power spectrum is of importance. Power concentration in the low frequencies indicates coarse texture whereas power concentration in the high frequencies indicates fine texture. The radial distribution of the power spectrum can be described by the integral:

$$P_r(r) = \int_0^{2\pi} P_{ff}(r, \psi) d\psi \qquad (6.6.30)$$

This integral can be approximated by the following sum, by splitting the spectral domain into concentric areas:

$$P_r(r) = \sum_{r_1^2 \leq \sqrt{\omega_1^2 + \omega_2^2} < r_2^2} |F(\omega_1, \omega_2)|^2 \qquad (6.6.31)$$

where $r_1 < r < r_2$. An algorithm for the calculation of the angular and radial power spectrum distributions is shown in Figure 6.6.6. The input of subroutine psRtheta() is the power spectrum of an image. A square spectrum matrix $N \times N$ is assumed. The input spectrum matrix is shifted circularly in such a way that the center of frequency coordinates (0,0) is located at the center $(N/2, N/2)$ of the input matrix. The image is scanned in a row-wise manner in order to calculate $P_r(r)$, $P_\phi(\phi)$. Only the points lying at distance less than $N/2$ from the center of the square are taken into account in the calculation.

```
int psRtheta(ps,psr,pstheta,N,M,NS)
matrix ps;
vector psr,pstheta;
int N,M,NS;
/* Subroutine to calculate angular and radial distribution of
power spectrum.
    ps: Magnitude of power spectrum.
    psr: Vector containing radial ps distribution.
         Its domain is [0,...,NMAX/2-1].
    pstheta: Vector containing angular ps distribution.
             Its domain corresponds to [-pi,...,pi].
    N,M: size of power spectrum buffer
    NS: size of vectors psr, pstheta */
{
  int y,x,rows2,cols2;
  int nr,nc;
  unsigned char r,theta;
  float coef,temp;
  rows2=(int)N/2;
  cols2=(int)M/2;
  coef=(float)((float)(NS)/6.28);
```

316   IMAGE SEGMENTATION ALGORITHMS

```
/* Rearrange input matrix so that DC coefficient (0,0)
   is at position (N/2,M/2). Rearrangement is
   performed in place. */
for(nr=0; nr<rows2; nr++)
 { for(nc=0; nc<cols2; nc++)
     {
       temp=ps[nr+rows2][nc+cols2];
       ps[nr+rows2][nc+cols2]=ps[nr][nc];
       ps[nr][nc]=temp;
     }
   for(nc=cols2; nc<M; nc++)
     {
       temp=ps[nr+rows2][nc-cols2];
       ps[nr+rows2][nc-cols2]=ps[nr][nc];
       ps[nr][nc]=temp;
     }
 }

/* Initialize psr, pstheta vectors */
for (y=0;y<NS;y++)
 { psr[y]=0.0F;pstheta[y]=0.0F; }
   /* Scan the entire image. Calculate psr, pstheta by taking
      into account only power spectrum points lying at
      maximal distance N/2-1 from central point (N/2, M/2).*/
   for (y=0;y<N;y++)
    for (x=0;x<M;x++)
      {
        r=(unsigned char)sqrt((y-rows2)*(y-rows2)+
                                    (x-cols2)*(x-cols2));
        if ((int)r<=(N/2-1))
          {
            if (abs(x-cols2)>=1)
              theta=(unsigned char)((atan2((double)(y-rows2),
                       (double)(x-cols2))+3.14159)*coef);
            else
              {
                if((y-rows2)>0) theta=
                            (unsigned char)(1.570796*coef);
                else if((y-rows2)<0) theta=
                            (unsigned char)(1.5*3.14159*coef);
              }
            psr[r]+=ps[y][x];
            if((x-cols2)!=0||(y-rows2)!=0)
              pstheta[theta]+=ps[y][x];
          }
```

```
      }
  return(0);
}
```

Fig. 6.6.6 Algorithm for the angular and radial power spectrum distributions.

The normalized angular and radial power spectrum distributions for image Baboon are shown in Figure 6.6.7. The horizontal axes are scaled versions of 2D DFT radius and angle. It is clearly seen that the spectral energy at or close to the DC component is strong. Furthermore, the peaks in the angular distribution reveal that a large number of linear features exist in this image. These remarks are supported by the image of the power spectrum of Baboon that is shown in Figure 2.7.1.

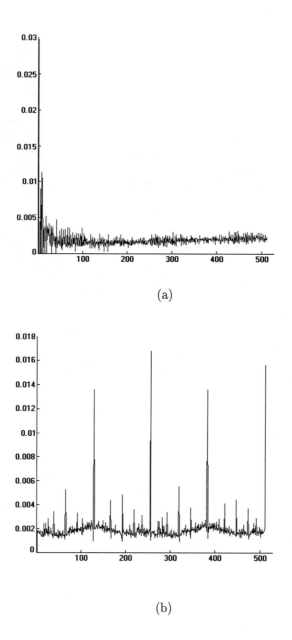

Fig. 6.6.7 (a) Radial power spectrum distribution of image Baboon; (b) angular power spectrum distribution.

References

BAL82. D.H. Ballard, C.M. Brown, *Computer vision*, Prentice-Hall, 1982.

BOR99. A.G. Bors, I. Pitas, "Object Classification in 3-D Images Using Alpha-Trimmed Mean Radial Basis Function Network," *IEEE Trans. on Image Processing*, vol. 8, no. 12, pp. 1744-1756, Dec. 1999.

CHO72. C.K. Chow, T. Kaneko, "Automatic boundary detection of the left ventricle from cineangiograms," *Computers and Biomedical Research*, vol. 5, no. 4, pp. 388-410, Aug.1972.

CYP89. R. Cypher, J.L.C. Sanz, "SIMD architectures and algorithms for image processing and computer vision," *IEEE Transactions on Acoustics, Speech, and Signal Processing*, vol. ASSP-37, no. 12, pp. 2158-2173, Dec. 1989.

GAL75. M.M. Galloway, "Texture classification using gray level run lengths," *Computer Graphics and Image Processing*, vol. 4, no. 2, pp. 172-179, June 1975.

GON87. R.C. Gonzalez, P. Wintz, *Digital image processing*, Addison-Wesley, 1987.

HAR79. R.M. Haralick, "Statistical and structural approaches to texture," *Proc. IEEE*, vol. 67, no. 5, pp. 786-804, May 1979.

HEI94. F. van der Heijden, *Image based measurement systems*, Wiley, 1994

HOR74. S.L. Horowitz, T. Pavlidis, "Picture segmentation by a directed split-and-merge procedure," *Proc. 2nd Int. Joint Conf. on Pattern Recognition*, pp. 424-433, 1974.

JAIN89. A.K. Jain, *Fundamentals of digital image processing*, Prentice-Hall, 1989.

KOH90. T.K. Kohonen, "The self-organizing map," *Proceedings of the IEEE*, vol. 78, no. 9, pp. 1464-1480, September 1990.

KOH97. T. Kohonen, *Self-organizing maps*, Springer Verlag, 1997.

KOT93CAIP. C. Kotropoulos, and I. Pitas, "A variant of learning vector quantizer based on split-merge statistical tests," in *Lecture Notes in Computer Science: Computer Analysis of Images and Patterns* (D. Chetverikov, and W.G. Kropatsch, Eds.), pp. 822-829, Springer Verlang, 1993.

KOT98IJCM. C. Kotropoulos, N. Nikolaidis, A. Bors, I. Pitas, "Robust and adaptive training techniques in self-organizing neural networks," *Int. Journal of Computer Mathematics*, vol. 67, pp. 183-200, 1998.

LEV72. S. Levialdi, "On shrinking binary input patterns," *Communications of the ACM*, vol. 15, no. 1, pp. 7-10, 1972.

LEV81. M.D. Levine, S.L. Shaheen, "A modular computer vision system for picture segmentation and interpretation," *IEEE Transactions on Pattern Analysis and Machine Intelligence*, vol. PAMI-3, no. 5, pp. 540-558, Sept. 1981.

LEV85. M.D. Levine, *Vision in man and machine*, McGraw-Hill, 1985.

LIM90. J.S. Lim, *Two-dimensional signal and image processing*, Prentice-Hall, 1980.

NAG73. R.N. Nagel, A. Rosenfeld, "Steps towards handwritten signature verification," *Proc. 1st Int. Joint Conf. on Pattern Recognition*, pp. 55-66, 1979.

NAG80. P.A. Nagin, B. Schwartz, "Approaches to image analysis of the optic disc," *Proc. 5th Int. Conf. on Pattern Recognition*, pp. 948-956, 1980.

NAS80. D. Nassimi, S. Sahni, "Finding connected components and connected ones on a mesh connected parallel computer," *SIAM Journal of Computation*, vol. 9, no. 4, pp. 744-757, 1980.

PEL78. S. Peleg, A. Rosenfeld, "Determining compatibility coefficients for curve enhancement relaxation processes," *IEEE Transactions on Systems, Man and Cybernetics*, vol. SMC-8, no. 7, pp. 548-555, July 1978.

PIT90. I. Pitas, A.N. Venetsanopoulos, *Nonlinear digital filters: Principles and applications*, Kluwer Academic, 1990.

PIT92PR. I. Pitas, and C. Kotropoulos, "A texture-based approach to the segmentation of seismic images," *Pattern Recognition*, vol. 25, no. 9, pp. 929-945, 1992.

PIT94CDS. I. Pitas, C. Kotropoulos and A.N. Venetsanopoulos, "Knowledge-based signal/image processing for geophysical interpretation," in *Control and Dynamic Systems* (Prof. C.T. Leondes, ed.), vol. 77, pp. 1-48, Academic Press, 1996.

PIT96TIP. I. Pitas, C. Kotropoulos, N. Nikolaidis, R. Yang and M. Gabbouj, "Order Statistics Learning Vector Quantizer," *IEEE Trans. on Image Processing*, vol. 5, no. 6, pp. 1048-1053, 1996.

ROS76. A. Rosenfeld, R.A. Hummel, S.W. Zucker, "Scene labeling by relaxation operators," *IEEE Transactions on Systems, Man and Cybernetics*, vol. SMC-6, no. 6, pp. 420-533, June 1976.

ROS83. A. Rosenfeld, "Parallel image processing using cellular arrays," *IEEE Computer*, pp. 14–20, Jan. 1983.

SAM90. H. Sammet, *The design and analysis of spatial data structures*, Addison-Wesley, 1990.

SIM90. P.K. Simpson, *Artificial neural systems*, Pergamon Press, 1990.

WAH87. F.M. Wahl, *Digital image signal processing*, Artech, 1987.

WES76. J.S. Westa, C.R. Dyer, A. Rosenfeld, "A comparative study of texture measure for terrain classification," *IEEE Transactions on Systems, Man and Cybernetics*, vol. SMC-6, no. 4, pp. 369-385, Apr. 1976.

UMB98. S.E. Umbaugh, *Computer vision and image processing*, Prentice-Hall, 1998.

7
Shape Description

7.1 INTRODUCTION

Shape description is an important issue both in image analysis and in image synthesis (graphics). Shape representations obtained by image analysis techniques can be used for object recognition applications. In graphics applications, two- and three-dimensional object representations, obtained through object modeling, are used for digital image synthesis. In the following, we shall describe various two-dimensional shape representation schemes and their applications in image analysis, computer vision and pattern recognition.

Two-dimensional shapes can be described in two different ways. The first method is to use the object boundary and its features (e.g. boundary length, curvature, signature, Fourier descriptors). This method is directly connected to edge and line detection. The resulting description schemes are called *external* representations. They enjoy a certain popularity because they produce compact shape representations. The second method is to describe the region occupied by the object on the image plane. This method is linked to the region segmentation techniques described in Chapter 6. The resulting representation schemes are called *internal* representations. Some segmentation techniques (e.g. region splitting) lead directly to object representation schemes (quadtrees). Region features (e.g. area, moments, skeletons) have been used extensively in object recognition applications.

Shape representation schemes must have certain desirable properties [PAV78]:

1. **Uniqueness.** This is of crucial importance in object recognition, because each object must have a unique representation.

2. **Completeness.** This refers to unambiguous representations.

3. **Invariance under geometrical transformations.** Invariance under translation, rotation, scaling and reflection is very important for object recognition applications.

4. **Sensitivity.** This is the ability of a representation scheme to reflect easily the differences between similar objects.

5. **Abstraction from detail.** This refers to the ability of the representation to represent the basic features of a shape and to abstract from detail. This property is directly related to the noise robustness of the representation.

In the following, we shall first describe some external representation schemes and then continue with the description of internal representations. An introduction to morphological representation schemes will conclude this chapter.

7.2 CHAIN CODES

Let us suppose that a boundary of an object is represented by a connected path of 1s in a binary image. Both 4- and 8-connected paths are considered. The path can be considered to consist of line segments connecting adjacent pixels as can be seen in Figure 7.2.1. Each segment possesses a specific direction if the boundary is traversed in a clockwise manner. The chain directions can be coded in the way shown in Figure 7.2.2 for 4-connected and 8-connected chains. The boundary path can be followed in a clockwise manner. The codes of the successive boundary segments form the boundary *chain code*. The 4-connected chain code of the boundary shown in Figure 7.2.1 is 00300333212232211011, when path following starts from the upper leftmost boundary corner. It is clear that the chain code depends on the start point of boundary following. This is not a problem, because a chain code can be considered as a circular list in the case of closed boundaries. If uniqueness is needed for object recognition applications, this list is shifted circularly until it forms an integer of minimum magnitude [GON87]. An advantage of chain code is that it is translation invariant. Scale invariance can be obtained by changing the size of the sampling grid. However, such a resampling seldom produces exactly the same chain code. Rotation invariance is obtained by using the *difference chain code* and is constructed in the following way. Let $x_1 x_2 \ldots x_N$ be a chain code. The difference code is $d_1 d_2 \ldots d_N$, where d_i is given by:

$$d_i = \begin{cases} diff(x_i, x_{i-1}) & \text{if } i \neq 1 \\ diff(x_i, x_N) & \text{if } i = 1 \end{cases} \quad (7.2.1)$$

CHAIN CODES 325

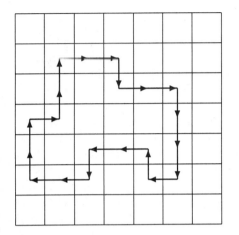 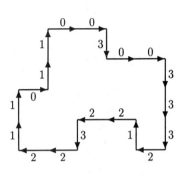

Fig. 7.2.1 Chain code of the digital boundary.

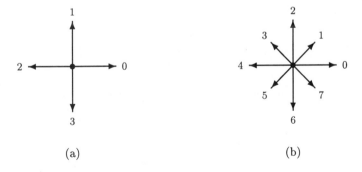

Fig. 7.2.2 Directions of boundary segments of a chain code for (a) a 4-connected chain; (b) an 8-connected chain.

The difference $diff(x_i, x_{i-1})$ is computed by counting, in a counterclockwise manner, the number of left-hand turns (of angle $\pi/2$ or $\pi/4$) that separate code element x_i from its preceding element x_{i-1}. The preceding element of x_1 is supposed to be the last list element x_N, under the assumption that the list is circular. The difference chain code represents direction differences. Therefore, it is rotation invariant (at least for angles that are multiples of $\pi/2$). Rotation at arbitrary angles on a rectangular grid may change the boundary shape, thus altering the resulting difference chain code.

Chain codes provide a good compression of boundary description, because each chain code element can be coded with 2 bits only (for a 4-connected chain code), instead of the 2 bytes required for the storage of the coordinates (x, y) of each boundary pixel. Chain codes can also be used to calculate certain boundary features. The boundary perimeter P is equal to N for a 4-connected chain code of length N, because all its segments have length 1. In the case of an 8-connected chain code, the even-numbered segments have length 1 whereas the odd-numbered segments have length $\sqrt{2}$. Thus, the boundary perimeter T is given by:

$$T = \sum_{i=1}^{N} n_i \qquad (7.2.2)$$

$$n_i = \begin{cases} 1 & \text{if } x_i \bmod 2 = 0 \\ \sqrt{2} & \text{if } x_i \bmod 2 = 1 \end{cases} \qquad (7.2.3)$$

The object width w and height h can also be easily calculated from the chain code. In the case of a 4-connected chain code, segments having code 0 (or 2) contribute to object width, whereas segments having code 1 (or 3) contribute to object height. Therefore the object width and height are given by:

$$w = \sum_{i=1}^{N} w_i \qquad (7.2.4)$$

$$h = \sum_{i=1}^{N} h_i \qquad (7.2.5)$$

where

$$w_i = \begin{cases} 0 & \text{if } x_i = 1, 2, 3 \\ 1 & \text{if } x_i = 0 \end{cases} \qquad (7.2.6)$$

$$h_i = \begin{cases} 0 & \text{if } x_i = 0, 2, 3 \\ 1 & \text{if } x_i = 1 \end{cases} \qquad (7.2.7)$$

Similarly, in the case of 8-connected chain codes, the direction codes $0, 1, 7$ (or $3, 4, 5$ equivalently) contribute to object width by 1 and the direction

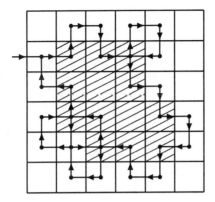

Fig. 7.2.3 Turtle procedure in binary object boundary following.

codes $1, 2, 3$ (or $5, 6, 7$ equivalently) contribute to object height. Therefore, the object width and height are given by (7.2.4–7.2.5), where:

$$w_i = \begin{cases} 1 & \text{if } x_i = 0, 1, 7 \\ 0 & \text{otherwise} \end{cases} \qquad (7.2.8)$$

$$h_i = \begin{cases} 1 & \text{if } x_i = 1, 2, 3 \\ 0 & \text{otherwise} \end{cases} \qquad (7.2.9)$$

Finally, chain codes can be used in the calculation of object area by an algorithm described in [FAI88].

The boundaries of binary objects can be easily followed by employing an algorithm similar to *Papert's turtle* [PAP73], [DUD73]. The algorithm operates on 4-connected neighborhoods. The *turtle* starts from a boundary point: if the current pixel value is 1, it turns left and advances one pixel; if the current pixel value is 0, it turns right and advances one pixel. The algorithm terminates when the turtle returns to the start boundary point. The procedure is illustrated in Figure 7.2.3. It is clearly seen that certain loops occur during boundary following. These loops can be removed by postprocessing the boundary sequence. The turtle algorithm can be easily modified to create the 4-connected chain code of the boundary, as can be seen in Figure 7.2.4.

```
int chain_code(a, ch, is, js, TMAX, N1, M1, N2, M2)
image    a;
CHAIN    *ch;
int      *is, *js;
int TMAX,N1,M1,N2,M2;
```

```
/* Subroutine to perform chain coding
   a:     input image
   ch:    output chain code (linked list)
   is,js: output starting points of image
   TMAX: maximal allowable number of chain nodes
   N1,M1: start coordinates
   N2,M2: end coordinates */
{
    int     i, j, c;
    /* typedef unsigned char byte; */
    byte    d=0;
    /* Find starting point. */
    for (i=N1;i<N2;i++)
     for (j=M1;j<M2;j++)
      if ((int)a[i][j]!=0) { *is=i; *js=j; goto cont; }
    cont:
    /* Initializations */
    i = *is; j = *js;
    /* Create chain head and tail */
    (*ch).start = (*ch).end = getchnode();

    /* Main loop. Border is followed. */
    c = 0;
    do { /* Follow border. */
         if (a[i][j]!=0) d=(d+1)%4; else d=(d-1)%4;
         switch(d)
           { /* Translate direction to pixel. */
             case 0: j++;break;
             case 1: i--;break;
             case 2: j--;break;
             case 3: i++;break;
           }
         /* Borders of ROI have been violated. */
         if(i<N1 || i>N2 || j<M1 || j>M2) return(-93);
         /* Add direction to end of chain. */
         (*ch).end = addchnode(d, (*ch).end);
         /* Check if turtle follower is running wild. */
         if (c++ > TMAX) return(-94);
         /* while not back at start point, continue. */
       } while(i!=*is||j!=*js);
    if ((*ch).start == NULL || (*ch).end == NULL) return(-95);
    return(0);
}
```

Fig. 7.2.4 Turtle algorithm for boundary following.

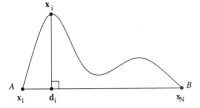

Fig. 7.3.1 Error definition for the approximation of a curve by a linear segment.

Subroutine chain_code() produces a chain whose elements are of the form:

typedef struct CHNODE {unsigned char num; struct CHNODE *next;}
    CHNODE;

The chain has a pointer both to its start and to its end:

typedef struct CHAIN {CHNODE *start; CHNODE *end;} CHAIN;

Subroutine addnode() is used in chain_code() to add new elements to the output list [AMM87].

7.3 POLYGONAL APPROXIMATIONS

Digital boundaries carry information which may be superfluous for certain applications. Boundary approximations can be sufficient in such cases. *Linear piecewise (polygonal) approximations* are the most frequently used ones and they produce a *polygon* which closely resembles the original line. The polygon vertices are a representation of the digital boundary. Error criteria, which measure the quality of fitness of the polygon to the digital curve, can be used in order to obtain good polygonal approximations. Let us suppose that a digital curve from point A to point B is approximated by the straight line segment AB, as shown in Figure 7.3.1. Let $\mathbf{x}_1, \mathbf{x}_2, \ldots, \mathbf{x}_N$ denote curve pixel coordinates and $\mathbf{x}_i - \mathbf{d}_i$, $i = 2, \ldots, N-1$, denote vectors that start from $\mathbf{x}_i$, $i = 2, \ldots, N-1$, and are perpendicular to the straight line segment AB, that is, to the vector $\mathbf{x}_N - \mathbf{x}_1$:

$$(\mathbf{x}_i - \mathbf{d}_i)^T (\mathbf{x}_N - \mathbf{x}_1) = 0 \qquad (7.3.1)$$

The distance $|\mathbf{x}_i - \mathbf{d}_i|$ is the approximation error of the curve AB by the line segment for a particular curve pixel $\mathbf{x}_i$, $i = 2, \ldots, N-1$. The fitness criterion can be either the mean square error E_2 or the maximal error E_{max}:

$$E_2 = \sum_{i=2}^{N-1} |\mathbf{x}_i - \mathbf{d}_i|^2 \qquad (7.3.2)$$

330 SHAPE DESCRIPTION

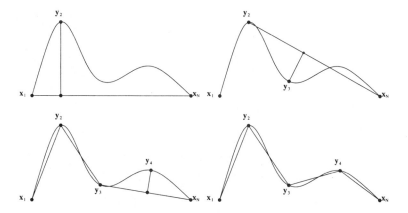

Fig. 7.3.2 Splitting method for polygonal approximations.

$$E_{max} = \max_{2 \le i \le N-1} | \mathbf{x}_i - \mathbf{d}_i | \qquad (7.3.3)$$

The same fitness criteria can be used when a curve is approximated by a polygon. The optimal linear piecewise approximation can be obtained by choosing the polygon vertices in such a way that the overall approximation error is minimized. A solution to this problem is not trivial and may result in iterative search algorithms with heavy computational complexity. However, there exist alternative techniques based on the split and merge approach that are fast and relatively efficient in most applications. Splitting techniques divide a curve segment recursively into smaller segments, until each curve segment can be approximated by a linear segment within an acceptable error range. Mean square error or maximal error can be used. The process is illustrated in Figure 7.3.2. The curve from $\mathbf{x}_1$ to $\mathbf{x}_N$ must be approximated by a polygon. If the maximal error measure is used, the point $\mathbf{y}_2$ is found which possesses the maximal error E_{max} over the entire curve $\mathbf{x}_i$, $i = 1, \ldots, N$. The curve is now approximated by the line segments $\mathbf{x}_1\mathbf{y}_2$ and $\mathbf{y}_2\mathbf{x}_N$. This process is repeated until the resulting error E_{max} for all line segments is below a certain threshold. The algorithm is described in Figure 7.3.3. Subroutine curve_split() is a recursive one and its output is a list of pixel coordinates of the form:

```
typedef struct PIXEL {int x; int y;} PIXEL;
typedef struct LNODE {PIXEL num; struct LNODE *next;} LNODE;
```

The list is double indexed, that is, it has one pointer at its start and one pointer at its end respectively:

```
typedef struct LIST {LNODE *start; LNODE *end;} LIST;
```

The recursion in this subroutine is organized in such a way that the polygon vertex coordinates are produced in a serial order in the output list. The end polygon vertex is not added by this subroutine but must be added to the end of the output list externally. The number of segments needed for a curve representation depends heavily on the error threshold.

```
int curve_split(x, p, n, T)
PIXEL *x;
LIST *p;
int n;
double T;

/* Subroutine for polygon approximation of curve by splitting.
    x: array of PIXELs
    p: output list of polygon vertices
    n: length of pixel array
    T: threshold */
{
    double *d, dmax;
    int i, ok=1, imax;

    /* Append starting point x[1] to the list. */
    if (p->start==p->end) p->end=add_list(x[1],p->end);
    /* Allocate memory for local distance buffer.
       This buffer will contain the distances of each curve
       pixel from a straight line. */
    d = (double *) calloc(n+1, sizeof( double ));
    /* Calculate distances. */
    for (i=2; i < n; i++)
      d[i]=dist(x[1].x, x[1].y, x[i].x, x[i].y, x[n].x, x[n].y);
    /* Check threshold violation. */
    for (i=3; i < n-1; i++)
      if (d[i] > T) { ok = 0; goto cont; }
    cont: if (ok == 1) return(0);
    /* Find max distance. */
    dmax = 0.0;
    for (i=3; i < n-1; i++)
      if (d[i] > dmax) { dmax=d[i]; imax=i; }
    free(d);
    /* Add point to list. Perform split on two subcurves. */
    curve_split( x, p, imax, T);
    p->end = add_list( x[imax], p->end );
    curve_split( x+imax, p, n-imax, T);
}

double dist(x1, y1, x2, y2, x3, y3)
int x1, y1, x2, y2, x3, y3;
```

```
/*  Subroutine to calculate the distance of point (x2,y2)
    from the straight line (x1,y1),(x3,y3).
    x1,y1: line   start
    x2,y2: point
    x3,y3: line end */
{
    int xb, yb, xc, yc;
    double  eg, par, angle, x;
    xb = x2-x1; xc = x3-x1; yb = y2-y1; yc = y3-y1;
    eg = abs((double)(xb*xc + yb*yc));
    par = sqrt( xb*xb + yb*yb ) * sqrt( xc*xc + yc*yc );
    if ((par == 0) || (eg/par > 1)) return(0.0);
    angle = asin( eg/par );
    x = cos( angle ) * sqrt( xb*xb + yb*yb );
    return(x);
}
```

Fig. 7.3.3 Splitting algorithm for polygonal approximation.

If the error threshold is small, the representation contains a large number of linear segments. If the curve is a closed one, the start and end points x_1, x_N are chosen to lie on two opposite remote parts of the curve, as shown in Figure 7.3.4. These points split the curve into two parts. The splitting algorithm can be applied to the two parts independently, as shown in Figure 7.3.4. A basic advantage of the splitting approach is that it can detect the inflection points on a curve and can use them in curve representation.

Merge techniques in the polygonal approximation operate in the opposite way. We start from the curve point x_1 and traverse the curve in a clockwise or counterclockwise manner. At each point x_i we update the error from x_1 to x_i. If the error exceeds a certain threshold, we declare the current curve point x_i as a polygon vertex and repeat the same procedure with x_i as the new start

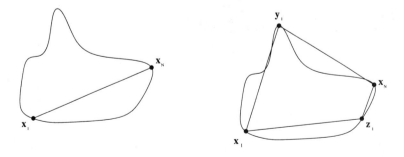

Fig. 7.3.4 Splitting method for the linear piecewise approximation of a closed curve.

point. If the curve is closed, we choose x_1 to be a prominent inflection point on the curve. The merge algorithm is shown in Figure 7.3.5.

```
int curve_merge(x, p, n, T)
PIXEL *x;
LIST *p;
int n;
double T;

/* Subroutine for polygonal approximation of curve by merging.
    x: array of PIXELs
    p: output list of polygon vertices
    n: length of pixel array
    T: threshold */
{
    double   *d, dmax;
    int start=1, end=1, i;
    /* Append starting point x[1] to the list. */
    if (p->start==p->end) p->end=add_list(x[1],p->end);
    /* Allocate memory for local distance buffer.
        This buffer will contain the distances of each curve
        pixel from a straight line. */
    d = (double *) calloc(n+1, sizeof( double ));
    /* Move one pixel forward in the curve. Calculate maximal
        distance until threshold is violated. */
    do { end++; dmax=0.0;
        for (i=start; i<end; i++)
            { d[i] = dist(x[1].x, x[1].y,
                        x[i].x, x[i].y, x[end].x, x[end].y);
                if (d[i] > dmax) dmax = d[i];
            }
        } while (dmax <= T && end != n);
    free(d);

    /* Append new vertex to pixel list. Start merging from
        the new start point. */
    if (end < n)
    { p->end = add_list(x[end], p->end); start = end;
        curve_merge(x+start, p, n-start, T);
    }
    else { p->end = add_list(x[end], p->end); return( 0 ); }
}
```

Fig. 7.3.5 Merge algorithm for polygonal approximation.

334 SHAPE DESCRIPTION

Subroutine curve_merge() is recursive and produces an output polygon vertex list of the type LIST. Both start and end vertices are included in this list. Both curve_merge() and curve_split() subroutines use the add_list() function to add nodes to the output list [AMM87]. The primary disadvantage of the merge algorithm is that polygon vertices do not coincide with curve inflection points. This disadvantage is a serious problem when the curves to be represented are close to linear piecewise models. This problem can be solved by combining split and merge techniques. However, such an approach tends to produce less concise curve representations than the split method. The algorithms described in this section have been tested in the image shown in Figure 7.3.6. Both split and merge algorithms produce good linear piecewise approximations of the object boundary. However, it can be seen that the merge algorithm produces vertices that do not coincide with curve corners.

7.4 FOURIER DESCRIPTORS

If a closed curve is traversed, its shape can be described by the curve coordinates $x(t), y(t)$, as shown in Figure 7.4.1. The resulting waveforms $x(t), y(t)$ are periodic with period 2π. These waveforms can be sampled and combined

Fig. 7.3.6 (a) Test image; (b) result of split algorithm; (c) result of the merge algorithm.

FOURIER DESCRIPTIONS

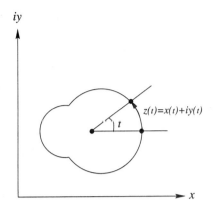

Fig. 7.4.1 Parametric curve representation.

to produce a complex periodic waveform with period N:

$$z(n) = x(n) + iy(n) \qquad n = 0, 1, \ldots, N-1 \qquad (7.4.1)$$

It is well known that such a signal can be represented by its discrete Fourier transform coefficients $Z(k)$, also called *Fourier descriptors*:

$$Z(k) = \sum_{n=0}^{N-1} z(n) \exp\left(-\frac{i2\pi nk}{N}\right) \qquad (7.4.2)$$

$$z(n) = \frac{1}{N} \sum_{k=0}^{N-1} Z(k) \exp\left(\frac{i2\pi nk}{N}\right) \qquad (7.4.3)$$

Fourier representation (7.4.2–7.4.3) has some very interesting properties. The coefficient $Z(0)$ represents the *center of gravity* of the curve. Fourier coefficients $Z(k)$ represent slowly and rapidly varying shape trends for small and large indices k, respectively. A translation in curve coordinates by z_0:

$$z_t(n) = z(n) + z_0 \quad z_0 = x_0 + iy_0 \qquad (7.4.4)$$

affects only the term $Z(0)$ of the representation:

$$Z_t(0) = Z(0) + z_0 \qquad (7.4.5)$$

A rotation of the curve coordinates by angle θ:

$$z_r(n) = z(n)e^{i\theta} \qquad (7.4.6)$$

results in a phase shift of the transform coefficients by an equal amount:

$$Z_r(k) = Z(k)e^{i\theta} \qquad (7.4.7)$$

A scaling operation by a factor a, with respect to a coordinate system having origin the center of gravity of the curve, results in a scaling of Fourier coefficients by an equal amount:

$$z_s(n) = az(n) \tag{7.4.8}$$

$$Z_s(k) = aZ(k) \tag{7.4.9}$$

Finally, a change in the starting point of curve traversal:

$$z_t(n) = z(n_1 - n_0) \tag{7.4.10}$$

produces *modulation* of the Fourier descriptors:

$$Z_t(k) = Z(k)e^{-i2\pi n_0 k/N} \tag{7.4.11}$$

Therefore, Fourier descriptors can provide useful invariant shape descriptions. The coefficient magnitude $|Z(k)|$, $k = 0, \ldots, N-1$, is rotation invariant. The magnitude of the $N-1$ coefficients $|Z(k)|$, $k = 1, \ldots, N-1$, is translation invariant as well. Finally, phase information $arg(Z(k))$, $k = 0, \ldots, N-1$, is invariant to scaling. Therefore, Fourier descriptors have interesting invariance properties that can be used in object recognition applications. The magnitude mean square error:

$$E = \sum_{k=0}^{N-1} (|Z_1(k)| - |Z_2(k)|)^2 \tag{7.4.12}$$

can be used as an error measure for matching two curves $z_1(n)$, $z_2(n)$. This error is small (ideally zero) if $z_2(n)$ is a rotated version of curve $z_1(n)$.

If N is chosen to be a power of 2, the Fourier descriptors can be easily calculated by using the radix-2 fast Fourier transform [OPE89]. If N is not a power of 2, other Fourier transform algorithms can be used (e.g. the prime factor algorithm) [NUS81]. The above-mentioned properties are valid for uniformly sampled curves $z(n)$. This is not the case if the 8-connected neighborhood is used for curve following, because the sampling intervals may be 1 or $\sqrt{2}$. Phase information suffers particularly from non-uniform curve sampling. Thus, if 8-connected curve traversing is used on a rectangular grid, only the transform magnitude possesses reliable information.

Fourier descriptors can be used to regenerate the contour by employing the inverse transform (7.4.3). However, if transform coefficients are processed (e.g. truncated, quantized), the inverse DFT may no longer produce accurate representations of the contour. Fourier coefficient modifications may result in curves that are not closed.

7.5 QUADTREES

In this section we shall describe *quadtrees* for binary region representation. We have already encountered them in region segmentation (Chapter 6), because

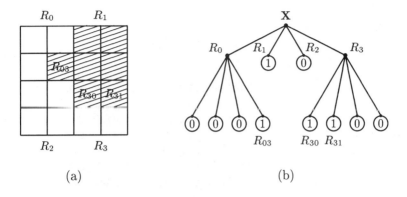

Fig. 7.5.1 (a) Binary image; (b) quadtree representation.

they are closely connected to region splitting algorithms. Quadtrees are based on the following recursive approach: if a binary image region of size $2^n \times 2^n$ consists of both 0s and 1s, it is declared inhomogeneous and is split into four square subregions R_0, R_1, R_2, R_3 having size $2^{n-1} \times 2^{n-1}$ each. This procedure continues until all subregions are homogeneous (i.e. they consist solely of 0s or 1s). The resulting representation is a quadtree (i.e. each node has four children). The root represents the entire binary image. Tree leaves represent homogeneous regions denoted by 0 or 1, for regions consisting of 0s or 1s respectively. Non-terminal nodes represent inhomogeneous image regions. A quadtree representation is shown in Figure 7.5.1. Let us suppose that the original image X has size $2^n \times 2^n = 4^n$. The quadtree has maximum $n+1$ levels, $k = 0, \ldots, n$ (including the root level). Each level has a maximum of $2^k \times 2^k = 4^k$ nodes. Therefore, the maximal number of nodes for quadtree representation is:

$$N = \sum_{k=0}^{n} 4^k \simeq \frac{4}{3} 4^n \qquad (7.5.1)$$

This means that the maximal quadtree memory requirements are approximately equal to 4/3 of the image size. Each quadtree node can be uniquely labeled. Let us suppose that each region R is split into four subregions having labels $0, 1, 2, 3$. A node at level k, $k = 1, \ldots, n$, can be labeled by a number of the form $n_1 n_2 \ldots n_k$, where n_i takes values from the set $\{0, 1, 2, 3\}$. A binary object can be represented by the list of labels of all its square subregions. For example, the representation of the binary object in Figure 7.5.1 is $R = R_1 \cup R_{03} \cup R_{30} \cup R_{31}$. An algorithm for the construction of the quadtree representation of a binary image is shown in Figure 7.5.2. The quadtree nodes have the format:

```
typedef struct QNODE { unsigned char color;
```

338 SHAPE DESCRIPTION

```
         struct QNODE *father;
         struct QNODE *quad[4];
     } QNODE;
```

The color of each node denotes node status: GRAY nodes are non-terminal nodes; terminal nodes have color 0 or 1. The quadtree has $\log_2 N + 1$ levels $[0,\ldots,\log_2 N]$, where $N \times N$ is the size of the binary image to be encoded. The subroutine construct_quadtree() that constructs the quadtree is recursive. At each tree level, it goes one level down and calls itself four times, one for each subregion. It does this recursively until level 0 (tree bottom). The Morton traverse scheme is used [SAM90]. If all children have the same color, the subroutine returns. If the children have different color, a father node is created and is assigned color GRAY=2. The existing child nodes are attached to the father node. Any new child nodes created are attached to the father node as well.

```
int image_to_quadtree(a,N1,M1,N2,M2,root)
image a;
int N1,N2,M1,M2;
QNODE **root;
/* Procedure to build a quadtree from a square image stored
   in array a. The procedure returns a pointer to the root of
   the quadtree. The image must be binary.
   a: array of the binary image
   N1,M1: start coordinates
   N2,M2: end   coordinates
   root: Pointer to the root of the quadtree which the
   routine returns. */
{
  unsigned char col,level=0;
  QNODE *p;
  int divizor,status;

  /* Find the level of the quadtree. If the dimension of
     the image is not a power of 2, return -33 */
  do
    { level=0; divizor=(N2-N1);
      while (divizor!=1)
        { if ((divizor%2)!=0) return(-33);
          divizor=divizor/2;
          level++;
        }
    } while ((int)pow(2.0,(double)level)!=N2);
  /* Construct the quadtree. If construct_quadtree returns NULL,
     the image is homogeneous and can be represented
```

```
      by one node. */
  status=construct_quadtree(a,&p,level,&col,N1,M1,N2,M2);
  if (status!=0) {*root=NULL; return(status);}
  if (p==NULL) p=create_node(NULL,0,col);
  *root=p;
  return(0);
}

int construct_quadtree(a,ptr,level,col,N1,M1,x,y)
image a;
QNODE **ptr;
unsigned char level,*col;
int x,y,N1,M1;
/* Recursive routine which builds the quadtree which corresponds
   to the image of dimensions 2^level x 2^level.
   ptr : pointer to the subtree to be constructed
   level: New level of the quadtree. (Level is 0 for
          nodes representing pixels.)
   col  : Node color to be returned.(0,1,GRAY for non-leaf nodes)
   N1,M1: starting point of the region of interest in the image
   x,y  : The coordinates of the south-most, east-most pixel of
          the square subregion which will be visited.
   image: Pointer to the array which is the image stored. */
{
 unsigned char t[4];
 int l,d,status;
 static unsigned char xf[4]={1,0,1,0}, yf[4]={1,1,0,0};
 QNODE *p[4],*q;
 int GRAY=2;
 int NW=0, NE=1, SW=2, SE=3;
 /* If we are at the bottom level of the subroutine, assign
    the pixel value to color and return. */
 if (level==0)
  { *col=a[N1+x-1][M1+y-1]; *ptr=NULL; return(0); }
 else
  { /* Go down one level. Call construct_quadtree four times
       (one for each subregion). Morton traverse is used. */
    level--;
    d=(unsigned char)pow(2.0,(double)level);
    status=construct_quadtree(a,&p[NW],
           level,&t[0],N1,M1,x-xf[0]*d,y-yf[0]*d);
    if(status!=0) return(status);
    status=construct_quadtree(a,&p[NE],
           level,&t[1],N1,M1,x-xf[1]*d,y-yf[1]*d);
    if(status!=0) return(status);
```

```
            status=construct_quadtree(a,&p[SW],
                   level,&t[2],N1,M1,x-xf[2]*d,y-yf[2]*d);
            if(status!=0) return(status);
            status=construct_quadtree(a,&p[SE],
                   level,&t[3],N1,M1,x-xf[3]*d,y-yf[3]*d);
            if(status!=0) return(status);
            /*If the region is homogeneous calculate color and return.*/
            if (t[NW]!=GRAY && t[NW]==t[NE] &&
                t[SW]==t[SE] && t[NW]==t[SW])
            { *col=t[NW]; *ptr=NULL; return(0) ; }
            else
            { /* If the region is inhomogeneous, create a father node.
                  Create children nodes or attach existing children
                  nodes. GRAY color is assigned to the father node. */
              q=create_node(NULL,0,GRAY);
              for (l=0;l<4;l++)
                {
                  if (p[l]==NULL) p[l]=create_node(q,l,t[l]);
                  else { q->quad[l]=p[l]; p[l]->father=q; }
                }
              *col=GRAY; *ptr=q;
              return(0);
            }
         }
     }
}
QNODE *create_node(father,t,col) QNODE *father; unsigned char
t,col; {
 /* Function to create a node of a quadtree as a child of node
     father with orientation t and color col.
     father: pointer to the father of the node to be created
     t     : the orientation of the node to be created
     col : the color of the node     */

 QNODE *pt; int ii;
 pt=(QNODE *)malloc(sizeof(QNODE));
 if (pt==NULL) return(NULL);
 if (father!=NULL) father->quad[t]=pt;
 pt->father=father; pt->color=col;
 for (ii=0;ii<4;ii++) pt->quad[ii]=NULL;
 return(pt);
}
```

Fig. 7.5.2 Algorithm to create a quadtree representation of a binary object.

The opposite procedure is shown in Figure 7.5.3. Subroutine quadtree_to_image() transforms a quadtree to a binary image. First, the number of levels of the binary tree is found by using the recursive procedure find_level() and then the quadtree is traversed by using the subroutine make_image(). At each terminal node we assign output image pixels by the appropriate color 0,1. Thus, a binary image is produced.

```
int quadtree_to_image(pt,a,N1,M1,N2,M2)
QNODE *pt;
image a;
int N1,M1,N2,M2;
{
  /* Routine for loading a binary image quadtree
     to an image buffer.
     pt    : Pointer to the root of the quadtree.
     a     : output image buffer.
     N1,M1 : Start point on the output buffer.
     N2,M2 : End point on the output buffer. */

  QNODE *p;
  int dim,lev,level; lev=0; p=pt;
  /* Find the number of quadtree levels.*/
  level=find_level(p,&lev);
  dim=(int)pow(2.0,(double)level);
  p=pt;
  /* Construct the binary image. */
  make_image(p,a,level,N1,M1,dim,dim);
  return(0);
}

int make_image(ptr,a,level,N1,M1,x,y)
QNODE *ptr;
image a;
int level,x,y,N1,M1;
{
  /* Recursive routine which constructs the binary image
     from a quadtree.
     ptr  : Pointer to the node of the quadtree which is visited
     a    : output image buffer
     level: The level of the quadtree.
            The size of the image is 2^level x 2^level.
     N1,M1: Destination point on the output image buffer.
     x,y  : The coordinates of the south-most, east-most square
            subregion of the image.        */
```

```
int d,i,j,l;
static int xf[4]={1,0,1,0}; static int yf[4]={1,1,0,0};
static unsigned char pix;
int GRAY=2;
/* If we are at level 0 of the quadtree, assign value
   to the corresponding output image pixel. */
if (level==0) a[N1+x-1][M1+y-1]=ptr->color;
else
  { /* If we are at GRAY node, we go one level down by
       calling make_image() */
    if (ptr->color==GRAY)
      { level--; d=(int)pow(2.0,(double)level);
        for (l=0;l<4;l++)
          make_image(ptr->quad[l],a,
                     level,N1,M1,x-xf[l]*d,y-yf[l]*d);
      }
    else
      { /*If we are at leaf node, we assign corresponding output
           image pixels with node color. */
        d=(int)pow(2.0,(double)level); pix=ptr->color;
        /* The region which the leaf node represents
           is homogeneous */
        for (i=x-d;i<x;i++)
          for (j=y-d;j<y;j++)
            a[N1+i][M1+j]=pix;
      }
  }
}
```

Fig. 7.5.3 Algorithm to find the binary image that corresponds to a quadtree representation.

7.6 PYRAMIDS

Multiresolution representations employ several copies of the same image at different resolutions. Thus, multiresolution algorithms can easily obtain abstraction from image details. They have been proven to be very efficient in certain image analysis applications (e.g. region segmentation, edge detection and correlation matching). Multiresolution techniques applied to grayscale or binary images lead to the so-called *image pyramids*. An image pyramid is a series $f_k(i,j)$, $k = 0, \ldots, n$, of image arrays, each having size $2^k \times 2^k$, as shown in Figure 7.6.1 [CAN86], [CYP89]. The top array has size 1 and represents the entire image. The bottom array is the entire image of size $2^n \times 2^n$. The in-between levels are image representations at resolutions $2^k \times 2^k$, $k = 1, \ldots, n-1$. The intermediate-level images are created in a bottom-up approach. The pixel $f_k(i,j)$ at level k is created from its four neighbors $f_{k+1}(i',j')$ at level $k+1$:

$$f_k(i,j) = g(f_{k+1}(2i,2j), f(2i, 2j+1), f(2i+1, 2j),$$
$$f(2i+1, 2j+1)) \quad i,j = 0, \ldots, 2^k - 1 \qquad (7.6.1)$$

where $g(\cdot)$ is a *mapping function*. This procedure is illustrated in Figure 7.6.1b. A simple mapping results by using the local arithmetic mean [TAN75]:

$$f_k(i,j) = \frac{1}{4} \sum_{l=0}^{1} \sum_{m=0}^{1} f_{k+1}(2i+l, 2j+m) \qquad (7.6.2)$$

Alternatively, the local maximum, the local minimum or local morphological transformations [PIT90] can be used as a mapping function. Pyramids can be constructed for binary images as well. The mapping function can be local AND or OR operations. It is clearly seen that the mapping function from one level to the next is a low-pass operator. Therefore, high frequencies (and image details) are gradually suppressed. This fact is useful when we want to start image analysis from rough object characteristics and to move gradually to object details. Pyramidal techniques enjoy a certain popularity for image analysis and compression applications, because they offer abstraction from image details [TAN86]. Binary image pyramids can be used in multiresolution edge detection and region segmentation. Image edges and regions can be displayed at different resolutions. Edges that are broken and regions that

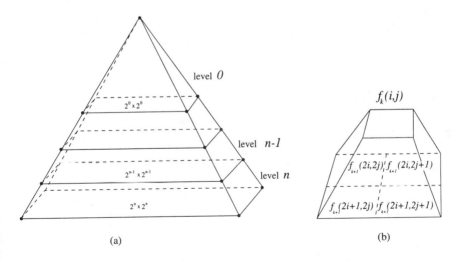

Fig. 7.6.1 (a) Image pyramid; (b) mapping from one pyramid level to the next level.

are disjoint at the bottom of the pyramid bottom may join at intermediate pyramid levels. Therefore, edge following can benefit from a multiresolution approach.

As was mentioned in the case of quadtrees, the total space required for the storage of a pyramid (and of a quadtree) is $\frac{4}{3}2^n \times 2^n$, where $2^n \times 2^n$ is the size of the original image. Thus, all levels $k = 0, \ldots, n-1$ of the pyramid can be stored in another image array of size $2^n \times 2^n$. The resulting algorithm is shown in Figure 7.6.2.

```
int pyramid_1(x,y,N,M)
image x,y;
int N,M;
/* Subroutine that creates the pyramid of input binary image x
   and stores pyramid levels (except the lowest one)
   to output image y.
   x: input image buffer
   y: output image buffer
   N,M: sizes of input and output image      */
{
 int NS=0,NSOLD,k,l;
 for (k=0;k<N;k++)
  for (l=0;l<N;l++)
   y[k][l]=0;
 N/=2;
 /* OR binary operations are used in the 2x2 image neighborhood */
 for (k=0;k<N;k++)
  for (l=0;l<N;l++)
   y[k][l]=(x[2*k][2*l]|x[2*k+1][2*l]|
           x[2*k][2*l+1]|x[2*k+1][2*l+1]);
 do { NSOLD=NS; NS+=N;N/=2;
      for (k=0;k<N;k++)
       for (l=0;l<N;l++)
        y[k][l+NS]=(y[2*k][2*l+NSOLD]|y[2*k+1][2*l+NSOLD]
                   |y[2*k][2*l+NSOLD+1]|y[2*k+1][2*l+NSOLD+1]);
    } while (N!=1);
}
```

Fig. 7.6.2 Storage of a pyramid on an image array.

Subroutine pyramid_1() uses local OR operations as a mapping function. This subroutine can be slightly modified to store the pyramid on $n+1$ image arrays of size $2^k \times 2^k$, $k = 0, \ldots, n$. The resulting subroutine is shown in Figure 7.6.3.

```
int pyramid_fill(im,x,N)
image im;
pyramid x;
int N;
/* Subroutine that creates the pyramid of input binary image im
   and stores pyramid levels to an output pyramid structure.
   im: input image buffer
   x: output pyramid
   N: size of input image      */
{
  int k,l,i,m=N,n,nn;
  /* Calculate pyramid levels. */
  for(n=0,nn=1;nn<N;n++) nn *=2;
  /* Load bottom pyramid level. */
  for (k=0;k<N;k++)
    for (l=0;l<N;l++)
      x[n][k][l]=im[k][l];
  /* Calculate other pyramid levels using OR operations. */
  for (i=n-1;i>=0;i--)
  { m/=2;
    for (k=0;k<m;k++)
      for (l=0;l<m;l++)
        x[i][k][l]=(x[i+1][2*k][2*l]|x[i+1][2*k+1][2*l]
                   |x[i+1][2*k][2*l+1]|x[i+1][2*k+1][2*l+1]);
  }
}
```

Fig. 7.6.3 Algorithm for the construction of a pyramid.

This subroutine employs an array of images of the form:

```
typedef image * pyramid;
```

The local OR operation is again used as a mapping function. Let us suppose that such a pyramid stored in the arrays x_i, $i = 0, \ldots, n$, has already been calculated. It is easy to perform multiresolution edge detection by applying an edge detector operator at each pyramid level. Alternatively it is possible to apply the edge detector at an appropriate pyramid level k, $0 \leq k \leq n$, where the edges of interest are visible at the desired resolution. The image regions where the edge detector output is strong enough can be refined recursively at higher pyramid levels i, $k < i < n$, until the pyramid bottom is reached. The resulting algorithm is a variation of the one presented in [TAN75], [BAL82] and is described in Figure 7.6.4. Subroutine `pyramid_edge_detection()` uses the recursive subroutine `refine()` for the refinement of image edges at higher pyramid levels.

```
int pyramid_edge_detection(x,e,N,k,T)
pyramid x,e;
int N,k;
int T;
/* Subroutine for edge detection using a pyramid structure
   x: input pyramid
   e: output pyramid containing the edge detection results
   N: size of lowest level image
   k: start level for edge detection
   T: threshold                                           */
{
 int i,m,l,s,n,nn;
 for (i=0,s=1;i<k;i++) s*=2;
 /* Calculate pyramid levels. */
 for(n=0,nn=1;nn<N;n++) nn *=2;
 /* Perform edge detection and refinement. */
 if (k<=n && k>=0)
   for (m=1;m<s-1;m++)
     for (l=1;l<s-1;l++)
     { e[k][m][l] = (unsigned char)edge_detector(x[k],m,l);
       if (e[k][m][l]>T)  refine(x,e,N,k,T,m,l);
     }
 return(0);
}


int edge_detector(a,i,j)
image a;
int i,j;
/* Subroutine for edge detection at pixel (i,j)
     a:   input image buffer
     i,j: pixel coordinates                    */
{
 int cc,c;
 cc=-(int)a[i-1][j-1]-2*(int)a[i-1][j]-(int)a[i-1][j+1];
 cc += (int)a[i+1][j-1]+2*(int)a[i+1][j]+(int)a[i+1][j+1];
 c=abs(cc);
 cc=-(int)a[i-1][j-1]-2*(int)a[i][j-1]-(int)a[i+1][j-1];
 cc += (int)a[i-1][j+1]+2*(int)a[i][j+1]+(int)a[i+1][j+1];
 c += abs(cc) ;
```

```
    if(c>255) c=255;
    return(c);
}

int refine(x,e,N,k,T,m,l)
pyramid x,e;
int N,k;
int T;
int m,l;
/* Recursive routine that refines edge detection
   from upper to lower pyramid levels.
   x: input pyramid
   e: output pyramid
   N: size of lowest level image
   k: start level for edge detection
   T: threshold
   m,l: current pixel coordinates for refinement   */
{
  int i,j,n,nn;
  /* Calculate pyramid levels. */
  for(n=0,nn=1;nn<N;n++) nn *=2;
  if (k<n)
    for (i=0;i<=1;i++)
      for (j=0;j<=1;j++)
      { e[k+1][2*m+i][2*l+j]=edge_detector(x[k+1],2*m+i,2*l+j);
        if (e[k+1][2*m+i][2*l+j]>T)
           refine(x,e,N,k+1,T,2*m+i,2*l+j);
      }
}
```

Fig. 7.6.4 Hierarchical edge detection based on pyramids.

The above-mentioned algorithms are illustrated in Figure 7.6.5. The original binary image and the resulting pyramid are shown in Figures 7.6.5a and 7.6.5b, respectively. The edges detected by subroutine pyramid_edge_detection() are shown in Figure 7.6.5c and the resulting edge pyramid is illustrated in 7.6.5d. The edge search started from level $k = 5$. Iterative edge refinement produced relatively good edges without excessive noise. Multiresolution image processing techniques based on pyramids have been extensively used in image coding and computer vision. An example of such an application is face detection in 'face and shoulder' images [YAN94], [KOT97].

Fig. 7.6.5 (a) Original binary image; (b) image pyramid; (c) output of the pyramid edge detector; (d) edge pyramid.

7.7 SHAPE FEATURES

Geometrical shapes possess certain features (e.g. perimeter, area, moments) that carry sufficient information for some object recognition applications. Such features can be used as object descriptors, resulting in a significant data compression, because they can represent the geometrical shape by a relatively small feature vector. Shape features can be derived either from a raw binary image or from shape representation schemes that have already been presented in this chapter. For example, it has been shown in a previous section how to calculate object perimeter, width and height from the boundary chain code. Shape features can be grouped in two large classes: *boundary features* and *region features*. In the following, we shall describe several boundary and region features and methods for their calculation.

Object *perimeter* is given by the following integral:

$$T = \int \sqrt{x^2(t) + y^2(t)} dt \qquad (7.7.1)$$

for the parametric boundary representation shown in Figure 7.4.1. The perimeter can be a by-product of boundary-following algorithms. If $\mathbf{x}_1, \ldots, \mathbf{x}_N$ is a boundary coordinate list, the object perimeter is given by:

$$T = \sum_{i=1}^{N-1} d_i = \sum_{i=1}^{N-1} |\mathbf{x}_i - \mathbf{x}_{i+1}| \qquad (7.7.2)$$

The distances d_i are equal to 1 for 4-connected boundaries and to 1 or $\sqrt{2}$ for 8-connected boundaries.

Corners are boundary locations where boundary *curvature magnitude* $|k(t)|$ takes large values or even becomes unbounded:

$$|k(t)|^2 \triangleq \left(\frac{d^2x}{dt^2}\right)^2 + \left(\frac{d^2y}{dt^2}\right)^2 \qquad (7.7.3)$$

The curvature magnitude can be approximated by employing numerical differentiation:

$$|k(n)| = \frac{1}{\Delta^2}\sqrt{[x(n-1) - 2x(n) + x(n+1)]^2 + [y(n-1) - 2y(n) + y(n+1)]^2} \qquad (7.7.4)$$

where Δ denotes the sampling interval. Another definition of the curvature is given by [SCH89]:

$$k(s) = \frac{d\phi(s)}{ds}$$
$$ds = \sqrt{dx^2 + dy^2} \qquad (7.7.5)$$

Based on this relation, an approximation of the local curvature $k(n)$ is given by:

$$k(n) \simeq \frac{x_n - x_{n-1}}{L(x_n) + L(x_{n-1})} \qquad (7.7.6)$$

where

$$L(x_i) = \begin{cases} \frac{1}{2} & \text{for } x_i \text{ even} \\ \frac{\sqrt{2}}{2} & \text{for } x_i \text{ odd} \end{cases} \qquad (7.7.7)$$

Relation (7.7.6) employs the boundary chain code $x_1 x_2 \ldots x_N$ described previously in this chapter. Each code element x_i is supposed to take values from the set $0, \ldots, m-1$. For 8-connected chains, m is equal to 8. The numerator of (7.7.6) denotes a local change in tangent direction and the denominator approximates the local curve length around point n. Another boundary feature related to curvature is its *bending energy* [CAN70], [YOU79]:

$$E = \frac{1}{T}\int_0^T |k(t)|^2 dt \qquad (7.7.8)$$

The bending energy at the point n, $1 \leq n \leq N$, can be found by combining (7.7.8), (7.7.6):

$$E = \frac{1}{T}\sum_{i=0}^{n-1} |k(i)|^2 \qquad (7.7.9)$$

Bending energy can also be easily calculated from boundary Fourier descriptors $Z(k)$ [JAI89]:

$$E = \sum |Z(k)|^2 \left(\frac{2\pi k}{T}\right)^4 \qquad (7.7.10)$$

The circle possesses the minimal bending energy:

$$E = \left(\frac{2\pi}{T}\right)^2 \qquad (7.7.11)$$

among shapes having equal perimeter T. Unfortunately, bending energy is perimeter-dependent, as can be seen in (7.7.9). Therefore, a normalization of bending energy leads to the following feature that takes values in the range [0,1] [LEV85]:

$$E_N = 1 - \frac{E_{circle}}{E_{object}} = 1 - \frac{4\pi^2}{T\sum_{i=0}^{n}|k(i)|^2} \qquad (7.7.12)$$

E_N is particularly useful in biomedical applications, where cells have approximately circular shapes.

Object *area* is a region-based feature given by:

$$A = \int\int_R dx\,dy \qquad (7.7.13)$$

where R is the object region. Area can also be described in terms of the object boundary ∂R by the well-known differential geometry formula:

$$A = \int_{\partial R} \left(y(t)\frac{dx}{dt} - x(t)\frac{dy}{dt}\right) dt \qquad (7.7.14)$$

The elementary area $dx\,dy$ in (7.7.13) is an object pixel. Therefore, in digital images, the object area can be found by counting object pixels. Area measurement can be a by-product of connected component labeling algorithms, because such algorithms scan every object pixel. Such an algorithm using the grassfire concept has already been presented in Chapter 6. Shape complexity can be measured in terms of its *compactness* or *circularity* [GRA71]:

$$\gamma = \frac{T^2}{4\pi A} \qquad (7.7.15)$$

where T, A are object perimeter and area respectively. Compactness attains its minimum value $\gamma=1$ for a circle. All other shapes that have more complicated boundaries have larger compactness. The square has compactness $\gamma = 4/\pi$. A normalized version of compactness can be used instead of (7.7.15):

$$\gamma_N = 1 - \frac{4\pi A}{T^2} \qquad (7.7.16)$$

Normalized compactness equals zero for circles and tends to unity for complex shapes.

The object *width* and *height* are given by the relations:

$$w = \max_t x(t) - \min_t x(t) \qquad (7.7.17)$$

$$h = \max_t y(t) - \min_t y(t) \qquad (7.7.18)$$

where (x, y) are the *major* and *minor* axes of the object. The major axis is the straight line joining the two object points lying furthest apart from each other. The minor axis is perpendicular to the maximal axis. A minimal box having sides parallel to the maximal and minimal axes is called the *basic rectangle* of the object. The ratio of its sides w/h is called object *eccentricity* or *elongation*. Sometimes elongation is defined as the ratio [LEV85]:

$$\ell = \frac{|w - h|}{h} \qquad (7.7.19)$$

This ratio is equal to zero for squares and circles and approaches unity for elongated objects. Finally, the diameter of the object:

$$D = \max_{\mathbf{x}_k, \mathbf{x}_l \in R} d(\mathbf{x}_k, \mathbf{x}_l) \qquad (7.7.20)$$

and the direction of the line segment $\mathbf{x}_k\mathbf{x}_l$ are important object descriptors. The maximization is performed over any points $\mathbf{x}_k, \mathbf{x}_l$ belonging to object region R. The distance $d(\mathbf{x}_k, \mathbf{x}_l)$ used in (7.7.20) can be any distance measure, for example, Euclidean, city-block, or chessboard distance [SER82], [BOR86], [ZHO88].

Topological descriptors can give useful global information about an object. Two important topological features are the *holes* and the *connected components* of an object. Let us suppose that an object X is a subset of an image plane E. Connected component labeling algorithms, described in Chapter 6, give the number C of the connected components as a byproduct. The same algorithms can be used to find the holes contained in the object X, provided that X is the only object contained in the image plane E. Object holes are connected components of the object *complement* X_E^c. If the image E has no background, i.e. $X_E^c \cap \partial E = \emptyset$, all connected components of X_E^c are holes of X. Otherwise, one of them corresponds to the object background and must not be counted as a hole. Let C and H be the number of connected components of X and the number of its holes, respectively. The *Euler number E* [BAL82]:

$$E = C - H \qquad (7.7.21)$$

is an important topological descriptor. The objects shown in Figure 7.7.1 have Euler numbers 0, −1 and 1, respectively.

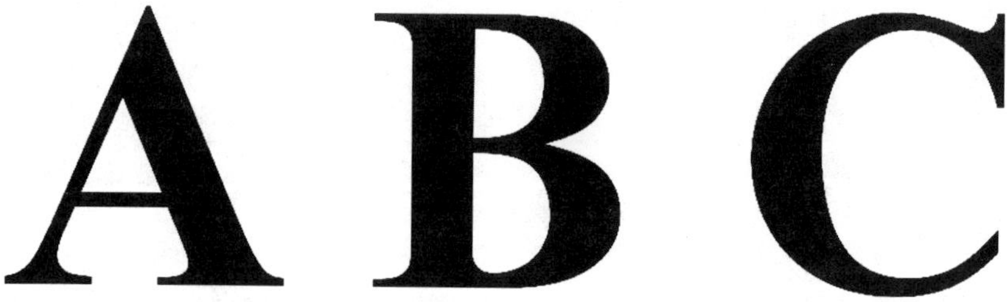

ABC

Fig. 7.7.1 Letters A,B,C have Euler numbers 0, −1,1, respectively.

7.8 MOMENT DESCRIPTORS

A very useful and practical set of shape descriptors is based on the theory of moments. The *moments* of an image $f(x,y)$ are given by:

$$m_{pq} = \int_{-\infty}^{\infty} \int_{-\infty}^{\infty} x^p y^q f(x,y) dx dy \quad p,q = 0,1,2,\ldots \quad (7.8.1)$$

The *center of gravity* of the object can be given by using object moments:

$$\bar{x} = \frac{m_{10}}{m_{00}} \qquad \bar{y} = \frac{m_{01}}{m_{00}} \quad (7.8.2)$$

The center of gravity is used in the definition of the *central moments*:

$$\mu_{pq} = \int_{-\infty}^{\infty} \int_{-\infty}^{\infty} (x-\bar{x})^p (y-\bar{y})^q f(x,y) dx dy \quad p,q = 0,1,2,\ldots \quad (7.8.3)$$

In the case of discrete images, the moment relations are given by [HU61], [HU62]:

$$m_{pq} = \sum_i \sum_j i^p j^q f(i,j) \quad (7.8.4)$$

$$\mu_{pq} = \sum_i \sum_j (i-\bar{x})^p (j-\bar{y})^q f(i,j) \quad (7.8.5)$$

It must be noted that indices i, j correspond to coordinate axes x, y respectively. If image $f(i,j)$ is binary, the moment calculations are given by:

$$m_{pq} = \sum_i \sum_j i^p j^q \quad (7.8.6)$$

$$\mu_{pq} = \sum_i \sum_j (i-\bar{x})^p (j-\bar{y})^q \quad (7.8.7)$$

Moment m_{00} represents the binary object area. Object moments can be calculated very easily. Such an algorithm for binary central moment calculation is shown in Figure 7.8.1.

```
float find_moment(x,p,q,N1,M1,N2,M2,_x,_y)
image x;
int p,q,N1,M1,N2,M2;
float _x,_y;
/* Subroutine to calculate sum sum [(j-_x)^p * (i-^_y)^q]
   over the ROI of the binary image x
   x: input image buffer
   p,q: moment indices
   N1,M1: start coordinates
   N2,M2: end   coordinates
   _x,_y: central point coordinates */
{
 int i,j,k;
 float a=0,h;
 for (i=N1,a=0.;i<N2;i++)
  for (j=M1;j<M2;j++)
   { if (x[i][j]==1)
      {
       for (k=0,h=1;k<p;k++) h*=(j-_x);
       for (k=0;k<q;k++) h*=(i-_y);
       a+=h;
      }
   }
 return (a);
}
```

Fig. 7.8.1 Algorithm for central moment calculation for binary objects.

It can be easily proven that the central moments are given by the following relations (up to the third-order moments) [HU61], [REE81], [WON78], [LEV85]:

$$\begin{aligned}
\mu_{00} &= m_{00} = \mu \\
\mu_{10} &= \mu_{01} = 0 \\
\mu_{20} &= m_{20} - \mu \bar{x}^2 \\
\mu_{11} &= m_{11} - \mu \bar{x} \bar{y} \\
\mu_{02} &= m_{02} - \mu \bar{y}^2 \\
\mu_{30} &= m_{30} - 3 m_{20} \bar{x} + 2\mu \bar{x}^3 \\
\mu_{21} &= m_{21} - m_{20}\bar{y} - 2m_{11}\bar{x} + 2\mu \bar{x}^2 \bar{y} \\
\mu_{12} &= m_{12} - m_{02}\bar{x} - 2m_{11}\bar{y} + 2\mu \bar{x} \bar{y}^2 \\
\mu_{03} &= m_{03} - 3 m_{02}\bar{y} + 2\mu \bar{y}^3
\end{aligned} \qquad (7.8.8)$$

The normalized central moments given by [GON87]:

$$\eta_{pq} = \mu_{pq}/\mu_{00}^{\gamma}$$

$$\gamma = \frac{p+q}{2} + 1$$

can be employed to produce a set of invariant moments. The seven low-order invariants $\phi_1, \ldots, \phi_7$ are given in terms of the second- and third-order central moments [HU61], [HU62]:

$$\begin{aligned}
\phi_1 &= \eta_{20} + \eta_{02} \\
\phi_2 &= (\eta_{20} - \eta_{02})^2 + 4\eta_{11}^2 \\
\phi_3 &= (\eta_{30} - 3\eta_{12})^2 + (3\eta_{21} - \eta_{03})^2 \\
\phi_4 &= (\eta_{30} + \eta_{12})^2 + (\eta_{21} + \eta_{03})^2 \\
\phi_5 &= (\eta_{30} - 3\eta_{12})(\eta_{30} + \eta_{12})[(\eta_{30} + \eta_{12})^2 - 3(\eta_{21} + \eta_{03})^2] \\
&+ (3\eta_{21} - \eta_{03})(\eta_{21} + \eta_{03})[3(\eta_{30} + \eta_{12})^2 - (\eta_{21} + \eta_{03})^2] \\
\phi_6 &= (\eta_{20} - \eta_{02})[(\eta_{30} + \eta_{12})^2 - (\eta_{21} + \eta_{03})^2] \\
&+ 4\eta_{11}(\eta_{30} + \eta_{12})(\eta_{21} + \eta_{03}) \\
\phi_7 &= (3\eta_{21} - \eta_{03})(\eta_{30} + \eta_{12})[(\eta_{30} + \eta_{12})^2 - 3(\eta_{21} + \eta_{03})^2] \\
&- (\eta_{30} - 3\eta_{12})(\eta_{21} + \eta_{03})[3(\eta_{30} + \eta_{12})^2 - (\eta_{21} + \eta_{03})^2]
\end{aligned} \quad (7.8.9)$$

All moments $\phi_1, \ldots, \phi_7$ are scale, rotation and translation invariant. Moments $\phi_1, \ldots, \phi_6$ are reflection invariant as well. The magnitude of ϕ_7 is reflection invariant. However, its sign changes under reflection.

Several object descriptors are based on moments. The center of gravity given by (7.8.2) is such a feature. An alternative method for the calculation of the center of gravity is the following. If the area of an image $f(i,j)$ is N pixels, the coordinates of the center of gravity are given by:

$$\bar{x} = \frac{1}{N} \sum_{(i,j) \in R} i \qquad (7.8.10)$$

$$\bar{y} = \frac{1}{N} \sum_{(i,j) \in R} j$$

Object *orientation* θ is the angle between the major axis of the object and axis x as can be seen in Figure 7.8.2. It can be derived by minimizing the function:

$$S(\theta) = \sum\sum_{(i,j) \in R} [(i - \bar{x})\cos\theta - (j - \bar{y})\sin\theta]^2 \qquad (7.8.11)$$

The minimization gives the following result:

$$\theta = \frac{1}{2} \arctan\left(\frac{2\mu_{11}}{\mu_{20} - \mu_{02}}\right) \qquad (7.8.12)$$

Object approximation by an ellipse has found extensive applications in face detection [SOB98]. It uses moments for finding facial ellipse parameters and

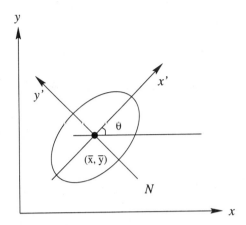

Fig. 7.8.2 Definition of object orientation.

face orientation. Face modeling by ellipse can also be used for face normalization prior to face recognition or verification [KOT99PR].

Object *eccentricity* can be measured by using [LEV85]:

$$\epsilon = \left[\frac{\mu_{02} \cos^2 \theta + \mu_{20} \sin^2 \theta - \mu_{11} \sin 2\theta}{\mu_{02} \sin^2 \theta + \mu_{20} \cos^2 \theta + \mu_{11} \cos 2\theta} \right]^{1/2} \quad (7.8.13)$$

An alternative definition of eccentricity is given by [JAI89]:

$$\epsilon = \frac{(\mu_{02} - \mu_{20})^2 + 4\mu_{11}}{A} \quad (7.8.14)$$

where A is the object area. Finally the object *spread* or *size* can be also measured in terms of the central moments [HU62]:

$$S = \mu_{02} + \mu_{20} \quad (7.8.15)$$

An algorithm for the calculation of the center of gravity, orientation, eccentricity and spread of a binary object is shown in Figure 7.8.3. It must be noted that the indices i, j used in subroutine find_moment() correspond to coordinates y, x, respectively.

```
int calc_features_1(x,N1,M1,N2,M2,area,xmean,ymean,orientation,
                    eccentricity1,eccentricity2,spread)
image x;
int N1,M1,N2,M2;
float *area,*xmean,*ymean,*orientation,*eccentricity1,
  *eccentricity2,*spread;
/* Subroutine that calculates moment-related features
```

```
   x: input image buffer
   N1,M1: start coordinates
   N2,M2: end   coordinates
   area: object area
   xmean, ymean: center of gravity coordinates
   orientation: object orientation
   eccentricity1, eccentricity2: object eccentricities
   spread: object spread */
{
 float m00,m10,m01,mi11,mi20,mi02,th,nom,denom;
 m00=find_moment(x,0,0,N1,M1,N2,M2,0.,0.);
 m10=find_moment(x,1,0,N1,M1,N2,M2,0.,0.);
 m01=find_moment(x,0,1,N1,M1,N2,M2,0.,0.);
 (*area)=m00;(*xmean)=m10/m00;(*ymean)=m01/m00;
 mi11=find_moment(x,1,1,N1,M1,N2,M2,(*xmean),(*ymean));
 mi20=find_moment(x,2,0,N1,M1,N2,M2,(*xmean),(*ymean));
 mi02=find_moment(x,0,2,N1,M1,N2,M2,(*xmean),(*ymean));
 th=(mi11!=0||mi02-mi20!=0)?(1./2)*atan2(2*mi11,mi20-mi02):0;
 nom=mi02*cos(th)*cos(th)+mi20*sin(th)*sin(th)-mi11*sin(2*th);
 denom=mi02*sin(th)*sin(th)+mi20*cos(th)*cos(th)+mi11*cos(2*th);
 (*orientation)=th;
 (*eccentricity1)=pow(nom/denom,0.5);
 (*eccentricity2)=((mi02-mi20)*(mi02-mi20)+4*mi11)/m00;
 (*spread)=mi02+mi20;
 return(0);
}
```

Fig. 7.8.3 Algorithm for center of mass and orientation calculation of a binary object.

7.9 THINNING ALGORITHMS

Many shapes, especially the elongated ones, are sufficiently well described by their thinned versions consisting of connected lines that, ideally, run along the medial axes of object limbs [DAV81]. These unit-width lines are closely related to object *skeleton*. A mathematical definition of skeleton and the appropriate algorithms will be given later on in this chapter. No precise mathematical definition exists for the thinning process. This fact partially explains the multitude of thinning algorithms existing in the literature [DAV81], [TAM78]. *Thinning* can be defined heuristically as a set of successive erosions of the outermost layers of a shape, until a connected unit-width set of lines (skeleton) is obtained [DAV81]. Thus, thinning algorithms are iterative algorithms that 'peel off' border pixels, that is, pixels lying at $0 \rightarrow 1$ transitions in a binary

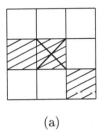 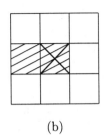

p_8	p_1	p_2
p_7	p_0	p_3
p_6	p_5	p_4

(a) (b) (c)

Fig. 7.9.1 (a) Border pixel whose removal may cause discontinuities; (b) border pixel whose removal will shorten an object limb; (c) local pixel notation used in connectivity check.

image. Connectivity is an important property that must be preserved in the thinned object. Therefore, border pixels are deleted in such a way that object connectivity is maintained. Thinning algorithms satisfy the following two constraints:

1. They maintain connectivity at each iteration. They do not remove border pixels that may cause discontinuities, for example, the ones shown in Figure 7.9.1a.

2. They do not shorten the end of thinned shape limbs. Therefore, they do not remove border pixels similar to the one shown in Figure 7.9.1b.

An algorithm which satisfies these two constraints is shown in Figure 7.9.2. It is a modification of the algorithm described in [FAI88].

```
int thin_1(x,N1,M1,N2,M2)
image x;
int N1,M1,N2,M2;
/* One pass thinning algorithm
   x:input and output image buffer
   N1,M1: start coordinates
   N2,M2: end    coordinates    */
{int k,l,i,j,count=0,y[9],trans=0,m,OK=1;
 do
   { OK=1;
     for (k=N1+1;k<N2-1;k++)
       for (l=M1+1;l<M2-1;l++)
         if (x[k][l]==1)
```

```
          { /* Count number of 1s in 3 x 3 window. */
            count=0;
            for (i=-1;i<=1;i++)
             for (j=-1;j<=1;j++)
              if (x[k+i][j+1]==1) count++;
            if ((count>2)&&(count<8))
             { /* Count number of transitions. */
               y[0]=x[k-1][l-1];y[1]=x[k-1][l];y[2]=x[k-1][l+1];
               y[3]=x[k][l+1];y[4]=x[k+1][l+1];y[5]=x[k+1][l];
               y[6]=x[k+1][l-1];y[7]=x[k][l-1];y[8]=x[k-1][l-1];
               trans=0;
               for (m=0;m<=7;m++)
                if (y[m]==0 && y[m+1]==1) trans++;
               /* If the number of transitions is 1,
                  delete current pixel. */
               if (trans==1) {x[k][l]=0;OK=0;}}
           }
     } while(OK==0);
}
```

Fig. 7.9.2 Simple thinning algorithm.

The subroutine thin_1() operates in a local 3×3 neighborhood. First, it checks if the number of object pixels $N(p_0)$ within the 3×3 window is less than or equal to 2. If so, no action is taken because the window includes the end of an object limb. If the number $N(p_0)$ is larger than 7 no action is taken, because object erosion would result. If more than two object pixels are inside the 3×3 window, it is checked if the removal of the central pixel would break object connectivity. This is done by using the local pixel notation shown in Figure 7.9.1c. The pixel sequence $p_1 p_2 p_3 \ldots p_8 p_1$ is formed. If the number of $0 \rightarrow 1$ transitions in this sequence is 1, only one connected component exists in the perimeter of the 3×3 square. In this case, it is allowed to remove the central pixel having value 1, because removal does not affect the local connectivity of the rest of the pixels in the local window. The basic disadvantage of this algorithm is that it does not thin the image object symmetrically. If the image is scanned in a row-wise manner from the left to the right, the thinned object lines are located at the South, East borders of the image objects, because the North, West border points are removed first. Furthermore, this subroutine does not perform in a satisfactory way when the binary image objects are relatively large and convex.

A more elaborate thinning algorithm that requires two successive iterative passes is described in [ZHA84], [GON87]. At step 1, a logical rule P_1 is applied locally in a 3×3 neighborhood to flag border pixels that can be deleted. These pixels are only flagged (not deleted) until the entire image is scanned. Deletion of all flagged pixels is performed afterwards. At step 2, another logical rule P_2

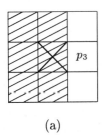 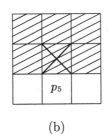 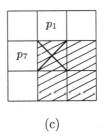

Fig. 7.9.3 Central window pixels belonging to: (a) an East boundary; (b) a South boundary; (c) a North-West corner point.

is applied locally in a 3×3 window to flag border pixels for deletion. When the entire image has been scanned, the flagged pixels are deleted. This procedure is applied iteratively, until no more image thinning can be performed. Let:

$$N(p_0) = \sum_{i=1}^{8} p_i \qquad (7.9.1)$$

denote the number of object pixels ($p_i = 0, 1$, $i = 1, \ldots, 8$) in the 3 × 3 window (excluding the central pixel) and $T(p_0)$ denote the number of $0 \to 1$ transitions in the pixel sequence $p_1 p_2 p_3 \ldots p_8 p_1$. The logical rules P_1, P_2 used in the two steps of the algorithm have the following form:

$$P_1 : (2 \leq N(p_0) \leq 6)\&\&(T(p_0) = 1)\&\&(p_1 \cdot p_3 \cdot p_5 = 0)\&\&(p_3 \cdot p_5 \cdot p_7 = 0) \qquad (7.9.2)$$

$$P_2 : (2 \leq N(p_0) \leq 6)\&\&(T(p_0) = 1)\&\&(p_1 \cdot p_3 \cdot p_7 = 0)\&\&(p_1 \cdot p_5 \cdot p_7 = 0) \qquad (7.9.3)$$

The products of the form $p_i \cdot p_j \cdot p_k$ denote logical conjunctions of the corresponding pixels. The first condition of both logical predicates states that the central pixel can be deleted if it possesses at least one and at most six 8-connected neighbors in a 3 × 3 window. If the central pixel has only one neighbor, it cannot be deleted, because the skeleton limb will be shortened. If the central pixel has more than six neighbors, central pixel deletion is not permitted because it will cause object erosion. The second predicate $T(p_0) = 1$ checks if the pixels of the perimeter of the 3 × 3 neighborhood form only one connected component. The third and fourth predicates ($p_1 \cdot p_3 \cdot p_5 = 0$) and ($p_3 \cdot p_5 \cdot p_7 = 0$) in (7.9.2) are satisfied if $p_3 = 0$ or $p_5 = 0$, or if ($p_1 = 0$ and $p_7 = 0$). Examples of these three cases are shown in Figure 7.9.3. The central pixel belongs to an East boundary ($p_3 = 0$), to a South boundary ($p_5 = 0$), or to a North-West object corner ($p_1 = 0$, $p_7 = 0$). In the first pass, the algorithm removes pixels belonging to one of these three cases. In the second

360   SHAPE DESCRIPTION

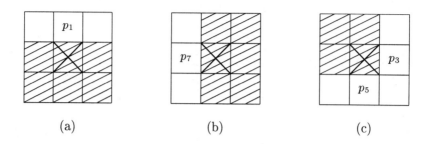

Fig. 7.9.4 Central window pixels belonging to: (a) a North boundary; (b) a West boundary; (c) a South-East corner.

pass, the thinning algorithm removes pixels having $p_1 = 0$ or $p_7 = 0$ or ($p_3 = 0$ and $p_5 = 0$), as can be seen by inspecting (7.9.3). In the second pass, pixels lying at North boundaries, or West boundaries, or South-East corner points are removed, as can be seen in Figure 7.9.4. A description of this two-pass thinning algorithm is shown in Figure 7.9.5.

```
int thin_2(x,N1,M1,N2,M2)
image x;
int N1,M1,N2,M2;
/* Subroutine for two-pass thinning algorithm
   x:input and output image buffer
   N1,M1: start coordinates
   N2,M2: end   coordinates    */
{
 int k,l,i,j,count=0,trans=0,m,OK=1,turn=0;
 int y[9];
 image z;  /* z is a flag buffer */
  z=matuc2(N2-N1,M2-M1);
 do
  { OK=1;turn=(turn+1) %2;
    /* Clear flag buffer. */
    for (k=1;k<N2-N1-1;k++)
      for (l=1;l<M2-M1-1;l++) z[k][l]=0;
    for (k=N1+1;k<N2-1;k++)
      for (l=M1+1;l<M2-1;l++)
        if (x[k][l]==1)
         { /* Count number of 1s in a 3 x 3 window. */
           count=0;
           for (i=-1;i<=1;i++)
```

```
            for (j=-1;j<=1;j++)
              if (x[k+i][j+1]==1) count++;
            if ((count>2)&&(count<8))
            { /* Count number of transitions. */
              y[0]=x[k-1][1-1]; y[1]=x[k-1][1];
              y[2]=x[k-1][1+1]; y[3]=x[k][1+1];
              y[4]=x[k+1][1+1]; y[5]=x[k+1][1];
              y[6]=x[k+1][1-1]; y[7]=x[k][1-1];
              y[8]=x[k-1][1-1];
              trans=0;
              for (m=0;m<=7;m++)
                if (y[m]==0&&y[m+1]==1) trans++;
              /* Flag current pixel to be deleted. */
              if (trans==1)
                if(turn==0&&(y[3]==0||y[5]==0||(y[1]==0&&y[7]==0)))
                  { z[k-N1][1-M1]=1; OK=0; }
                else
                  if(turn==1&&(y[1]==0||y[7]==0||(y[3]==0&&y[5]==0)))
                    { z[k-N1][1-M1]=1;OK=0; }
            }
         }
      /* Delete flagged pixels. */
      for (k=N1+1;k<N2-1;k++)
        for (l=M1+1;l<M2-1;l++)
          if (z[k-N1][1-M1]==1) x[k][l]=0;
    } while(OK==0);
    mfreeuc2(z,N2-N1);
}
```

Fig. 7.9.5 A two-pass thinning algorithm.

The results of the thinning algorithms are illustrated in Figure 7.9.6. Figure 7.9.6a is the result of a Sobel edge detector. The thresholded image is shown in Figure 7.9.6b. The result of the one-pass algorithm is shown in Figure 7.9.6c. It can be seen that the thinned object lines are not located in the 'middle' of the image objects. On the contrary, the output of the two-pass algorithm shown in Figure 7.9.6d is very good. The thinned object lines are located close to object skeletons.

7.10 MATHEMATICAL MORPHOLOGY

Mathematical morphology is a powerful tool for geometrical shape analysis and description. It uses a set theoretic approach to image analysis. A bi-

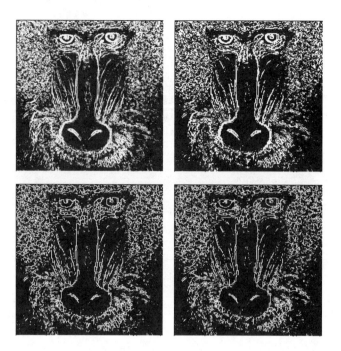

Fig. 7.9.6 (a) Sobel edge detector output; (b) binary image; (c) output of the one-pass thinning algorithm; (d) output of the two-pass thinning algorithm.

nary image object X and its subparts are described by sets in mathematical morphology. Subpart interrelationships describe object structure [SER82]. According to mathematical morphology principles, an image possesses no information if it is not observed by an observer. It is the observer who examines the object and extracts the information needed for structure description and analysis. Observer–object interaction is obtained by using a simple set (e.g. circle, square) that is called a *structuring element*. The mode of interaction is a morphological transformation $\Psi(X)$. Its result can be a measure $m(\Psi(X))$ that describes aspects of the object structure. Information about the object size, shape, connectivity, smoothness and orientation can be obtained by using various structuring elements and morphological operators. The morphological transformations must possess the following properties:

1. **Translation invariance**

$$\Psi(X_z) = [\Psi(X)]_z \qquad (7.10.1)$$

where X_z is a translation of X by a vector z.

2. **Scale invariance**

$$\Psi_\lambda(X) = \lambda\Psi(\lambda^{-1}X) \qquad (7.10.2)$$

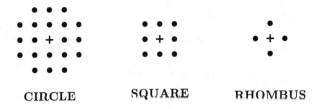

Fig. 7.10.1 Simple symmetric structuring elements.

3. **Local knowledge.** Transformation $\Psi(X)$ must require only information within a local neighborhood for its operation.

4. **Semicontinuity.** The morphological transformation $\Psi(X)$ must possess certain continuity properties. More analytical descriptions of the semicontinuity property can be found in [PIT90], [SER82].

Let $E = R^2$ or Z^2 be the two-dimensional Euclidean space or Euclidean grid, respectively. The binary image object X is a subset of E. Let us also describe by B a structuring element (set) and by B^s its symmetric set with respect to the origin (0,0):

$$B^s = \{-b : b \in B\} \tag{7.10.3}$$

B^s is obtained by rotating B 180 degrees. Simple structuring elements are shown in Figure 7.10.1. The basic morphological transformations are *dilation*:

$$X \oplus B^s = \bigcup_{b \in B} X_{-b} = \{z \in E : B_z \cap X \neq \emptyset\} \tag{7.10.4}$$

and *erosion*:

$$X \ominus B^s = \bigcap_{b \in B} X_{-b} = \{z \in E : B_z \subset X\}$$

Erosion $X \ominus B^s$ is a shrinking operator, because it contains only the points $z \in E$ such that the translates B_z are subsets of X. These points form a subset of X. Thus, erosion shrinks an object, shortens its limbs and expands its holes, as can be seen in Figure 7.10.2c. Dilation $X \oplus B^s$ includes all translates B_z of B that intersect with X. It is an expanding operator and tends to shrink object holes and to expand object limbs, as can be seen in Figure 7.10.2d. Erosion and dilation are special cases of *Minkowski set addition* $X \oplus B$ and *Minkowski set subtraction* $X \ominus B$:

$$X \oplus B = \bigcup_{b \in B} X_b \tag{7.10.5}$$

364 SHAPE DESCRIPTION

Fig. 7.10.2 Basic morphological transformations. (a) Original image; (b) thresholded image; (c) erosion by the structuring element SQUARE; (d) dilation by SQUARE.

$$X \ominus B = \bigcap_{b \in B} X_b \qquad (7.10.6)$$

Erosion, dilation, Minkowski set addition and subtraction have the following interesting properties:

1. **Commutativity**
$$A \oplus B = B \oplus A \qquad (7.10.7)$$

2. **Associativity**
$$A \oplus (B \oplus C) = (A \oplus B) \oplus C \qquad (7.10.8)$$

3. **Translation invariance**
$$A_z \oplus B = (A \oplus B)_z \qquad (7.10.9)$$
$$A_z \ominus B = (A \ominus B)_z \qquad (7.10.10)$$
$$A \ominus B_z = (A \ominus B)_z \qquad (7.10.11)$$

4. **Increasing property**
$$A \subseteq B \Rightarrow A \oplus D \subseteq B \oplus D \qquad (7.10.12)$$

Fig. 7.10.3 (a) Opening by the structuring element SQUARE; (b) closing by SQUARE.

$$A \subseteq B \Rightarrow A \ominus D \subseteq B \ominus D \qquad (7.10.13)$$

5. **Distributivity**

$$\begin{align}
(A \cup B) \oplus C &= (A \oplus C) \cup (B \oplus C) & (7.10.14)\\
A \oplus (B \cup C) &= (A \oplus B) \cup (A \oplus C) & (7.10.15)\\
A \ominus (B \cup C) &= (A \ominus B) \cap (A \ominus C) & (7.10.16)\\
(A \cap B) \ominus C &= (A \ominus C) \cap (B \ominus C) & (7.10.17)\\
A \ominus (B \oplus C) &= (A \ominus B) \ominus C & (7.10.18)
\end{align}$$

In the general case, erosion and dilation are non-reversible operations. An erosion followed by a dilation does not generally recover the original object X. Instead, it defines a new morphological transformation called *opening* X_B:

$$X_B = (X \ominus B^s) \oplus B = \cup \{B_z : B_z \subset X\} \qquad (7.10.19)$$

and consists of the union of those translations of B that lie inside X. Its dual transformation is *closing* X^B:

$$X^B = (X \oplus B^s) \ominus B = \cap \{B_z^c : B_z \subset X^c\} \qquad (7.10.20)$$

It consists of the intersection of the complements of the translations of B that lie outside X. Opening smooths the contours of X, cuts the narrow isthmuses, suppresses the small islands and the sharp capes of X, as can be seen in Figure 7.10.3a. Closing blocks up narrow channels, small holes and thin gulfs of X. Opening and closing have some interesting properties:

1. **Duality**

$$(X^B)^c = (X^c)_B \qquad (7.10.21)$$

$$(X_B)^c = (X^c)^B \qquad (7.10.22)$$

2. Extensivity and antiextensivity

$$X_B \subset X \qquad (7.10.23)$$

$$X^B \supset X \qquad (7.10.24)$$

3. Increasing property

$$X_1 \subset X_2 \Rightarrow (X_1)_B \subset (X_2)_B \qquad (7.10.25)$$

$$X_1 \subset X_2 \Rightarrow (X_1)^B \subset (X_2)^B \qquad (7.10.26)$$

4. Idempotence

$$(X_B)_B = X_B \qquad (7.10.27)$$

$$(X^B)^B = X^B \qquad (7.10.28)$$

An important property of the morphological operations is their computational simplicity. The definition of binary dilation:

$$X \oplus B^s = \{z \in E : B_z \cap X \neq 0\} \qquad (7.10.29)$$

can be used for its computation in the following way. The entire image plane E is scanned by the set (template) B. The pixel $z \in E$ belongs to the dilation $X \oplus B^s$, if the translated template B_z intersects with X. Such an algorithm for the calculation of dilation is shown in Figure 7.10.4 for the structuring elements SQUARE and RHOMBUS.

```
int bdilate(a,b, ki, N1,M1,N2,M2)
image a,b;
int ki;
int N1,M1,N2,M2;
/* Subroutine for 2-d 3 x 3 binary dilation
   a,b: input and output buffers
   ki: parameter denoting the filter window
   N1,M1: start coordinates
   N2,M2: end    coordinates          */
{ int k,l, i,j, s;
  /* SQUARE structuring element */
  if(ki==1)
  for(k=N1+1; k<N2-1; k++)
    for(l=M1+1; l<M2-1; l++)
    { s=a[k][l];
      for(i=0; i<3; i++)
        for(j=0; j<3; j++)
          if(s<a[k+i-1][l+j-1]) s=a[k+i-1][l+j-1];
      b[k][l]=s;
    }
```

```
/* RHOMBUS structuring element */
if(ki==2)
  for(k=N1+1; k<N2-1; k++)
    for(l=M1+1; l<M2-1; l++)
    { s=a[k][l];
      if(s<a[k-1][l]) s=a[k-1][l];
      if(s<a[k+1][l]) s=a[k+1][l];
      if(s<a[k][l-1]) s=a[k][l-1];
      if(s<a[k][l+1]) s=a[k][l+1];
      b[k][l]=s;
    }
  return(0);
}
```

Fig. 7.10.4 Algorithm for binary dilation.

The definition of binary erosion:

$$X \ominus B^s = \{z \in E : B_z \subset X\} \tag{7.10.30}$$

suggests the following way for its calculation. The image plane E is scanned by B. The pixel $z \in E$ belongs to the erosion $X \ominus B^s$, if the translated template B_z is a proper subset of X. The corresponding subroutine berode() is similar to the subroutine bdilate() shown in Figure 7.10.4. The above-mentioned dilation and erosion routines can be used for the calculation of binary closing and opening. If $N(B)$ is the number of points in the structuring element B, both erosion and dilation require $N(B) - 1$ elementary operations at each image pixel. All operations involved require information within a local image neighborhood. Thus, *cellular* computer architectures can be used efficiently for the calculation of morphological operations [PIT90], [PIT91].

An alternative way for the calculation of binary erosion and dilation is given by:

$$X \oplus B^s = \bigcup_{b \in B} X_{-b} \tag{7.10.31}$$

$$X \ominus B^s = \bigcap_{b \in B} X_{-b} \tag{7.10.32}$$

where X_{-b} denotes the translation of the image object X by a vector $-b$, $b \in B$. Dilation and erosion can be calculated by shifting the image X and by calculating the union or intersection of the translations X_{-b} by simple binary OR, AND operations. If $N(B)$ denotes the number of pixels in the structuring element B, $N(B) - 1$ binary OR, AND operations per image point are required. The computation of erosion and dilation by using (7.10.31–7.10.32) is preferred, when image translation and binary operations can be performed in an easy way. This is the case if cellular computer architectures are used.

368 SHAPE DESCRIPTION

Fig. 7.10.5 Decomposition of the structuring element CIRCLE.

The associativity (7.10.8) and the distributivity property (7.10.18) of the morphological transformations are very useful in pipelining morphological operations. Let us decompose the structuring element B^s as follows:

$$B^s = B_1^s \oplus B_2^s \oplus \ldots \oplus B_k^s \qquad (7.10.33)$$

The dilation $X \oplus B^s$ and the erosion $X \ominus B^s$ can be implemented in a pipeline having k stages:

$$X \oplus B^s = (\ldots((X \oplus B_1^s) \oplus B_2^s) \ldots \oplus B_k^s) \qquad (7.10.34)$$

$$X \ominus B^s = (\ldots((X \ominus B_1^s) \ominus B_2^s) \ldots \ominus B_k^s) \qquad (7.10.35)$$

If $N(B_i)$ is the number of points in each structuring element B_i, the computational complexity of the pipeline implementation is:

$$\sum_{i=1}^{k} N(B_i) - k \ll N(B) - 1 \qquad (7.10.36)$$

which is usually much less than the computational complexity without using structuring element decomposition. An example of structuring element decomposition is shown in Figure 7.10.5. The CIRCLE contains 21 pixels. Therefore, the computation of binary erosion and dilation by using (7.10.31–7.10.32) requires 20 image translations and unions/intersections. The structuring element decomposition shown in Figure 7.10.5 allows the use of a three-stage pipeline that requires two, two and four translations at each stage. Therefore, the decomposition implementation has more than 50% computational savings. The savings are even larger when large structuring elements are employed. The decomposed implementation of morphological operations has been used in dedicated morphological image processors, for example, in *Texture Analyzer* [KLE72] and *Cytocomputer* [STE83].

The classical morphological operators erosion and dilation are very sensitive to noise. More robust morphological operators can be based on fuzzy set operations. The Generalized Fuzzy Mathematical Morphology (GFMM) [CHA99] is based on a novel definition of a Fuzzy Inclusion Indicator, which is a fuzzy set defined as a measure of the inclusion of a fuzzy set into another. Fuzzy Inclusion Indicator obeys a set of axioms, which are extensions of the known inclusion indicator axioms, and which correspond to the desirable properties of any mathematical morphology operations. The classical binary and grayscale mathematical morphologies can be considered as special cases of GFMM. GFMM is a very powerful and flexible tool for morphological operations. Its properties can be handled by using various types of fuzzy structuring elements.

7.11 GRAYSCALE MORPHOLOGY

Morphological operations can be extended to functions and, therefore, to grayscale images [SER82], [PIT90], [MAR90]. The tools for grayscale morphological operations are simple functions $g(x)$ having domain G. They are called *structuring functions*. Their symmetric counterparts are given by:

$$g^s(x) = g(-x) \qquad (7.11.1)$$

The grayscale dilation and erosion of a function $f(x)$ by $g(x)$ are defined by:

$$[f \oplus g^s](x) = \max_{z \in D, z-x \in G} \{f(z) + g(z-x)\} \qquad (7.11.2)$$

$$[f \ominus g^s](x) = \min_{z \in D, z-x \in G} \{f(z) - g(z-x)\} \qquad (7.11.3)$$

where D is the domain of $f(x)$. A subroutine for the implementation of grayscale dilation is given in Figure 7.11.1. The subroutine for grayscale erosion is very similar.

```
int imdilate(a,b, g, NW,MW, N1,M1,N2,M2)
image a,b;
image g;
int NW,MW;
int N1,M1,N2,M2;
/* Subroutine for 2-d grayscale dilation
   a,b: input and output buffers
   g:   structuring function
   NW,MW: window size
   N1,M1: start coordinates
   N2,M2: end    coordinates        */
{ int k,l, i,j, NW2,MW2;
  unsigned char * * z;
  z=matuc2(NW,MW);
  NW2=NW/2; MW2=MW/2;
```

```
    for(k=N1+NW2; k<N2-NW2; k++)
     for(l=M1+MW2; l<M2-MW2; l++)
      { for(i=0; i<NW; i++)
         for(j=0; j<MW; j++)
          z[i][j]=a[k+i-NW2][l+j-MW2]+g[i][j];
        b[k][l]=z[0][0];
        for(i=0; i<NW; i++)
         for(j=0; j<MW; j++)
          if(b[k][l]<z[i][j]) b[k][l]=z[i][j];
      }
   mfreeuc2(z,NW);
   return(0);
}
```

Fig. 7.11.1 Algorithm for grayscale dilation.

Grayscale opening and closing is another set of dual operations:

$$f_g(x) = [(f \ominus g^s) \oplus g](x) = [f(x) \ominus g(-x)] \oplus g(x) \quad (7.11.4)$$

$$f^g(x) = [(f \oplus g^s) \ominus g](x) = [f(x) \oplus g(-x)] \ominus g(x) \quad (7.11.5)$$

Opening has a very interesting graphical representation shown in Figure 7.11.2. It is essentially a *rolling ball transformation*. The shape of the 'ball' is determined by the structuring function $g(x)$. The rolling ball traces the smooth contours and deletes the protruding (positive) impulses. When negative impulses are encountered, they are enhanced by the rolling ball transformation. On the contrary, grayscale closing enhances positive impulses and deletes negative ones. Therefore, both opening and closing operations have low-pass

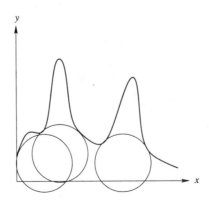

Fig. 7.11.2 Opening as a rolling ball transformation.

characteristics. Grayscale dilation, erosion, opening and closing have interesting properties that are detailed in [PIT90], [SER82]. An important property is grayscale morphological pipelining that is used for the implementation of morphological operations. Let us suppose that the structuring function $g(x)$ can be decomposed as follows:

$$g = g_1 \oplus g_2 \oplus \ldots \oplus g_k \qquad (7.11.6)$$

In this case, grayscale dilation and erosion can be implemented in pipeline:

$$f \oplus g = (\ldots((f \oplus g_1) \oplus g_2) \oplus \ldots \oplus g_k) \qquad (7.11.7)$$

$$f \ominus g = (\ldots((f \ominus g_1) \ominus g_2) \ominus \ldots \ominus g_k) \qquad (7.11.8)$$

Let us suppose that the structuring function $g(x)$ is zero and that its domain is given by $G = [-\nu, \ldots, 0, \ldots, \nu]$. In this case, grayscale dilation and erosion are called *dilation* and *erosion of a function by a set*:

$$[f \oplus G^s](x) = [f \oplus g^s](x) = \max\{f(i-\nu), \ldots, f(i), \ldots, f(i+\nu)\} \qquad (7.11.9)$$

$$[f \ominus G^s](x) = [f \ominus g^s](x) = \min\{f(i-\nu), \ldots, f(i), \ldots, f(i+\nu)\} \qquad (7.11.10)$$

Dilation and erosion of a function by a set are essentially local max and min filtering operations respectively. Thus, grayscale morphological operations are closely related to filters based on order statistics that have been described in Chapter 3 [PIT90].

Opening and closing of a function by a set are two-step operations defined as:

$$f_G(x) = [(f \ominus G^s) \oplus G](x) \qquad (7.11.11)$$

$$f^G(x) = [(f \oplus G^s) \ominus G](x) \qquad (7.11.12)$$

They can be used to define morphological filters [STE87]. Two such filters, the *close-opening* (CO) and the *open-closing* (OC) filters:

$$y = [(f^G)_G](x) \qquad (7.11.13)$$

$$y = [(f_G)^G](x) \qquad (7.11.14)$$

are four-step morphological operations that have been proposed for impulsive filtering [STE87], [PIT90]. Morphological operators can be used as edge detectors. The erosion $[f \ominus B](x)$ tends to decrease the area of high-intensity image plateaus. The changes are more profound around the object edges. Therefore, the following algebraic difference:

$$y = f(x) - [f \ominus B](x) \qquad (7.11.15)$$

has a large output close to image edges and can be used as an edge detector [GOE80]. The edge orientation is determined by the shape of B. An extension of this edge detector is given by:

$$y = f(x) - [f \ominus nB](x) \qquad (7.11.16)$$

where
$$nB = B \oplus B \oplus \ldots \oplus B \quad (n \text{ times}) \tag{7.11.17}$$

The parameter n controls edge thickness. Opening f_{nB} is a low-pass nonlinear filter, because it destroys the high-frequency content of the signal. Therefore, the algebraic difference:
$$y = f(x) - f_{nB}(x) \tag{7.11.18}$$
is a nonlinear high-pass filter, called the *top-hat transformation* [MEY79]. For sufficiently large structuring element size nB, the opening f_{nB} erases the peaks of the signal f. The difference $f - f_{nB}$ contains only these peaks and no background at all. Thus, the top-hat transformation can be used as a peak detector. Its application in detecting and counting stars in astronomical images has been reported. Another recent application is in the detection of cracks in paintings for conservation applications. Cracks have usually small luminance. Therefore, their negatives appear as local peaks and ridges. Thus, they can be detected very well by top hat transformation of the negative of the painting luminance [GIA98].

7.12 SKELETONS

Object *skeleton* is an important topological descriptor of a two-dimensional binary object and has been used extensively in various applications [LEV85], [SER82], [PIT90]. Shape description by employing its skeleton was originated by Blum [BLU67], who used the term *medial axis*. The medial axis can be obtained by setting up *grassfires* simultaneously at $t = 0$ along the object boundary. The fire wavefronts are allowed to propagate to the object interior by using Huygen's principle and by assuming uniform speed. The medial axis consists of those points where fire wavefronts intersect and extinguish. Their arrival times constitute the *medial axis function*. The process is illustrated in Figure 7.12.1. Another definition of skeleton is based on the notion of maximal disks. The skeleton $SK(X)$ of an object X consists of the centers of the maximal inscribable disks in X. This definition is illustrated in Figure 7.12.1b. A maximal disk is not contained properly in any other disk totally included in X. It touches the object contour in at least two points. Each skeleton point has an associated value of the radius r of the corresponding maximal disk. These values form the *skeleton function*. In the following, we shall use mathematical morphology to describe the skeletonization process [SER82]. Other mathematical descriptions may lead to slightly different definitions of skeleton. Let us denote by $S_r(x)$ the subset of skeleton $SK(X)$, which corresponds to maximal inscribable disks of radius r. The morphological skeleton is given by [PIT90], [SER82]:

$$SK(X) = \bigcup_{r>0} S_r(X) = \bigcup_{r>0} [(X \ominus rB) - (X \ominus rB)_{drB}] \tag{7.12.1}$$

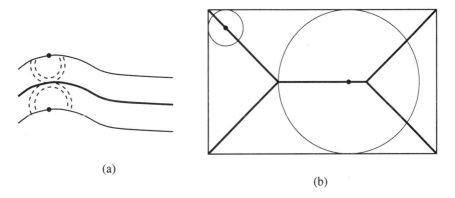

Fig. 7.12.1 (a) Grassfire propagation model of medial axis; (b) maximal disk definition of skeleton.

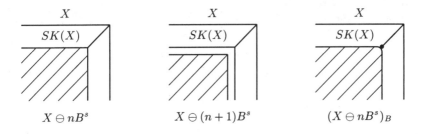

Fig. 7.12.2 Illustration of morphological skeletonization.

where rB denotes a disk of radius r and drB denotes a disk of very small radius dr. In the case of Euclidean grid $E = Z^2$ the disk drB can be approximated by a small structuring element B. The disk rB can be approximated by a structuring element nB of discrete radius n given by:

$$nB = B \oplus B \oplus \ldots \oplus B \quad (n \text{ times}) \qquad (7.12.2)$$

The skeleton is described by:

$$SK(X) = \bigcup_{n=0}^{N} S_n(X) = \bigcup_{n=0}^{N} [(X \ominus nB^s) - (X \ominus nB^s)_B] \qquad (7.12.3)$$

where $N + 1$ is the number of skeleton subsets and $-$ denotes set difference. This definition gives an iterative procedure for skeleton calculation that is illustrated in Figure 7.12.2. The procedure is based on the following observation: the set $(X \ominus nB^s)_B = [X \ominus (n+1)B^s] \oplus B$ differs from the set $X \ominus nB^s$ only close to protruding points of its boundary (e.g. its corners). Therefore,

374 SHAPE DESCRIPTION

the set difference $(X \ominus nB^s) - (X \ominus nB^s)_B$ contains only those protruding boundary points that belong to the object skeleton. The object X can be reconstructed very easily from its skeleton:

$$X = \bigcup_{n=0}^{N} [S_n(x) \oplus nB] \qquad (7.12.4)$$

Partial object reconstructions can be obtained, if only $N - k$ skeleton subsets $n = k, \ldots, N$ are used:

$$X' = \bigcup_{n=k}^{N} [S_n(x) \oplus nB] = X_{kB} \qquad (7.12.5)$$

The result of the partial object reconstruction is the opening X_{kB} of X by kB. Skeletons are translation invariant. However, only the skeletons defined by (7.12.1) over the Euclidean space R^2 are scale invariant. Their major disadvantage is that they are very sensitive to boundary noise. A fast algorithm for the calculation of skeletons based on the serial decomposition of (7.12.3) is proposed in [MAR86]:

$$S_0(X) = X - [X \ominus B^s] \oplus B \qquad (7.12.6)$$

$$S_1(X) = (X \ominus B^s) - [(X \ominus B^s) \ominus B^s] \oplus B \qquad (7.12.7)$$

$$\vdots$$

$$S_N(X) = (X \ominus NB^s) - [(X \ominus NB^s) \ominus B^s] \oplus B \qquad (7.12.8)$$

Object reconstruction is based on a similar procedure:

$$X = [\ldots [[S_N(x) \oplus B \cup S_{N-1}(x)] \oplus B] \cup S_{N-2}(x) \ldots] \oplus B \cup S_0(x) \qquad (7.12.9)$$

The fast skeletonization and reconstruction procedures are illustrated in Figure 7.12.3. The corresponding skeletonization algorithm is shown in Figure 7.12.4. Subroutines clear() and copy() clear an image buffer and copy a source buffer to a destination buffer respectively. Subroutine berode() is similar to subroutine bdilate() shown in Figure 7.10.4 and performs binary erosion. Finally, subroutines or(), xor() perform binary OR, XOR operations on image buffers.

```
int skeleton(a,b,c,d, ki, N1,M1,N2,M2)
image a,b,c,d;
int ki;
int N1,M1,N2,M2;
/* Subroutine for binary image skeletonization
    a,d: input and output buffers
    b,c: buffer for intermediate results
```

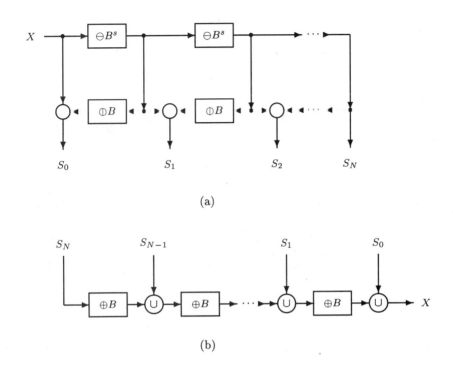

Fig. 7.12.3 (a) Fast skeletonization algorithm; (b) fast object reconstruction from skeleton subsets.

```
  ki: parameter denoting the filter window
  N1,M1: start coordinates
  N2,M2: end    coordinates            */
{ int k,l;
  clear(b,0,N1,M1,N2,M2); clear(c,0,N1,M1,N2,M2);
  clear(d,0,N1,M1,N2,M2);
  for( ; ; )
  { berode(a,b,ki,N1,M1,N2,M2);
    for(k=N1; k<N2; k++)
     for(l=M1; l<M2; l++)
      if(b[k][l]==1) goto L1;
    goto L2;
    L1:  bdilate(b,c,ki,N1,M1,N2,M2);
    xor(a,c,a,N1,M1,N2,M2);
    or(a,d,d,N1,M1,N2,M2);
    copy(b,a,N1,M1,N1,M1,N2,M2);
  }
```

```
L2: or(d,a,d,N1,M1,N2,M2);
    return(0);
}
```

Fig. 7.12.4 Skeletonization algorithm.

Skeletons can be extended to grayscale images. Let $ng(x)$ be an n-fold Minkowski addition of the structuring function $g(x)$:

$$ng = g \oplus g \oplus \ldots \oplus g \qquad (7.12.10)$$

The skeleton subfunctions $S_n(f)$ are given by the following algebraic difference:

$$S_n(f) = (f \ominus ng^s) - (f \ominus ng^s)_g \qquad (7.12.11)$$

Given the skeleton subfunctions $S_n(f)$, there are two ways to define grayscale image skeletons:

$$SK_{sum}(f)(x) = \sum_{n=0}^{N} [S_n(f)](x) \qquad (7.12.12)$$

$$Sk_{max}(f)(x) = \max_{n=0}^{N} [S_n(f)](x) \qquad (7.12.13)$$

Skeletons have found extensive applications in biological shape description, pattern recognition, industrial inspection, quantitative metallography and image coding [PIT90], [MAR86].

Classical skeleton definition is very sensitive to noise. Robust skeletons can be devised by using fuzzy extensions of mathematical morphology, e.g. Generalized Fuzzy Mathematical Morphology (GFMM) [CHA99]. Extensive simulations showed that the reconstruction of noisy objects from their skeletal subsets, that can be achieved by using the GFMM is better than by using the classical BMM in most cases. Besides, the use of the GFMM for skeletonization and shape decomposition preserves the shape and the location of the skeletal subsets and spines, and therefore, can be efficiently used for object representation especially in cases of impulsive noise.

7.13 SHAPE DECOMPOSITION

A complex object X can be decomposed into a union of 'simple' subsets $X_1, \ldots, X_n$, thus providing an intuitive object description scheme called shape decomposition. It must use simple geometrical primitives in order to conform with our intuitive notion of simple shapes (e.g. squares, circles, polygons). The derivation of the decomposition must be well-defined and object-independent. Uniqueness, translation, scaling and rotation invariance are very important

```
inscribe(X,L,Z,B,n)
X,L,Z,B: set;
n:   integer;
{ n=0; for( ; ; ) { L=X ⊖ B^S;
if(L== ∅) {L=X; break;}
n++; X=L;}
Z=L ⊕ nB}
```

Fig. 7.13.1 Algorithm to obtain the maximal inscribable disk in an object.

properties for pattern recognition applications. The complexity of the decomposition must be small compared with the original description of X. Finally, a small noise sensitivity is desirable.

In the following, *morphological shape decomposition* will be presented [PIT90PAMI]. It employs morphological techniques to decompose an object X into a union of non-overlapping subsets $X_1, \ldots, X_K$:

$$X = \bigcup_{i=1}^{K} X_i \qquad (7.13.1)$$

The shape of the structuring element B is the geometrical primitive of the morphological shape decomposition of an object X. In the two-dimensional case it can be circle, triangle, square, and so forth. In the three-dimensional object representation it can be a cube, sphere or other simple 3-d primitive. The maximal inscribable element B in X has the form:

$$X_1 = X_{n_1 B} = (X \ominus n_1 B^s) \oplus n_1 B \qquad (7.13.2)$$

and size n_1. It can be found by eroding object X by B n_1 times. Object X is successively eroded by structuring elements nB of ever increasing size n. The procedure stops when the output of the erosion $L = X \ominus nB^s$ becomes the empty set. The size of the maximal inscribable element B is the size n_1 at this last step. This procedure is described in C-like format in Figure 7.13.1. The process can be repeated for $X - X_1$. The whole procedure can be described by a recursive relation:

$$\begin{aligned} X_i &= (X - X'_{i-1})_{n_i B} \\ X'_i &= \bigcup_{j=1}^{i} X_j \\ X'_0 &= \emptyset \end{aligned} \qquad (7.13.3)$$

Stopping condition: $(X - X'_K) \ominus B^S = \emptyset$

The resulting algorithm is shown in Figure 7.13.2.

378 SHAPE DESCRIPTION

```
int msd(a,b,c,d, ki, N,N1,M1,N2,M2)
image a,b,c,d;
int ki,N;
int N1,M1,N2,M2;
/* Subroutine for the morphological representation
   of binary images
   a,d: input and output buffers
   b,c: buffer for intermediate results
   ki: parameter denoting the filter window
   N: maximum number of components
   N1,M1: start coordinates
   N2,M2: end   coordinates          */

{ int k,l,OK,OK2,n,i,j=0;
  clear(d,0,N1,M1,N2,M2);
  do
    { OK2=0; n=0; j++;
      copy(a,b,N1,M1,N1,M1,N2,M2);
      clear(c,0,N1,M1,N2,M2);
      /* Find the center(s) of maximal inscribable element. */
      do
        { OK=0; n++;
          berode(b,c,ki,N1,M1,N2,M2);
          for(k=N1; k<N2; k++)
           for(l=M1; l<M2; l++)
             if(c[k][l]==1) OK=1;
          if(OK==1)
          for(k=N1; k<N2; k++)
           for(l=M1; l<M2; l++)
             b[k][l]=c[k][l];
        } while(OK==1);
      /* Create the maximal inscribable element(s). */
      for(i=1;i<n;i++)
        {
          bdilate(b,c,ki,N1,M1,N2,M2);
          for(k=N1; k<N2; k++)
           for(l=M1; l<M2; l++)
             b[k][l]=c[k][l];
        }
      /* Update output buffer. */
      or(c,d,d,N1,M1,N2,M2);
      /* Subtract current maximal elements from original image.*/
      xor(a,c,a,N1,M1,N2,M2);
      for(k=N1; k<N2; k++)
       for(l=M1; l<M2; l++)
```

```
        if(a[k][l]==1) OK2=1;
    } while(OK2==1 &&j<N);
    return(0);
}
```

Fig. 7.13.2 Algorithm for the morphological shape decomposition of an object.

Subroutine msd() uses subroutines berode(), bdilate() to perform binary erosion and dilation. If the structuring element used is symmetric with respect to the origin bdilate() can be used to implement $X \oplus B$. Morphological object decomposition decomposes an object X into a series of simple objects $X_1, \ldots, X_K$. The object X is reconstructed by using (7.13.1). The simple objects X_i are of the form:

$$X_i = L_i \oplus n_i B \qquad (7.13.4)$$

where L_i is either a point or a line. Objects of the form (7.13.4) are called *Blum ribbons* [ROS86]. They consist of a *spine* L_i and a structuring element $n_i B$ of size n_i which is translated along the spine. An example of the morphological shape description is shown in Figure 7.13.3. The SQUARE structuring element has been used to perform this decomposition. Morphological shape decomposition has successfully been used for object recognition applications [PIT92]. One disadvantage of morphological shape decomposition is that it is susceptible to boundary noise. Another disadvantage is that the representation produced is not close to human shape perception if the object consists of unions, intersections and differences of various geometrical primitives. This disadvantage can be alleviated by combining morphological techniques with *constructive solid geometry* (CSG) [FOL90], [HEA86], [REQ80]. The resulting representation scheme is called *morphological shape representation*. One of the main advantages of CSG over skeleton representation or morphological

Fig. 7.13.3 (a) Original binary image; (b) first 16 components of its morphological shape decomposition.

```
msr1(X,S,X_k, k=1,2,...)
X,X_k set;
S set of N structuring elements;
{L_j, Z_j set;
k=1;
do
{for (B_j ∈ S, j=1,...,N)
inscribe(X,L_j,Z_j,B_j,n_j);
find Z_i such that M(X-Z_i) is minimum;
X_k=Z_i; X=X-X_k;
k++; }
while(X ≠ ∅);}
```

Fig. 7.13.4 Algorithm for the morphological shape representation of an object.

shape decomposition is that it uses a multitude of geometrical primitives (e.g. squares, circles, triangles) instead of one. This fact not only enhances the descriptive power of CSG but also conforms to our intuitive notion of simple geometrical shapes. *Gestalt psychology* suggests that we organize our perception of shapes in the simplest possible form. Furthermore, the use of many geometrical primitives increases the conciseness of the representations. In the following, we shall describe such a morphological representation scheme which is unique and employs many geometrical primitives [PIT92PR].

Let $S = \{B_1, \ldots, B_N\}$ be a set of structuring elements, each describing a geometric primitive. An object is assumed to be a simple geometric object if it is described by (7.13.4) for any structuring element belonging to S. Usually objects have a 'body' of rather simple geometrical shape. The body can be fairly well described by a simple object $Z_j = L_j \oplus n_j B_j$, $j = 1, \ldots, N$, of the form (7.13.4). n_j is the size of the maximal inscribable structuring element B_j in X. The choice of Z_j, $j = 1, \ldots, N$, for any structuring element B_j can be done by using the routine shown in Figure 7.13.1. The same procedure can be repeated for every structuring element B_j in S. Let us also denote by $M(X)$ a measure of the object X, for example the area of X. The figure of merit for the evaluation of the approximation of the body of X by Z_j, $j = 1, \ldots, N$, is the measure $M(X - Z_j)$, $j = 1, \ldots, N$. If the simple object Z_i, $1 \leq i \leq N$, has the minimal measure $M(X - Z_i)$, it is chosen as the best representation of the 'body' of X. The best representation Z_i is subtracted from X and the same procedure is applied for the remainder $X - Z_i$. The whole procedure is an iterative one and can be described in C-like format as shown in Figure 7.13.4.

The binary object is represented by the series $X'_1, \ldots, X'_i, \ldots$ or by the series $X_1, \ldots, X_i, \ldots$. This representation is always finite for digitized images. The object representation X'_i is unique, unambiguous, translation and scale invariant. It increases monotonically and it is upper bounded by X.

```
enclose(X,L,Y,B,n)
X,Y,B: set;
n:   integer;
{n=0; L=∅;
do {L=E; n++; for (x ∈ X) { L&=(x ⊕ nB^s); if (L==∅) break; }}
while(L== ∅); Y=L ⊕ nB;}
```

Fig. 7.13.5 Algorithm to find the minimal enclosing element of an object.

```
msr3(X,S,X_k,  k=1,2,...)
X,X_k set;
S set of N structuring elements;
{L_j, M_j,Z_j, Y_j set;
k=1;
do
{for (B_j ∈ S, j=1,..,N)
{inscribe(X,L_j,Z_j,B_j,n_j); enclose(X,M_j,Y_j,B_j,N_j); }
find Z_i such that M(X-Z_i) is minimum;
find Y_i such that M(Y_i-X) is minimum;
if(M(Y_i-X)<M(X-Z_i)) { X_k=Y_i; X=X_k-X;}
else { X_k=Z_i; X=X-X_k;}
k++; }
while(X ≠ ∅);}
```

Fig. 7.13.6 Algorithm for the morphological shape representation of an object.

In certain cases the 'body' of an object X can be represented better by a minimal enclosing structuring element rather than by a maximal inscribable structuring element. A technique to find the minimal enclosing structuring element is presented as follows. Let x be a point of the object X and nB a structuring element of size n. The locus of the centers of the structuring elements nB that include x is the set $L_x = x \oplus nB^s$. Therefore, the locus (if any) of the structuring element nB that encloses X is the intersection of L_x for every x belonging to X. If we want to find the minimal enclosing structuring element, we can start this procedure for $n = 1$. If the locus is the empty set, we increase the structuring element size by one and we continue until the locus becomes a non-empty set. The procedure is described in Figure 7.13.5. $E \subset R^2$ is the entire image plane. If y is any point of L, $y \oplus B$ encloses X. The set $Y = L \oplus nB$ encloses X and is unique.

Morphological shape representation using enclosing simple sets can proceed as follows. First, the minimal enclosing simple sets $Y_j = L_j \oplus N_j B_j$, $j = 1, \ldots, N$, are found by using the above-mentioned algorithm. Each of them has size N_j, $j = 1, \ldots, N$. Then the maximal inscribable simple sets Z_j, $j =$

$1, \ldots, N$, are found by using the procedure described in Figure 7.13.2. The measure of 'fitness' of each enclosing set Y_j is the area $M(Y_j - X)$, $j = 1, \ldots, N$. The measure of 'fitness' of each inscribable set Z_j, $j = 1, \ldots, N$, is the area $M(X - Z_j)$. The first element in the representation is the one from $Y_j, Z_j, j = 1, \ldots, N$, which has the minimal measure. If an inscribable simple set Z_i fits better to the 'body' of the object, we continue the procedure iteratively with $X - Z_i$. If an enclosing simple set Y_i has the minimal measure, then the difference $Y_i - X$ is used in the second step. The whole procedure is repeated iteratively as shown in Figure 7.13.6. The above-mentioned representation scheme is unique, because the choice of both the enclosing and the inscribable structuring elements is unique. It is translation and scale invariant and more concise than the representation given by the algorithm of Figure 7.13.4 for objects which tend to have 'caves' at their border.

7.14 VORONOI TESSELATION

Voronoi tessellation is a basic problem in computational geometry [PRE85]. It is also a tool used in image analysis, such as object recognition [AHU83] and texture analysis [PIT89]. Several important problems can be solved by employing Voronoi tessellation, for example Delaunay triangulation, convex hull computation and object decomposition into simple components (triangles). The input to the tessellation procedure is an image containing a set of classified points (seeds or markers). The goal is to classify the rest of the image points to the nearest seed (in the geometrical sense). The nearest neighbor criterion is implemented by using a distance metric. The mathematical morphology dilation operation can be used to perform Voronoi tessellation. A set of structuring elements can be used to implement the various distance metrics on the discrete grid of a digitized image. The use of dilation in order to tessellate an image is discussed in [BOR86], [MAZ89] and methods that approximate Euclidean disk growing are discussed in [YE88], [SHI92]. An efficient approach to Euclidean disk growing can be found in [KOT93]. This work is also related to [SER82], [PIT90], [LAN80], [DAN80], [BOR84].

The basic idea of Voronoi tessellation is to find proximity properties of points in the Euclidean plane $\mathcal{R}^2$. The Voronoi diagram partitions the plane into regions (sometimes called *influence zones*). Let $X = \{x_1, x_2, \ldots, x_m\}$ be a set of m points in $\mathcal{R}^2$ (called *seeds*) as shown in Figure 7.14.1a. The points in the interior of a Voronoi region $V(i)$ possess the following property: they are the points of the plane $\mathcal{R}^2$ that are closer to the seed point $x_i \in X$ than to any other seed point x_j of $X, (j \neq i)$. As a result, the definition of the Voronoi regions $V(i)$ and Voronoi diagram $\text{Vor}(X)$ can be given by using the notion of a distance function $d(.)$ between two points:

$$V(i) = \{p \in \mathcal{R}^2 \ : \ d(p, x_i) \leq d(p, x_j) \, , \, j \neq i \, \} \qquad (7.14.1)$$

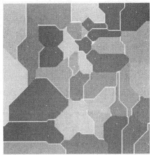

Fig. 7.14.1 Image domain containing seed points and a corresponding Voronoi diagram using the city block distance.

The union of Voronoi regions for all $x_i \in X$ is called *Voronoi diagram* of X (shown in Figure 7.14.1b):

$$\text{Vor}(X) = \bigcup_{i=1}^{m} V(i) \qquad (7.14.2)$$

The edges and the vertices of the Voronoi diagram are denoted by e_i and v_i, respectively. The process of constructing the Voronoi diagram of a set $X \subset \mathcal{R}^2$ is called *Voronoi tessellation* of $\mathcal{R}^2$. The 'borders' of the Voronoi regions are sometimes called *skeletons by influence zones (SKIZ)*. They contain the points that are equidistant from two seeds.

Voronoi tessellation is a classical topic in computational geometry. Several algorithms exist for this purpose. In earlier publications [PRE85], [TOU80], [LEE80], the Voronoi diagram is constructed using the following approach based on a divide and conquer strategy. First of all, the set X is separated into two sets equal in size. The Voronoi diagram of these subsets is recursively constructed by further splitting them. This continues until a trivial number of points remains in each subset. Then the Voronoi diagram is constructed for each of the subsets. This is done by using an algorithm for finding intersections of half planes [PRE85]. After the construction of each subdiagram, a merging must take place. This procedure needs a double connected edge list [PRE85]. When the algorithm ends, the vertices of the Voronoi polygons have been found and saved in a vertex list.

The notion of the distance between two points is fundamental in a number of geometrical problems, including Voronoi tesselation. The best known distance function between two points in $\mathcal{R}^2$ (continuous coordinates) is the Euclidean distance, defined as:

$$d_e(x, y) = \sqrt{(x_1 - y_1)^2 + (x_2 - y_2)^2}, \quad x, y \in \mathcal{R}^2 \qquad (7.14.3)$$

where $x = (x_1, x_2) \in \mathcal{R}^2$ and $y = (y_1, y_2) \in \mathcal{R}^2$ are the Cartesian coordinates of the points x and y respectively. Two of the most popular definitions of distance functions for discrete images $\mathcal{Z}^2$ are the following:

$$d_4(x,y) = |x_1 - y_1| + |x_2 - y_2| \quad \text{(city block distance)}$$
$$d_8(x,y) = \max\{|x_1 - y_1|, |x_2 - y_2|\} \quad \text{(chessboard distance)}$$

where $x = (x_1, x_2) \in \mathcal{Z}^2$ and $y = (y_1, y_2) \in \mathcal{Z}^2$. Such distances are directly related to the morphological dilation operation:

$$d_h(x,y) = \inf\{n : x \oplus B(n) \uparrow y\}, \quad x,y \in \mathcal{Z}^2, \quad n = 0, 1, 2, \ldots \quad (7.14.4)$$

where $B(n)$ defines a structuring element of size n that is created by dilating B n times. By using a RHOMBUS or the SQUARE structuring elements, it can easily be proven that the city block or the chessboard distance function are calculated, respectively.

There are many serial and parallel algorithms for computing the morphological Voronoi tesselation. The fastest though, are those that use *propagation fronts* (also called grassfires or wavefronts) to process each point in $\mathcal{Z}^2$ only once, when a zone growth reaches it. These propagation fronts are implemented by a data structure (queue, list, stack, array) that stores all points in the edge of the influence zone. In the following, we shall refer to that structure as a propagation front, without specifically stating how it is implemented. Sometimes, it is possible to implement variations of the propagation algorithm without lists, just by storing the growth information and labels on an image buffer [KOT93].

These algorithms have the following steps:

1. Label the points of the Voronoi region of each seed S_i with a unique label in a label image. Labels are employed by all algorithms reported in the literature.

2. Put the points of each seed in the propagation front.

3. For each top point of each propagation front, find all adjacent points that are not labeled in the label image, put them in the propagation front and label them.

4. Remove each point from a propagation front after it has been processed, then go to step 3.

Region growing in such algorithms can be implemented in two, roughly (but not completely) equivalent ways [VER89]:

Read formalism: A new point is labeled according to its neighbors' labels.

Write formalism: A point labels all its unlabeled neighbors with its label.

Algorithms also differ in the way they handle region collisions. They either:

- Use the above definitions of Voronoi regions to define the skeleton points;
- Label no skeleton points;
- Define extra skeleton points so that the skeleton is always one point thick and no influence zone is adjacent to another.

However, the last two methods are not *consistent*, because their result depends on the order by which the regions are processed. A performance analysis of such propagation algorithms can be found in [PIT98].

7.15 WATERSHED TRANSFORM

A discrete grayscale image f in domain $D \subset \mathcal{Z}^2$ can be decomposed into a series of binary images $f_i = \{\mathbf{p} \in D, f(\mathbf{p}) = i\}$ which are called *graylevels*. In the continuous case, the watershed transform treats the image as a relief and traces the flow of an inertialess liquid originating in a point $\mathbf{p}$, until it reaches a minimum of the relief. The result of this transformation is the segmentation of the image into *catchment basins*, one for every regional minimum of the image. At this point we need to define a regional minimum as a connected set of points in a specific graylevel surrounded only by points of higher graylevels. The absence of continuous gradients in digital images makes the computation of the watershed transform less obvious. However, it can be simulated by assuming that the flow from a point in a graylevel is directed towards the closest point belonging to a lower graylevel (according to the discrete geodesic distance within the current graylevel). From there, it continues likewise to the next closest point of a lower graylevel, until it reaches a regional minimum. By taking the above into account, we define the *flowdown distance* $dist_W(\mathbf{p}, \mathbf{p}')$ between point $\mathbf{p} \in f_i$ and point $\mathbf{p}' \in f_j$, where $i \leq j$, as the length of the non-decreasing path which connects $\mathbf{p}$ to $\mathbf{p}'$ that contains the minimal number of points in the highest graylevel and, progressively, the minimal number in the successively lower graylevels. If such a path does not exist, the flowdown distance is infinite, meaning that it is impossible for the liquid to flow from $\mathbf{p}$ to $\mathbf{p}'$. It must also be noted that the flowdown distance is not commutative. The flowdown distance between a point $\mathbf{p}$ and a set S is defined as the minimum distance between $\mathbf{p}$ and all points in S. If we suppose that there are n regional minima $M_1, \ldots, M_n$ in an image, the catchment basin of a minimum M_i is defined as:

$$CB_i = \{p \in D : \ dist_W(M_i, p) < dist_W(M_j, p), \ \forall j \neq i\}$$

The watershed points of the image are defined as:

$$WSHD = \{p \in D : \exists i, j \ dist_W(M_i, p) = dist(M_j, p)\}$$

It can be shown that the above definition of the watershed transform yields unnecessarily thick watershed regions. This problem can be solved in a similar way as in the case of the influence zones, that is, by excluding from consideration watershed points close to the minima during the computation of further distances.

The computation of the watersheds using the definition is impossible in practice, since all possible paths would have to be considered. Again, various algorithms have been proposed for the computation of watersheds. The most commonly used is the *immersion algorithm*, since almost all others are inexact and/or slow [VIN91I]. The immersion algorithm works by inverting the process of computing the watersheds. Instead of tracing the flow from the points to be classified to the minima, it 'floods' the image by computing the skeleton influence zones in each graylevel using as markers the minima of the image and the catchment basins computed in previous graylevels. It can be shown that the above procedure yields a result identical to the one that results from the definitions. Again, the algorithm can handle basin collision in different ways, consistently or not.

7.16 FACE DETECTION AND RECOGNITION

Face recognition has been an active research topic in computer vision for more than two decades. A critical survey of the literature on human and machine face recognition can be found in [CHE95]. The key tasks in this area are the following:

- Face detection (or localization). It concerns face(s) localization within an image.

- Facial feature extraction. The output are relevant key face features (e.g. eyes, mouth, chin) that can be used for recognition or tracking.

- Face recognition. It provides matching of a facial image to a reference image existing in a facial image data base.

- Face authentication or verification. It provides a negative/positive reply to the question whether a new facial image matches to a reference one.

Naturally, these tasks are interrelated.

Recently, the research on model-assisted coding schemes [ELE95] in addition to the need for multi-modal verification techniques in tele-services and tele-shopping applications [M2VTS96] has reinforced the interest on face detection algorithms. Several approaches have been used to this end. A very attractive approach for face detection is based on multi-resolution images (also known as *mosaic images*) attempting to detect a facial region at a coarse resolution and subsequently to validate the outcome by detecting facial features at the next resolution level [YAN94]. Towards this goal, the method employs a hierarchical knowledge-based pattern recognition system. A variant of this method has been proposed in [KOT97A]. The proposed algorithm can determine very fast an image region where the face is roughly located and to provide a search area for other algorithms that work at the resolution of pixels and yield more accurate results. Alternatively, face localization can be performed by color segmentation keeping only regions with skin-like colors [SOB98A]. A usual problem in this approach is the misclassification of irrelevant regions as facial ones due to inaccurate color segmentation. It is well known that accurate face localization is very important for further accurate extraction of facial features.

Face is usually modeled as an elliptic region. Best-fit ellipsis can be used for this purpose. A better approximation of the facial contour can be performed by using active contours. Geometrical, graylevel and anthropometric considerations can be used for facial feature extraction. For example, the vertical eye, nose and mouth position can be determined by examining the local minima of the vertical face profile [SOB98A]. Facial feature extraction can be performed by a fusion of techniques based both on graylevel template matching and on certain geometrical features. Fusion usually provides more stable results than using each technique separately. Following such a reasoning a mixture of complementary techniques that are based either on geometrical shape parameterization or on template matching or on dynamic deformation of active contours is proposed in [NIK98], [NIK99]. Figure 7.16.1 shows some results of face contour extraction using the active contour deformation method. The best-fit and inner ellipses of the face region are also denoted. Figure 7.16.2 demonstrates some results of correct and complete feature constellations obtained using the adaptive Hough transform, pattern matching and graylevel relief minima algorithms. The corresponding best-fit ellipse is also depicted in these figures.

The success of a specific technique depends on the nature of the extracted features. Moreover, the compensation of head rotation enhances further the feature extraction. By considering these remarks, modifications as well as novel design ideas were introduced in the principles described in [SOB98A] based on the face symmetry in frontal views and the biometric analogies that increase the flexibility and the robustness of the face localization and the facial feature extraction modules and reduce their computational requirements [TSEK98].

Automated face recognition has exhibited a tremendous growth for more than two decades. Many techniques for face recognition have been developed whose principles span several disciplines, such as image processing, pattern recognition, computer vision and neural networks [CHEL95]. Three primary approaches are currently employed for face recognition or authentication:

- Feature-based approach. It is based on extracted facial features (e.g. mouth, eyes). It is sensitive to errors during feature extraction.

- Eigenfaces. The facial image is decomposed to eigen images (called eigenfaces) by using Principal Component Analysis (PCA, Karhunen

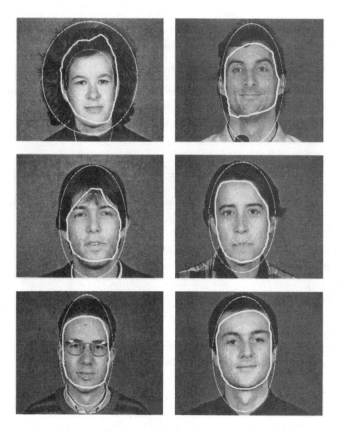

Fig. 7.16.1 Results of face contour extraction.

Loeve transform) [CHEL95]. The recognition method is based on matching the most important eigenfaces. Generally, it is a robust technique and provides very good recognition results.

- Elastic graph matching (Dynamic link architectures, DLA). It is a well-known approach that represents an object by projecting its image onto a rectangular elastic grid where a Gabor wavelet bank response is measured at each node [LAD93], [ZHA97].

The elastic graph matching algorithm is based on the analysis of a facial image region and its representation by a set of local descriptors extracted at the nodes of a sparse grid. The grid nodes are either evenly distributed over a rectangular image region or they are placed on certain facial features (e.g., nose, eyes, etc.) called fiducial points. In both cases, the elastic graph matching algorithm consists of the following steps:

Step 1 Build an information pyramid by using scale-space image analysis techniques. For example, the responses of a set of 2D Gabor filters tuned to different orientations and scales [LAD93], [ZHA97], [DUC97B] or the output of multiscale morphological dilation-erosion at several scales can be employed to form a local descriptor (i.e., a feature vector):

$$\mathbf{j}(\mathbf{x}) = \left(\hat{f}_1(\mathbf{x}), \ldots, \hat{f}_M(\mathbf{x})\right) \quad (7.16.1)$$

where $\hat{f}_i(\mathbf{x})$ denotes the output of a local operator applied to image f at the i-th scale or at the i-th pair (scale, orientation), $\mathbf{x}$ defines the pixel coordinates and M is feature vector dimensionality.

Step 2 Detect the face or the fiducial points on the reference image and place the grid nodes over the facial image region or the fiducial points, respectively. The method that represents the face as a labeled graph whose nodes are placed at different fiducial points is more difficult to be applied automatically, since the detection module has to find the precise coordinates of facial features. The detection of a rectangular facial region that encloses the face is generally an easier task.

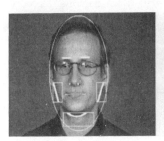

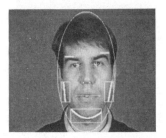

Fig. 7.16.2 Results of extracted feature constellations.

Step 3 Translate and deform an elastic graph comprised of local descriptors measured at variable pixel coordinates on the test image so that a cost function is minimized. The cost function is based on both the norm of the difference between the reference and test local descriptors and the distortion between the reference grid and the variable test graph. Let the superscripts t and r denote a test and a reference person (or grid), respectively. The L_2 norm between the feature vectors at the l-th grid node is used as a (signal) similarity measure, that is, $C_v(\mathbf{j}(\mathbf{x}_l^t), \mathbf{j}(\mathbf{x}_l^r)) = \|\mathbf{j}(\mathbf{x}_l^t) - \mathbf{j}(\mathbf{x}_l^r)\|$. Let us define by $\mathcal{V}$ the set of grid nodes that are considered as graph vertices. Let also $\mathcal{N}(l)$ denote the four-connected neighborhood of vertex l. The objective in elastic graph matching is to find the set of test grid node coordinates $\{\mathbf{x}_l^t, l \in \mathcal{V}\}$ that minimizes the cost function:

$$C(\{\mathbf{x}_l^t\}) = \sum_{l \in \mathcal{V}} \left\{ C_v(\mathbf{j}(\mathbf{x}_l^t), \mathbf{j}(\mathbf{x}_l^r)) + \lambda \sum_{\xi \in \mathcal{N}(l)} C_e(l, \xi) \right\} \quad (7.16.2)$$

where $C_e(l, \xi)$ penalizes grid deformations, i.e., $C_e(l, \xi) = \|(\mathbf{x}_l^t - \mathbf{x}_l^r) - (\mathbf{x}_\xi^t - \mathbf{x}_\xi^r)\|$, $\xi \in \mathcal{N}(l)$. Obviously, the cost function (7.16.2) defines a distance measure between two persons. One may interpret the optimization of (7.16.2) as a simulated annealing with additional penalties imposed by the grid deformations. Accordingly, (7.16.2) can be simplified to

$$D(t, r) = \sum_{l \in \mathcal{V}} C_v(\mathbf{j}(\mathbf{x}_l^t), \mathbf{j}(\mathbf{x}_l^r)) \text{ subject to } \mathbf{x}_l^t = \mathbf{x}_l^r + \mathbf{s} + \boldsymbol{\delta}_l, \ \|\boldsymbol{\delta}_l\| \leq \delta_{\max}$$
(7.16.3)

where $\mathbf{s}$ is a global transposition of the graph and $\boldsymbol{\delta}_l$ denotes a local perturbation of the grid coordinates. The choice of $\delta_{\max}$ controls the rigidity/plasticity of the graph.

It was found that elastic graph matching achieves a better performance than the eigenfaces [KIR90], [TUR91], auto-association and classification neural networks [COT90], [BOUR88] due to its robustness to lighting, varying face position and facial expression variations. Three major extensions to dynamic link architecture that allowed for handling larger galleries, tolerated larger variations in pose and increased its matching accuracy were introduced in [WIS97A]. Another variant that aims at increasing the robustness of dynamic link architecture in translations, deformations and changes in background was proposed in [WUR97TPAMI]. Recently, a variant of dynamic link architecture based on multiscale dilation-erosion, the so-called *morphological dynamic link architecture* (MDLA), was proposed and tested for face authentication [KOT97ICIP], [KOT98ICIP], [TEF99ICASSP]. Figure 7.16.3 demonstrates the image analysis step employed in morphological dynamic link

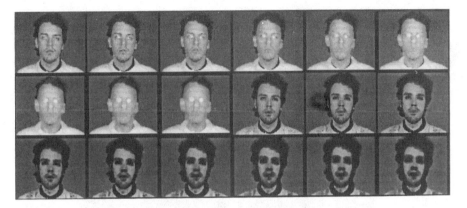

Fig. 7.16.3 The response of multiscale erosion dilation for scales $-9,\ldots,9$ used in MDLA.

architecture. The grids formed in the matching procedure of a test person with himself and with another person are depicted in Figure 7.16.4. MDLA has proven very efficient for face authentication applications during experiments on the M2VTS database [KOT98ICIP].

A recent approach to face recognition/authentication is based on fusing multiple modalities (e.g. frontal and profile images, speech and face movements). This multimodality method is proven to provide superior results than the corresponding single modality methods.

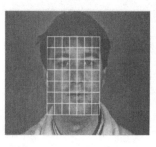
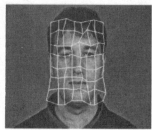
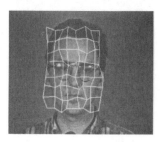

Fig. 7.16.4 Grid matching procedure in MDLA: (a) Model grid of the reference person; (b) best grid when the test person is identical to the reference one; (c) best grid when the test person is different from the reference one.

References

AHU83. N. Ahuja and B.J. Schachter, *Pattern Models*, New York: Wiley, 1983.

AMM87. L. Ammerdaal, *Programs and data structures in C*, Wiley, 1987.

BAL82. D.H. Ballard, C.M. Brown, *Computer vision*, Prentice-Hall, 1982.

BLU67. H. Blum, "A transformation for extracting new descriptors of shape," in *Models for the perception of speech and visual forms*, W. Wathen-Dunn (editor), pp. 362-380, MIT Press, 1967.

BOR84. G. Borgefors, "Distance Transformations in Arbitrary Dimensions," *Comput. Vision, Graphics, Image Proc.*, vol. 27, pp. 321-345, 1984.

BOR86. G. Borgefors, "Distance transformation in digital images," *Computer Vision Graphics and Image Processing*, vol. 34, pp. 344-371, 1986.

BOUR88. H. Bourlard and Y. Kamp, "Auto-association by multilayer perceptrons and singular value decomposition," *Biological Cybernetics*, vol. 59, pp. 291-294, 1988.

CAN70. P.B. Canham, "The minimum energy of bending as a possible explanation of the biconcave shape of the human red blood cell," *Journal of Theoretical Biology*, vol. 26, pp. 61-81, 1970.

CAN86. V. Cantoni, S. Levialdi (editors), *Pyramidal systems for computer vision*, Springer Verlag, 1986.

CHA99. V. Chatzis and I. Pitas, "A generalized fuzzy mathematical morphology approach and its application in robust 2D and 3D object representation," *IEEE Transactions on Image Processing*, accepted for publication, 1999.

CHE95. R. Chellapa, C.L. Wilson and S. Sirohey, "Human and machine recognition of faces: A survey," *Proceedings of the IEEE*, vol. 83, no. 5, pp. 705-740, May 1995.

CHEL95. R. Chellapa, C.L. Wilson and S. Sirohey, "Human and machine recognition of faces: A survey," *Proceedings of the IEEE*, vol. 83, no. 5, pp. 705-740, May 1995.

COT90. G.W. Cottrell and M. Fleming, "Face recognition using unsupervised feature extraction," in *Int. Neural Network Conf.*, vol. 1, pp. 322-325, Paris, July 1990.

CYP89. R. Cypher, J.L.C. Sanz, "SIMD architectures and algorithms for image processing and computer vision," *IEEE Transactions on Acoustics, Speech, and Signal Processing*, vol. ASSP-39, no. 12, pp. 2158-2174, Dec. 1989.

DAN80. P.E. Danielsson, "Euclidean Distance Mapping," *Computer Graphics and Image Processing*, vol. 14, pp. 227-248, 1980.

DAV81. E.R. Davies, A.P.N. Plummer, "Thinning algorithms: A critique and a new methodology," *Pattern Recognition*, vol. 14, no. 1-6, pp. 53-63, 1981.

DUC97B. B. Duc, S. Fischer and J. Big, "Face authentication with Gabor information on deformable graphs," *IEEE Transactions on Image Processing*, vol. 8, no. 4, pp. 504-516, April 1999.

DUD73. R.O. Duda, P.E. Hart, *Pattern recognition and scene analysis*, Wiley, 1973.

ELE95. E. Elefteriadis and A. Jacquin, "Automatic face location, detection and tracking for model-assisted coding of video teleconferencing sequences at low bit-rates," *Signal Processing: Image Communication*, vol. 7, no. 3, pp. 231-248, July 1995.

FAI88. M.G. Fairhurst, *Computer vision for robotic systems*, Prentice-Hall, 1988.

FOL90. J.D. Foley, A. van Dam, S.K. Feiner, J.F. Hughes, *Computer graphics: Principles and practice*, Addison-Wesley, 1990.

GIA98. I. Giakoumis, I. Pitas, "Digital restoration of painting cracks," *Proc. IEEE Intl. Symp. on Circuits and Systems (ISCAS98)*, pp. 357, 1998.

GOE80. V. Goetcherian, "From binary to gray level tone image processing using fuzzy logic concepts," *Pattern Recognition*, vol. 12, pp. 7-15, 1980.

GON87. R.C. Gonzalez, P. Wintz, *Digital image processing*, Addison-Wesley, 1987.

GRA71. S.B. Gray, "Local properties of binary images in two dimensions," *IEEE Transactions on Computers*, vol. C-20, no. 5, pp. 551-561, May 1971.

HEA86. D. Hearn, M.P. Baker, *Computer graphics*, Prentice-Hall, 1986.

HU61. M.K. Hu, "Pattern recognition by moment invariants," *Proc. IEEE*, vol. 49, no. 9, p. 1428, Sept. 1961.

HU62. M.K. Hu, "Visual pattern recognition by moment invariants," *IRE Transactions on Information Theory*, vol. 17-8, no. 2, pp. 179-187, Feb. 1962.

JAI89. A.K. Jain, *Fundamentals of digital image processing*, Prentice-Hall, 1989.

KAL94. E. Kalaitzis and I. Pitas, "Performance analysis of a morphological Voronoi tessellation algorithm," in *Mathematical Morphology and its applications*, J. Serra, P. Soille (editors) pp. 201-208, Kluwer Academic, 1994.

KIR90. M. Kirby and L. Sirovich, "Application of the Karhunen-Loeve procedure for the characterization of faces," *IEEE Trans. on Pattern Analysis and Machine Intelligence*, vol. 12, no. 1, pp. 103-108, January 1990.

KLE72. J.C. Klein, J. Serra, "The texture analyzer," *Journal of Microscopy*, vol. 95, pt. 2, pp. 349-356, Apr. 1972.

KOT93. C. Kotropoulos, I. Pitas, A. Maglara, "Voronoi Tessellation and Delaunay Triangulation Using Euclidean Disk Growing in Z^2," *IEEE Proc. Int. Conf. on Acoust. Speech and Signal Processing*, Minneapolis, pp. V29-V32, Apr. 1993.

KOT97. C. Kotropoulos and I. Pitas, "Rule-based face detection in frontal views," in *IEEE Proc. Int. Conf. on Acoustics, Speech, and Signal Processing*, vol. IV, pp. 2537-2540, Munich, Germany, 1997.

KOT97A. C. Kotropoulos and I. Pitas, "Rule-based face detection in frontal views," in *Proc. of IEEE Int. Conf. on Acoustics, Speech and Signal Processing (ICASSP 97)*, vol. IV, pp. 2537-2540, Munich, Germany, April 21-24, 1997.

KOT97ICIP. C. Kotropoulos and I. Pitas, "Face authentication based on morphological grid matching," in *Proc. of the IEEE Int. Conf. on Image Processing (ICIP-97)*, vol. I, pp. 105-108, October 1997.

KOT98ICIP. C. Kotropoulos, A. Tefas and I. Pitas, "Frontal face authentication using variants of dynamic link matching based on mathematical morphology," in *1998 IEEE Int. Conf. on Image Processing*, Chicago, vol. I, pp. 122-126, October 1998.

KOT99PR. C. Kotropoulos, A. Tefas and I. Pitas, "Morphological elastic graph matching applied to frontal face authentication under well-controlled and real conditions," *Pattern Recognition*, accepted for publication, 1999.

LAD93. M. Lades, J.C. Vorbrüggen, J. Buhmann, J. Lange, C.v.d. Malsburg, R.P. Würtz and W. Konen, "Distortion invariant object recognition in the dynamic link architecture," *IEEE Trans. on Computers*, vol. 42, no. 3, pp. 300-311, March 1993.

LAN80. C. Lantuejoul, "Skeletonization in Quantitative Metallography," in *Issues in Digital Image Processing* (R.M. Haralick and J.C. Simon, eds.), Maryland: Sijthoff and Noordhoff, 1980.

LAN84. C. Lantuejoul and F. Maisonneuve, "Geodesic Methods in Quantitative Image Analysis," *Pattern Recognition*, vol. 17, no. 2, pp. 177-187, 1984.

LEE80. D.T. Lee, "Two dimensional Voronoi diagram in the L_p-metric," *J.ACM 27*, pp. 604-618, 1980.

LEV85. M.D. Levine, *Vision in man and machine*, McGraw-Hill, 1985.

M2VTS96. M. Acheroy, C. Beumier, J. Bigün, G. Chollet, B. Duc, S. Fischer, D. Genoud, P. Lockwood, G. Maitre, S. Pigeon, I. Pitas, K. Sobottka, L. Vandendorpe, "Multi-modal person verification tools using speech and images," in *Proc. of the European Conference on Multimedia Applications, Services and Techniques*, pp. 747-761, Louvain-La-Neuve, Belgium, 1996.

MAR86. P. Maragos, R.W. Schafer, "Morphological skeleton representation and coding of binary images," *IEEE Transactions on Acoustics, Speech, and Signal Processing*, vol. ASSP-34, no. 5, pp. 1228-1244, Oct. 1986.

MAR90. P. Maragos, R.W. Schafer, "Morphological systems for multidimensional signal processing," *Proc. IEEE*, vol. 78, no. 4, pp. 690-710, Apr. 1990.

MAZ89. J.E. Mazille, "Mathematical Morphology and Convolutions," *Journal of Microscopy*, vol. 156, pp. 3-13, 1989.

MEY79. F. Meyer, "Iterative image transformations for an automatic screening of cervical smears," *J. Histoch. Cytochem.*, vol. 27, pp. 128-135, 1979.

NIK98. A. Nikolaidis and I. Pitas, "Facial Feature Extraction and Determination of Pose," in *Proc. 1998 NOBLESSE Workshop on Non-Linear Model Based Image Analysis (NMBIA'98)*, pp. 257-262, Glasgow, Scotland, July 1-3, 1998.

NIK99. A. Nikolaidis and I. Pitas, "Facial feature extraction and pose determination," *Pattern Recognition*, accepted for publication, 1999.

NUS81. H.J. Nussbaumer, *Fast Fourier transform and convolution algorithms*, Springer Verlag, 1981.

OPE89. A. Oppenheim, R.W. Schafer, *Discrete-time signal processing*, Prentice-Hall, 1989.

PAP73. S. Papert, "Uses of technology to enhance education," *Technical report 298*, AI Lab, MIT, 1973.

PAV78. T. Pavlidis, "A review of algorithms for shape analysis," *Computer Graphics and Image Processing*, vol. 7, no. 2, pp. 243-258, Apr. 1978.

PIT89. I. Pitas and C. Kotropoulos, "Texture Analysis and Segmentation of Seismic Images," *IEEE Proc. Int. Conf. on Acoust. Speech and Signal Processing*, pp. 1337-1340, Glasgow, 1989.

PIT90. I. Pitas, A.N. Venetsanopoulos, *Nonlinear digital filters: Principles and applications*, Kluwer Academic, 1990.

PIT90PAMI. I. Pitas, A.N. Venetsanopoulos, "Morphological shape decomposition," *IEEE Transactions on Pattern Analysis and Machine Intelligence*, vol. PAMI-12, no. 1, pp. 38-45, Jan. 1990.

PIT91. I. Pitas, "Parallel image processing algorithms and architectures," *Computing with parallel architectures: T.Node*, D. Gassiloud, J.C. Grossetie (editors), Kluwer Academic, 1991.

PIT92. I. Pitas, N. Sidiropoulos "Pattern recognition of binary image objects by using morphological shape decomposition," *Computer vision graphics and image processing*, L. Shapiro, A. Rosenfeld (editors), pp. 279-305, Academic Press, 1992.

PIT92PR. I. Pitas, A.N. Venetsanopoulos, "Morphological shape representation," *Pattern Recognition*, vol. 25, no. 6, pp. 555-565, 1992.

PIT98. I. Pitas, "Performance analysis of morphological Voronoi tessellation algorithms," in *Advances in digital and computational geometry*, R. Klette, A. Rosenfeld, F. Sloboda (editors), pp. 227-254, Springer, 1998.

PRE85. F.P. Preparata and M.I. Shamos, *Computational geometry*, Springer-Verlag, 1985.

REE81. A.P. Reeves, A. Rostampour, "Shape analysis of segmented objects using moments," *Proc. IEEE Conf. on Pattern Recognition and Image Processing*, pp. 171-174, 1981.

REQ80. A.A.G. Requicha, "Representations for rigid solids: Theory, methods and systems," *ACM Computing Surveys*, vol. 12, no. 4, pp. 436-464, Dec. 1980.

ROS86. A. Rosenfeld, "Axial representations of shape," *Computer vision graphics and image processing*, vol. 33, pp. 156-173, 1986.

SAM90. H. Sammet, *The design and analysis of spatial data structures*, Addison-Wesley, 1990.

SCH89. R.J. Schalkof, *Digital image processing and computer vision*, Wiley, 1989.

SER82. J. Serra, *Image analysis and mathematical morphology*, Academic Press, 1982.

SER94. J. Serra, P. Soille, *Mathematical morphology: Its applications to image processing*, Kluwer Academic, 1994.

SHI92. F.Y.C. Shin and O.R. Mitchell, "A Mathematical Morphology Approach to Euclidean Distance Transformation," *IEEE Trans. on Image Processing*, vol. 1, no. 2, pp. 197-204, April 1992.

SOB98. K. Sobottka, and I. Pitas, "A novel method for automatic face segmentation, facial feature extraction and tracking," *Signal Processing: Image Communication*, vol. 12, pp. 263-281, 1998.

SOB98A. K. Sobottka and I. Pitas, "A Novel Method for Automatic Face Segmentation, Facial Feature Extraction and Tracking," *Signal Processing: Image Communication*, vol. 12, no. 3, pp. 263-281, June 1998.

SOI99. P. Soille, *Morphological image analysis*, Springer, 1999.

STE83. S.R. Sternberg, "Biological image processing," *Computer*, vol. 16, no. 1, pp. 22-34, Jan. 1983.

STE87. R.L. Stevenson, G.R. Arce, "Morphological filters: Statistics and further properties," *IEEE Transactions on Circuits and Systems*, vol. CAS-34, pp. 1292-1305, Nov. 1987.

TAM78. H. Tammura, "A comparison of line thinning algorithms from a digital geometry viewpoint," *Proc. 4th Int. Conf. on Pattern Recognition*, pp. 715-719, Kyoto, 1978.

TAN75. S. Tanimoto, T. Pavlidis, "A hierarchical data structure for picture processing," *Computer Graphics and Image Processing*, vol. 4, no. 2, pp. 104-119, 1975.

TAN86. S.L.Tanimoto, "Paradigms for pyramid machine algorithms," *Pyramidal systems for computer vision*, V. Cantoni, S. Levialdi (editors), Springer Verlag, 1986.

TEF99ICASSP. A. Tefas, Y. Menguy, C. Kotropoulos, G. Richard, I. Pitas and P. Lockwood, "Compensating for variable recording conditions in frontal face authentication algorithms," in *Proc. of the IEEE Int. Conf. on Acoustics, Speech and Signal Processing (ICASSP-99)*, vol. 6, pp. 3561-3564, Phoenix, March 1999.

TOU80. G.T. Toussaint, "Pattern Recognition and Geometrical Complexity," *Proc. 5th International Conference on Pattern Recognition*, pp. 1324-1347, 1980.

TSEK98. S. Tsekeridou and I. Pitas, "Facial Feature Extraction in Frontal Views Using Biometric Analogies," in *IX European Signal Processing Conference (EUSIPCO'98)*, vol. I, pp. 315-318, Rhodes, Greece, September 8-11, 1998.

TUR91. M. Turk and A. Pentland, "Eigenfaces for recognition," *Journal of Cognitive Neuroscience*, vol. 3, no. 1, pp. 71–86, 1991.

VER89. B.J.H. Verwer, P.W. Verbeek and S.T. Dekker, "An Efficient Uniform Cost Algorithm Applied to Distance Transforms," *IEEE Trans. Pattern Anal. Machine Intel.*, vol. 11, no. 4, pp. 425-429, April 1989.

VIN91I. L. Vincent and P. Soille, "Watershed in Digital Spaces: An Efficient Based on Immersion Simulations," *IEEE Trans. Pattern Anal. Machine Intel.*, vol. 13, no. 6, pp. 583-598, June 1991.

VIN91II. L. Vincent, "Morphological Transformations of Binary Images with Arbitrary Structuring Elements," *Signal Processing*, vol. 22, pp. 3-23, 1991.

WIS97A. L. Wiskott, J.M. Fellous, N. Krüger and C.v.d. Malsburg, "Face recognition by elastic bunch graph matching," *IEEE Trans. on Pattern Analysis and Machine Intelligence*, vol. 19, no. 7, pp. 775-779, July 1997.

WON78. R.Y. Wong, E.L. Hall, "Scene matching with invariant moments," *Computer Graphics and Image Processing*, vol. 8, no.1, pp. 16-24, Aug. 1978.

WUR97TPAMI. R.P. Würtz, "Object recognition robust under translations, deformations, and changes in background," *IEEE Trans. on Pattern Analysis and Machine Intelligence*, vol. 19, no. 7, pp. 769-775, July 1997.

YAN94. G. Yang and T.S. Huang, "Human face detection in a complex background," *Pattern Recognition*, vol. 27, no. 1, pp.53-63, 1994.

YE88. Q.Z. Ye, "The Signed Euclidean Distance Transform and Its Applications," in *Proc. 9th Int. Conf. on Pattern Recognition*, pp. 495-499, Rome, Italy, 1988.

YOU79. I.T. Young, J.E. Walker, J.E. Bowie, "An analysis technique for biological shape," *Information and Control*, vol. 25, no. 4, pp. 357-370, Aug. 1979.

ZHA84. T.Y. Zhang, C.Y. Suen, "A fast parallel algorithm for thinning digital patterns," *Communications of the ACM*, vol. 27, no. 3, pp. 236-239, 1984.

ZHA97. J. Zhang, Y. Yan and M. Lades, "Face recognition: Eigenface, elastic matching and neural nets," *Proceedings of the IEEE*, vol. 85, no. 9, pp. 1423-1435, September 1997.

ZHO88. Z. Zhou, A.N. Venetsanopoulos, "Morphological skeleton representation and shape recognition," *Proc. IEEE Int. Conf. on Acoustics, Speech and Signal Processing*, pp. 948-951, 1988.

8

Digital Image Processing Lab Exercises Using EIKONA

8.1 INTRODUCTION

This book is accompanied by multimedia material (lab exercises, transparencies) for classroom use, distance learning or self experimentation. A variety of lab exercises based on image processing package EIKONA are given. The transparencies in PDF format that accompany the book highlight important topics covered in each book chapter. The book, together with its transparencies and the exercises, can be used as stand alone teaching tool in a digital image processing course, either in a traditional classroom or in a distance learning environment by videoconferencing over IP or over ISDN. In particular, the networked version of EIKONA can be used in an Internet environment to create a virtual image processing lab, where the student can execute his lab exercises at home at his own pace. The multimedia material (transparencies, exercises) can be found in the site *ftp://ftp.wiley.com/public/sci_tech_med/image_processing.* The demo, student and full version of EIKONA can be found in the site *http://www.alphatecltd.com.* In the following we give an overview of the educational material and its suggested use.

8.2 OVERVIEW

EIKONA is a powerful, but yet simple to use, digital image processing software package that runs under either Microsoft Windows or Unix systems and implements over 500 image processing routines in the following areas: image display, scanning and printing, image thresholding, clipping, addition, subtraction and multiplication of images, AND, OR, XOR bit-level operations between images, addition/multiplication of image, various image noise generators, two dimensional filters including adaptive and nonlinear filters, histogram and cdf histogram computation and equalization, matrix histogram and cdf matrix histogram computation, image enhancement and sharpening, region segmentation, edge detection, morphological filters, image transforms, color coordinate transformations, image mosaicing, registration and watermarking.

EIKONA's user interface is based completely on pull-down menus, dialog boxes and common user interface elements. As a consequence, it is extremely easy to operate even for users not very familiar with image processing. Images are stored on image buffers. All that is needed to apply an image processing function to an image is to specify the source and destination image buffer and the related parameters. Furthermore, implemented functions are grouped into categories according to the type of operation that they carry out, in order to facilitate the search for a specific task. The user can choose the image region where processing is to be performed. Multiple images can be displayed on the screen at the same time, a valuable feature when comparing the results of different processing functions on the same image. Being a pure 32-bit application, EIKONA fully exploits the flat memory model allowing efficient handling of big images.

EIKONA FOR WINDOWS can run on any 386, 486, Pentium/II/III computer under Microsoft Windows 95/98 and Windows NT. However, since many image processing applications are rather time consuming and since images, especially color images, require large amounts of memory, a minimum configuration of a Pentium machine with at least 64MByte of RAM is recommended. Also, a color monitor with true color graphics card is recommended in order to take advantage of the image display capabilities of EIKONA FOR WINDOWS.

EIKONA FOR UNIX can run on any Unix machine which supports the X-Windows system, release 4 or higher with a minimum configuration of 256MBytes of RAM. Also, a color monitor with a 24-bit (True Color) video card capable of displaying 16 million colors is required.

EIKONA supports TARGA, JPEG, BMP, TIFF, GIF and Postscript file formats, as well as binary (raw) images. Conversion from one image file format to another is possible. Processed images that are stored on image buffers can be saved on disk files.

8.3 STRUCTURE

The central concept in EIKONA is that of a *buffer*. EIKONA supports four types of buffers that are used by various image processing functions:

- **Black-and-white buffers** are used for storing grayscale images. Essentially they are two-dimensional arrays of unsigned characters. Their format is described in detail in Chapter 1. From now on, they will be referred to as BW image buffers.

- **Color buffers** are used for storing color images. They are nothing more than three BW buffers (one for each color channel) that EIKONA groups together. However the user can view and otherwise act on each channel separately as if it was and ordinary BW buffer. Furthermore three BW buffers can be merged into one color buffer.

- **Float matrices** are used to store the results of some functions implemented in EIKONA (most notable examples being the Fast Fourier Transform and the Discrete Cosine Transform).

- **Color processing matrices** used mainly for color space transformations and other color image processing functions.

Buffers are created and destroyed dynamically. They are also created dynamically when loading a new image. EIKONA can allocate as many buffers as required, the only limit being the memory capacity of the computer. Buffer dimensions are either specified by the user or determined by EIKONA according to the operation. The contents of a BW or Color image buffer can be displayed by clicking the Display command in Black & White or Color menus respectively.

EIKONA also allocates 2 one-dimensional float buffers (referred to as vectors) that are used internally in some operations (e.g. histogram calculation or matrix histogram calculation). The dimension of these vectors defaults to 256 elements (although it can be user-defined as described in the next section). Their contents can be saved to disk in decimal format in order to be further studied.

Color images are stored on color image buffers. In color image processing functions, three BW buffers are considered as one color buffer and are referred to by a single number. Red component occupies the first buffer, green the second and blue the third one.

The image buffer size as well as their maximum number that can be allocated by the user is specified, along with other initialization information, in the ASCII file EIKONA.PAR that resides in the directory of EIKONA. The user can modify this file to user-define the initial size and maximum number of those buffers. Typical contents of this file may be the following:

```
WindowXSize=15
WindowYSize=15
VectorSize=256
Xo_reference=1.0
Yo_reference=1.0
Zo_reference=1.0
```

These EIKONA.PAR file contents instruct EIKONA the maximal allowable window size in image filtering operations is 15 × 15, the vector size is 256 points and that the reference values for color space transformations are 1.0. EIKONA is not sensitive to the order of these definitions in the EIKONA.PAR file but it requires that every one of them is in a separate line.

Some functions implemented in EIKONA (e.g. image transforms, color coordinate system transformations) produce as an output two-dimensional float or integer matrices. Therefore, EIKONA can allocate and manipulate two-dimensional integer or float buffers. Any image processing algorithm implemented in EIKONA can be applied to an image stored in some buffer by simply selecting the corresponding option from the menu. The related dialog box is then presented on the screen, prompting the user to fill in the necessary parameters. Almost all operations require from the user to specify the input (source) buffer (or buffers) and the output (destination) buffer. A new buffer can be created dynamically as an output buffer. Selecting the input buffer to serve also as the output buffer is not prohibited. However, since many image processing algorithms cannot be executed "in-place", this is not recommended (unless the reader knows that the processing routine is a pixel-wise one).

By default, any image processing function is applied to the entire image buffer. However, the user can select a region of interest (ROI), that is, a part of the image to be processed, by clicking the right button over the upper left corner of the region of interest and keep it pressed while moving to the lower right corner. During the selection operation, a rectangle shows the selected area. The user can undo the region selection and re-select the entire image by the CTRL-A button or by a special 'Select Whole Image' button in the control window.

Various filtering operations require a float or integer coefficient mask. These masks, which are read by EIKONA without user intervention, are contained in the files WIN_F.PAR and WIN_INT.PAR that must reside in the directory of EIKONA. The user must edit these files prior to applying the filtering operation and fill in the filter coefficients. The contents of the WIN_F.PAR file could be the following:
3, 3
0.11111, 0.11111, 0.11111, 0.11111, 0.11111, 0.11111, 0.11111, 0.11111, 0.11111
The first two numbers show the column and row number of the floating point mask (3 × 3 in this case). Nine floating point numbers follow that correspond to the mask elements written in a row-wise manner. This mask, when used with convolution corresponds to the moving average filter. The contents of the WIN_UC.PAR file could be the following:
3, 3
0, 1, 0, 1, 1, 1, 0, 1, 0
The first two numbers show the column and row number of the integer mask (3 × 3 in this case). Nine integer numbers follow that correspond to the mask elements written in a row-wise manner.

The so-called Control Window can provide useful information for the currently active image buffer. It contains the current ROI size and position and the gray or RGB, XYZ, HLS, Lab value of the image pixel at the current mouse position. The coordinates of an image pixel and its value are displayed in the Control Window as well. EIKONA allows multiple images to be displayed at the same time. Each image buffer is displayed in a different window. Changing the contents of the buffer will automatically change the associated image window.

8.4 BW IMAGE PROCESSING

8.4.1 Black-and-White

This section contains all routines for the processing, analysis and display of black-and-white images. In most cases, the processing routines have one source BW image buffer, one destination BW image buffer and a number of parameters. The user can either specify an existing buffer as the destination buffer or ask for the creation of a new buffer. Processing is performed only within the region of interest (ROI). The display is automatically updated to reflect the result of the operation.

8.4.2 Basic

EIKONA can perform a number of basic image operations. These include clearing the contents of an BW buffer, copying the contents of a buffer to another buffer, bit-level operations between images (AND, OR, XOR), adding a constant to an image, multiplying an image by a constant, adding/subtracting two images, transforming an image to matrix and a matrix to image with normalization or truncation, image halftoning etc.

8.4.3 Processing

Several image processing operations can be performed, e.g. negation of an image, image sharpening, image zoom/decimation and also additive and multiplicative noise generation (Gaussian, Uniform, Laplacian) that can be used to corrupt an image for simulation purposes.

8.4.4 Analysis

Edge detection algorithms (Sobel, Laplace, etc.), line and point detection functions, as well as various algorithms for edge following and related algorithms, like forward and inverse Hough transform, can be performed. Region segmentation algorithms including thresholding and counting, texture description and shape analysis algorithms, along with thinning and pyramid methods exist in the analysis menu. Finally, image and matrix histogram, Cdf Histogram and Cdf Matrix Histogram are also included in this menu.

8.4.5 Transforms

This sub-menu includes the well-known Fast Fourier Transform (FFT) for one and two dimensions and the Discrete Cosine Transform (DCT), and also their inverses. Power Spectrum Density estimation, image convolution and periodogram are contained as well.

8.4.6 Filtering

This sub-menu contains some of the most widely used filters in Digital Image Processing. These include Histogram equalization, moving average, median, minimum, rank filter, L-filter and many more that are described in Chapter 3.

8.4.7 Nonlinear filtering

Some more nonlinear operators like morphological operators for grayscale and binary images (opening-closing, erosion-dilation), Signal Adaptive Median filter, hybrid, multistage, weighted, separable, median filters, homomorphic filter, harmonic filter are included in this menu.

8.5 COLOR IMAGE PROCESSING

It contains all routines for the processing, analysis and display of color images. In most cases, the processing routines have one color image buffer, one destination color image buffer and a number of parameters. The user can either specify an existing buffer as the destination buffer or ask for the creation of a new buffer. Processing is performed only within the region of interest (ROI).

8.5.1 Basic

Basic color image operations like bit-level operations (AND, OR etc.), buffer copy and clear operations, addition and multiplication of the three RGB components of an image with the same constant etc. are included in this sub-menu.

8.5.2 Processing

Noise generators for the three color image components, color image filters like marginal median, marginal minimum and also erosion-dilation operators are included in this sub-menu.

8.5.3 Analysis

Here we have Laplace, Prewitt and Sobel algorithms. Line, point detectors and segmentation algorithms can be found as well.

8.5.4 Color Representation

This sub-menu contains a number of color coordinate transformations. Many of the color coordinate systems that are included, require float color coordinates. Therefore, float buffers must be allocated for such transforms. A User Defined Color Transformation can be defined as well. The user specifies the mask of the transformation in the WIN_F.PAR file.

8.6 MODULES

EIKONA FOR WINDOWS is structured in a modular format for simplicity. In the following, we describe some existing image processing modules. User defined modules can be easily written as well, thus saving development time. The modules are DLLs linked with EIKONA FOR WINDOWS. They operate on EIKONA FOR WINDOWS buffers. Therefore, the user, who develops a new module, does not have to worry about image I-O, buffer allocation and image display, since all these capabilities already exist in the main EIKONA FOR WINDOWS menu. This leads to significant time and effort savings for new developments. In the Microsoft Windows operating system dynamic-link libraries (DLL) are modules that may contain code or data. The functions making up a DLL code can be of two kinds :

Internal functions, which can only be called by other functions defined in the same DLL.

Exported functions, which can be called by other modules.

A DLL can be loaded at run-time by other programs which can then call the DLL's exported functions. Thus, DLLs provide a way to modularize applications so that basic application functionality can be easily updated or extended (through the DLL's exported code). EIKONA makes full use of the benefits provided by DLLs. More specifically, EIKONA can dynamically load any DLL that conforms to a well-defined *interface*. This interface defines a set of rules that a DLL should follow in order to be loaded and used by

EIKONA. The rules specify mainly a set of functions that a DLL should export for EIKONA and a number of functions and services that EIKONA provides to attached DLL modules. Among others, the DLL can:

- Create a private menu under EIKONA's Modules menu and install a function to handle commands from this menu.

- Bypass default EIKONA message processing (if this is required) and provide its own message handler ...

- ... or just enjoy the readily available EIKONA GUI while concentrating on image processing issues.

The services provided by a DLL are accessed through the Modules menu of EIKONA. Clearly, the DLL should create an appropriate entry under the Modules menu and respond to choices from this menu accordingly. This is done in close cooperation with EIKONA through the EikonaDLLReturnMenu and EikonaDLLReturnCmdHandler function which every DLL should provide. EIKONA allows an attaching DLL to respond to standard window messages thus taking control of windows displayed by it. The DLL may install a special procedure to which window messages will be forwarded. The DLL may then process the message or pass it back to EIKONA, if it decides that it should not handle it. In its full generality, this model allows the DLL to take complete control of windows. This, however, requires considerable effort and good knowledge of the WIN32 API. In the following, examples of existing modules are described. The related theory is presented in previous book Chapters.

8.6.1 Arts module

Arts is a powerful EIKONA FOR WINDOWS module. It contains a mosaicing algorithm for automatic, semi-automatic or manual image mosaicing, and two image registration algorithms. The Arts module has been developed for analyzing digital paintings and other works of art. It can also be used in several other applications where image mosaicing or registration is needed, notably in remote sensing and in medical imaging.

8.6.1.1 Image Mosaic Image Mosaic is performed on a set of equal size images having overlapping parts in the horizontal and/or in the vertical direction, and creates an output image as the result of input image mosaicing. The images must have the same root filename and must be characterized by the number of the relevant scanning order. Image mosaicing is used in painting reflectography that employs infrared imaging. Infrared cameras have low resolution. Therefore, mosaicing of IR patches is extensively used for high resolution painting reflectography. Mosaicing is also used in remote sensing, when we want to create mosaics of aerial or even satellite images. Panoramic digital television applications employ mosaicing techniques. Finally, mosaicing can be used for creating panoramic digital X-ray images in dental and medical applications. Geometrical corrections or registration must be performed prior to mosaicing in many of the above mentioned applications.

Fig. 8.6.1 (a) Original image; (b) the mosaic. (*Courtesy of Fr. M. Politis, Skete of St. Andrew, Mt. Athos, Greece.*)

8.6.1.2 Image Registration Registration is performed on two source images contained in buffers of equal size, with the first image considered as the *reference* image. Registration produces an output image which is a scaled, translated and rotated version of the second source image, based on certain feature points specified by the user. User-defined feature points are used for calculating the registration output. Wrongly chosen feature points can be automatically rejected.

8.6.1.3 Combined Registration Combined Registration is an advanced version of registration. It differs from the basic version of registration in a number of ways. The most important is that the result is a combination of the two source images (hence the name), which can be computed in two ways:

- *Image merging.* It is performed by averaging the two images where they overlap and by keeping their values where they do not overlap.

- *Image subtraction.* It is performed by subtracting the images and adding a bias (in order to preserve the negative values) where the images overlap and by keeping their values where they do not overlap.

A second difference is the use of mask images to define arbitrarily shaped regions of interest. These masks are white in areas where there is image information, black where there is none and have intermediate values in border areas. Generally, for most images we only need to define an all-white mask. The third difference is that that the refined feature point rejection procedure has been improved. All the user has to do is to select the number of feature points he wants to keep. The best feature points will be kept and the rest will be rejected. A minor fourth difference is that the user has the ability to disable scaling and keep the size of the image unaltered. Image registration has important applications in painting conservation, when we want to register and superimpose various image modalities, notably X-ray, visible, infrared and ultraviolet images. It is also applied in remote sensing when we want to register satellite images and aerial images. Finally, it is used in medical imaging when we employ X-ray images to monitor the progress of a medical treatment or of a biological process. In this case we have to register images taken at various time instances and, sometimes, to subtract images after registration in order to highlight the differences (e.g. in subtractive angiography).

8.6.2 Crack Restoration

Many natural phenomena, like unfavorable weather conditions, cause frequently destructive effects on paintings. Wood is an anisotropic substance, exhibiting different degrees of hardness, toughness, which means that its physical characteristics in different directions are not similar. In dry environments, natural drying could lead to too rapid loss of water, resulting in non-uniform contraction and, eventually, in cracking. The cracks are frequent mostly in old paintings. The appearance of cracks on paintings deteriorates the perceived image quality. Thus, digital image processing techniques can be used in order to give the ability to the art historian to study the icon, as has been created by the painter.

Cracks have usually low luminance and can be considered as being local minima with rather elongated structural characteristics. Therefore, crack detector will be applied on a luminance image, where local minima will be identified. The detection of the cracks can be obtained with the implementation of the top-hat transformation described in Chapter 7. This high-pass filter can detect bright details in an image. The cracks have usually very small luminance. They look like dark details in a bright background. Thus, if we want to extract the cracks, we must negate the luminance image and then apply the top-hat transformation. The user can modify the size of the structuring element. The size of the element depends on the thickness of the crack to be detected. We must choose carefully the size of the structuring element, because, otherwise, it is possible to misidentify some thin brush strokes of small luminance.

The top-hat transformation produces a grayscale output image $t(k,l)$. If its value is large, the corresponding pixel belongs to a crack (or crack-like element). Otherwise, this pixel location corresponds to background. Therefore, a thresholding operation is required to separate cracks from the background. The threshold T can be chosen by the user. The thresholding is global, because T is chosen based on global information.

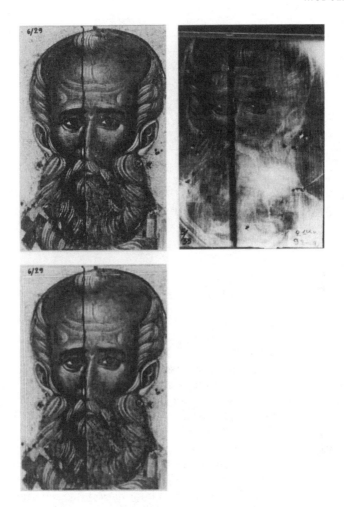

Fig. 8.6.2 (a) Original image of a Byzantine icon in the visible wavelength domain; (b) X-ray image; (c) the superimposed registered images. (*Courtesy of Fr. M. Politis, Skete of St. Andrew, Mt. Athos, Greece.*)

In some paintings there are thin dark brush strokes (e.g. in hair), which have almost the same features as cracks (e.g thickness, luminance). Therefore, it is possible that the top-hat transformation misclassifies these brush strokes as cracks. It would be better to separate these brush strokes, before the implementation of the crack filling procedure, in order to avoid any undesirable alterations to the original image.

A simple approach to the separation of cracks is to start from some pixels (seeds) in the thresholded output of the top-hat transformation, which represent distinct cracks. Then we *grow* them, until they cover all the cracks in the image. The pixel seeds are chosen by the user in a supervised mode. At

Fig. 8.6.3 (a) Original cracked image; (b) image after crack filling. (*Courtesy of Fr. M. Politis, Skete of St. Andrew, Mt. Athos, Greece.*)

least one seed per crack is chosen. The growth mechanism is quite simple. We must check if there are unclassified pixels with value 1 in the 8-neighborhood of each pixel of the crack. At the end of this process, the pixels in the binary image, which correspond to the brush strokes will be removed, since we have not determined seeds for them. An alternative automatic way for crack identification/separation is to use radial basis function (RBF) networks.

After identifying cracks, our aim is to restore them, possibly by filling (interpolating) image information content within cracks. The restoration method is implemented on each RGB channel independently and only on those pixels, which belong to cracks. We use the local information (neighboring pixels) for crack restoration and filling. Anisotropic diffusion has been used as an efficient nonlinear technique for crack filling. The classical anisotropic diffusion filter has been modified, by taking into account also the crack orientation.

8.6.3 Watermark module

The Watermark module for EIKONA FOR WINDOWS is an extension module containing an innovative algorithm for casting and detecting watermarks in digital images. Image watermarking is a technique for embedding a signal called *signature* or *watermark* to a digital image, so that unauthorized copying of the image is prevented. The watermark completely characterizes the person who applied it, and, therefore, proves the origin of the image. The embedding level specifies the robustness of the watermark. As the embedding level increases, the watermark becomes more robust under image modifications, such as JPEG compression and digital filtering. However, perceptually visible effects may arise. An example of watermarking is shown in Figure 8.6.4. The watermark is completely invisible.

Fig. 8.6.4 (a) Original image; (b) the watermarked image.

8.7 EIKONA SOURCE, LIBRARY/DLL

The routines presented in this book are simplified (e.g. they have few error checks only) in order to increase code readability. Their full versions are part of EIKONA SOURCE that contains more than 500 image processing routines altogether. Thus, EIKONA SOURCE forms a complete set of source code image processing routines. The routines are written in ANSI C and can be compiled by any C compiler. The 500 image processing routines that are included in EIKONA SOURCE have been compiled to static library EIKONA LIBRARY and the dynamic library EIKONA DLL. Either of them can be used to write stand alone image processing applications. Compilation has been performed by the MICROSOFT VISUAL C++ compiler. The EIKONA routines contained in EIKONA LIBRARY or EIKONA DLL van be easily called and linked to any calling environment. EIKONA LIBRARY versions for Unix workstations exist as well.

8.8 INSTRUCTIONS FOR USING THE EDUCATIONAL MATERIAL

The multimedia materials that accompany this book can be downloaded from *ftp://ftp.wiley.com/public/sci_tech_med/image_processing*. They can be used in the following way:

- The transparencies can be used for highlighting book chapter topics either in the classroom or at home or in a teleteaching environment through application sharing in videoconferencing over ISDN or over IP. They are in PDF format and are printable.

- The lab exercises can be executed either in an organized PC Lab under tutors' supervision or at home. They are in PDF format and are printable.

- The student or full version of EIKONA FOR WINDOWS or EIKONA FOR UNIX are required for performing the exercises. They can be found in *www.alphatecltd.com.*Virtual labs can be created over Internet, if the networked version of EIKONA FOR WINDOWS is employed. The networked version allows controlled execution of EIKONA FOR WINDOWS. The SSL certificates are used for access control and cryptography [WAG96]. Students can execute EIKONA and lab exercises only for a pre-selected time or on a pre-selected computer defined by its IP address. Virtual lab exercises can be performed with or without tutor's online assistance.

- EIKONA manual. It is in PDF format and is printable. It can be used for assistance for lab exercises. It can also be used for developing new EIKONA modules or stand alone applications using EIKONA SOURCE or EIKONA DLL or EIKONA LIB. They can be found in *www.alphatecltd.com.*

- Glossary. It can be used by beginners or by self-teaching students in a distance learning environment.

- Sample images. They can be used for executing lab exercises.

The multimedia material can be used by the following target student groups:

- Beginners in image processing. First they can study the book chapters of interest (or preferably in serial order) without putting efforts into understanding the C code. They can use the transparencies for highlighting the important topics. Glossary can be used for assistance. Finally they can perform lab exercises.

- Advanced users in image processing. They can study the book chapters of their interest trying to understand the related algorithms. They can execute lab exercises as well as their variations by using the sample images or images of their interest. They should try to experiment and understand the meaning and the effect of the various routine parameters.

- Programmers. They can thoroughly study the topic(s) and the routine(s) of their interest. They can thoroughly experiment with the related lab exercises. They can use EIKONA SOURCE, EIKONA LIB, EIKONA DLL for developing their own EIKONA module. Finally, they can modify their own EIKONA module and build their own stand alone application.

The teachers/tutors can use this book and the accompanying material in the following way:

- *Classroom use.* The transparencies can be used while teaching image processing theory and examples. The lab exercises can be used in a lab environment or at home with/without tutor's assistance.

- *Teleteaching.* The transparencies can be used as application sharing during videoconferencing. It is preferable that students have printed versions at home. The students can execute lab exercises at home by using the full EIKONA FOR WINDOWS. Alternatively, they can use the networked EIKONA FOR WINDOWS to perform a controlled execution of the lab exercises with/without tutor's assistance.

It is obvious that the use of the material is very flexible and can be adapted to the teacher and student needs.

References

PIT90. I. Pitas, A.N. Venetsanopoulos, *Nonlinear digital filters: Principles and applications*, Kluwer Academic, 1990.

PIT93. I. Pitas, editor, *Parallel algorithms for digital image processing, computer vision and neural networks*, Wiley, 1993.

EIK98. EIKONA manual v.4.0., 1998.

WAG96. D. Wagner and B. Schneier, "Analysis of the SSL 3.0 Protocol," in *Proceedings of the Second USENIX Workshop on Electronic Commerce*, pp. 29-40, USENIX Press, November 1996.

Index

AR model, 101
Binary operations, 14
Block convolution, 128
Chain codes, 324
Circular convolution, 54
Circular shift, 54
Clipping, 14
Closing, 365
Co-occurrence matrix, 311
Color transformations, 20
Connected component labeling, 300
Connectedness, 276
Constructive solid geometry, 379
Correlation, 56
Decimation, 177
Decision-directed filters, 159
Dilation, 363
Direct implementation of FIR filters, 122
Discrete cosine transform, 103
Discrete Fourier Transform, 52
Discrete wavelet transform, 113
Display, 18
Dithering, 172
Dynamic programming, 267
Edge-following algorithms, 257
Edge detection, 242
Edge templates, 243
Edge thresholding, 249
Elementary operations, 13
Entropy coding, 192

Erosion, 363
Face detection, 386
Face recognition, 386
FFT implementation of FIR filters, 125
Fourier descriptors, 334
Geometric transformations, 15
Gray-level difference, 306
Grayscale morphology, 369
Halftoning, 168
Heuristic edge search, 261
Histogram, 162
Histogram equalization, 164
HLS color space, 35
Hough transform, 249
HSI color space, 32
Huffman coding, 192
Image formation, 4
Image Mosaicing, 179
Image representation, 7
Image rotation, 15
Image storage, 12
Image watermarking, 180
Interpolation, 174
Inverse filter, 133
JPEG compression standard, 232
L-filters, 154
Laplace edge detector, 247
LZW compression, 205
Mathematical morphology, 361
Max/min filters, 150

Medial axis, 372
Median filter, 139
Memory constraints in FFT, 68
Merging algorithm, 286
MMSE filter, 157
Modified READ coding, 203
Moment descriptors, 352
Morphological shape decomposition, 376
Morphological shape representation, 379
Moving average filter, 122
Multistage median filter, 148
Neighborhood, 276
Noise generators, 38
Opening, 365
Overlap-add, 129
Overlap-save, 130
Papert's turtle, 327
Parseval's Theorem, 59
Periodogram, 96
Photoelectronic sensor, 6
Polygonal approximation, 329
Polynomial transform, 92
Power spectrum, 96
Predictive coding, 221
Prewitt edge detector, 242
Pseudocoloring, 166
Pyramids, 342
Quadtrees, 336
Quantization, 6

Range edge detector, 248
Rank-order filters, 149
Recursive median filter, 146
Region growing, 282
Region splitting algorithm, 290
Relaxation labeling, 297
Rolling ball transformation, 370
Row-column FFT, 59
Run-length calculation, 308
Run-length coding, 200
Running median algorithm, 143
Separable median filter, 145
Shape features, 348
Signal-adaptive median filter, 160
Skeleton, 372
Sobel edge detector, 242
Split and merge algorithm, 292
Texture description, 303
Thinning algorithms, 356
Thresholding, 14, 277
Top-hat transformation, 372
Transform image coding, 229
Trimmed mean filters, 153
Two-component model filtering, 159
Vector-radix FFT, 85
Voronoi tesselation, 382
Watershed transform, 385
Weighted median filter, 146
Wiener filter, 135
Zoom, 175